Disrupting Science

Princeton Studies in Cultural Sociology

SERIES EDITORS: Paul J. DiMaggio, Michèle Lamont, Robert J. Wuthnow, Viviana A. Zelizer

A list of titles in this series appears at the back of the book.

Disrupting Science

SOCIAL MOVEMENTS, AMERICAN SCIENTISTS, AND
THE POLITICS OF THE MILITARY, 1945–1975

Kelly Moore

PRINCETON UNIVERSITY PRESS
PRINCETON AND OXFORD

Copyright © 2008 by Princeton University Press
Published by Princeton University Press, 41 William Street,
Princeton, New Jersey 08540
In the United Kingdom: Princeton University Press, 3 Market Place,
Woodstock, Oxfordshire OX20 1SY

Library of Congress Cataloging-in-Publication Data
Moore, Kelly.
Disrupting science : social movements, american scientists, and the politics of the military,
1945–1975 / Kelly Moore.
   p.   cm.
Includes bibliographical reference and index.

ISBN 978-0-691-11352-4 (hardcover : alk. paper)
1. Science—Social aspects—United States—History—20th century. 2. Science—Political
aspects—United States—History—20th century. 3. Scientists—Political activity—
United States—History—20th century. I. Title.
Q127.U6M656   2008
509.73—dc22   2007019974

British Library Cataloging-in-Publication Data is available

This book has been composed in Sabon

Printed on acid-free paper. ∞

press.princeton.edu

Printed in the United States of America

10 9 8 7 6 5 4 3 2 1

Contents

Acknowledgments

Early in this project, four people influenced my thinking and ways of investigating the social world. Elisabeth S. Clemens encouraged my early interest in the sociology of science, provided excellent advice on how to structure historical arguments in the social sciences, and showed me how to work with archival evidence. Walter W. Powell's work on biotechnology shaped my thinking about how science has changed since 1980, and his work on organizations helped me to think about them as systems of meaning creation. David A. Snow played an especially important role in my thinking about symbols and identity creation. I owe a special debt to Doug McAdam. He inspired me to study how and why social life can be changed through collective action, and his ideas about states and social movements and the centrality of organizations in the political mobilization process have stimulated my own thinking about these issues.

I am especially grateful to those who read large parts or all of *Disrupting Science*. Lynn S. Chancer, Scott Frickel, Thomas F. Gieryn, David H. Guston, Daniel L. Kleinman, Annulla Linders, Rhys H. Williams, and Gilda Zwerman provided challenging and wise advice about my evidence and argument. I owe a special debt of gratitude to Daniel Kleinman, whose comments encouraged me to think more deeply about the relationship between scientists' authority and the authority of scientific knowledge and practices. Gilda Zwerman's insistence on linking my cases, and her insights about the politics of mid-twentieth-century America helped me to more fully develop my argument as I was finishing the book. I did not take all the advice that these readers provided, and so all errors are my own.

Conversations, and sometimes more heated debates, with Elisabeth S. Clemens; Elizabeth Bernstein; Lois Horowitz; Kerwin Kay; Laura Kay; Joel Kaye; Nicole Hala; Michael Lounsbury; Francesca Polletta; Benjamin H. Shepard; Marc Ventresca; members of the People, Power, and Politics Workshop, Department of Sociology, New York University; members of the Contentious Politics Workshop, Columbia University; audiences at meetings of the Society for Social Studies of Science and the American Sociological Association; and audiences at Cornell University, Princeton University, and Rensselaer Polytechnic Institute pushed me to sharpen my argument.

I gratefully acknowledge financial support from the American Institute of Physics, Columbia University, the Lilly Endowment, Inc., and the National Science Foundation (Grant # 9101175).

Without the excellent research assistance of Karen Elaine Bailey, Eileen Clancy, Nicole Hala, Jonathan Goldberg, Bari Meltzer, and James Park, this book would not have been written. I am grateful to have worked with such talented and dogged people.

Amy Crumpton of Archives of the American Association for the Advancement of Science, Charles Griefenstein of the American Philosophical Society, Barbara Truesdell at the Indiana University Center for the Study of History and Memory, Diana Franzusoff Peterson at the Haverford Library Quaker and Special Collections, and the staffs of the American Institute of Physics Niels Bohr Library, the Library of Congress, the MIT Archives and Special Collections, the Charles Deering Memorial Library Special Collections at Northwestern University, the Swarthmore College Peace Collection, the University of California Bancroft Library Regional Oral History Office, the University of Chicago Archives and Special Collections, and the Western Historical Manuscript Collection at the University of Missouri—St. Louis all provided valuable assistance in using their collections.

It has been a joy to work with Princeton University Press. Many thanks to Ian Malcolm for his support of this project, to Madeleine Adams for her outstanding work editing the manuscript, and to Natalie Baan for shepherding the book through the production process.

The University of Cincinnati Department of Sociology has provided me with a collegial and supportive environment in which to write. I am deeply appreciative.

In ways too numerous to identify, Elizabeth Bernstein, Lynn Chancer, Eileen Clancy, Steve Duncombe, Laura Kay, Ben Shepard, Caitlin von Schmidt, Gilda Zwerman, Kathleen, Bryan, Jane, and Chris have inspired and supported me. Thank you.

Rhys Williams deserves more than a simple thanks for all of the ways he has contributed to this book. My gratitude is great for his editing skill, for his willingness to discuss the biggest ideas and the smallest details, for his patience—his infinite patience!—for his love of ideas, and for his not complaining that our lives were on hold while I finished this book.

Abbreviations

AAAS	American Association for the Advancement of Science
AAG	Alternative Agriculture Group
AASW	American Association of Scientific Workers
ABM	antiballistic missile
AEC	Atomic Energy Commission
AFSC	American Friends Service Committee
AMRC	Army Mathematics Research Center
APS	American Physical Society
ARPA	Advanced Research Projects Agency
BSSRS	British Society for Social Responsibility in Science
CALC	Clergy and Laity Concerned
CBNS	Center for the Biology of Natural Systems
CNI	Greater St. Louis Citizens' Committee for Nuclear Information; later, Committee for Nuclear Information
CPP	Computer People for Peace
CRC	Community Relations Committee
CRG	Council for Responsible Genetics
CSPHW	Committee on Science in the Promotion of Human Welfare
CSRE	Committee for Social Responsibility in Engineering
CUNY	City University of New York
DOD	Department of Defense
ER	Edward Ramberg
FAO	Food and Agricultural Organization of the United Nations
FAS	Federation of Atomic Scientists; later, Federation of American Scientists
FCRC	federal contract research center
FLOC	Farm Labor Organizing Committee
FOIA	Freedom of Information Act
FOR	Fellowship of Reconciliation
HCUA	House Committee on Un-American Activities
IDA	Institute for Defense Analyses
IEEE	Institute of Electrical and Electronics Engineers
IGY	International Geophysical Year

ILGWU	International Ladies Garment Workers Union
LCI	Liquid Crystals Institute
MIRV	multiple independently targetable reentry vehicle
MLA	Modern Language Association
NAS	National Academy of Sciences
NCAI	National Committee on Atomic Information
NDEA	National Defense Education Act
NEPA	National Environmental Policy Act
NSF	National Science Foundation
NUC	New University Conference
NVAANW	Nonviolent Action against Nuclear Weapons
OTB	Otto Theodore Benfey
PHS	Port Huron Statement
PSAC	President's Science Advisory Committee
PT	*Physics Today*
R&D	research and development
SAC	Science Advisory Committee
SACC	Science Action Coordinating Committee
SAG	Scientists' Action Group
SANE	National Committee for a Sane Nuclear Policy
SDS	Students for a Democratic Society
SFN	Science for Nicaragua
SftP	Science for the People
SFV	Science for Vietnam
SIPI	Scientists' Institute for Public Information
SNCC	Student Nonviolent Coordinating Committee
SRI	Stanford Research Institute
SSG	Sociobiology Study Group
SSPA	Scientists for Social and Political Action
SSRS	Society for Social Responsibility in Science
TAP	Technical Assistance Program
UC	University of California
UCS	Union of Concerned Scientists
WILPF	Women's International League for Peace and Freedom
WITCH	Women's International Terrorist Conspiracy from Hell
WU	Washington University

Disrupting Science

CHAPTER 1

Introduction

In 1960, American scientists were *Time* magazine's "Men of the Year," described as superheroes whose powers and social contributions surpassed those of any other group in human history. The "true 20th century adventurers, the real intellectuals of the day," and the "leaders of mankind's greatest inquiries into life itself," scientists were "statesman and savants, builders and even priests" whose work shaped the "life of every human being on the planet."[1] In 1970, after a decade of criticism from environmentalists, antiwar activists, and members of the counterculture, *The Nation* declared that science had become a "war/space machine." As a result, some citizens had grown "hostile to science, identifying it with war, pollution, and every manner of evil."[2] Philip Abelson, the editor of *Science*, decried the growing "war on scientists," caused, he argued, by unrealistic demands for "relevant" scientific research.[3] Once lauded for their contributions to national security, scientists were now under fire for helping to perpetuate warfare. One of the most interesting aspects of the challenges to the relationship between scientists and the military was that these challenges were not simply waged by "outsiders." Scientists themselves filled the ranks of critics, charging their peers, the government, and industry with a failure to make good on the promise of science to improve human life. Although criticism of science and scientists and doubts about the benefits of technology have a long history in America, by 1970 the criticisms of science and of scientists were more vociferous and diverse than ever before.

Although it is tempting to treat scientists' self-criticism as an aberration, the historical record demonstrates quite the opposite. Throughout the twentieth century, American scientists were involved in varied and visible forms of public political action, especially in efforts against racism and war, often working closely with and inspired by activists who were not scientists. *Disrupting Science* examines the development of scientists' activism against the financial and political relationship between scientists and the military between 1945 and 1975. To do so, the book compares three episodes in which scientists formed organizations that articulated different public political roles for themselves and their peers. In the early 1950s, pacifist scientists formed the Society for Social Responsibility in

Science to convince other scientists to renounce all research that might contribute to war; in the late 1950s, scientists and citizens embroiled in the public debate over the wisdom of above-ground atomic testing developed a method of providing the public with information about the health effects of fallout through the formation of the Committee for Nuclear Information; and in the late 1960s, scientists formed Scientists for Social and Political Action (later known as Scientists and Engineers for Social and Political Action), which eventually came to call itself Science for the People. At first, Science for the People used direct action, public education, and other methods to call attention to and to discourage scientists' association with the military, racism, and sexism. Later, they used a variety of methods to put scientific knowledge into the hands of "the people." Each group represented a different vision of the place of science in public life, shaped by new arrangements between science and the state and by social movements of the day.

Scientists' roles in transforming the political meaning and uses of science raise three puzzles that are the central focus of *Disrupting Science*. Why did scientists engage in activism against the relationship between the military and science, the most radical of which undermined their privileged social position and the ideological foundation of their own work? What forms did their actions take, and why did they differ from one another? How did their efforts simultaneously contribute to buttressing the power of science in American political life and transforming it? The scientists who were involved in these debates grappled with the classical question Max Weber posed in "Science as a Vocation": "What is the value of science?"[4] In more specific form, they asked what the proper relationship between science and politics was and ought to be. None came up with the same answer, but all defined ideals and practices that they believed should govern the normative link between science and public politics.[5]

At the heart of this book are the vibrant efforts of scientists to redefine the relationships between fact and value, between politics and science, and between expert and citizen. Although the most active critics were a small minority of all scientists, they were drawn from many ranks, disciplines, and institutions. Some were highly visible members of prestigious universities and government agencies, and others worked in industry. They ranged in rank from graduate students to Nobel Prize winners. Their strategies for linking—and separating—were equally varied, including the disruption of scientific meetings, letter writing and public speaking, the provision of information to the public, and collaborating with like-minded groups of scientists and other activists. Whatever their tactics, scientists were above all engaged in thoughtful and earnest debates over how to best make good on the promise of science to provide the greatest benefit to the largest number of people. These efforts helped make one of the most important changes in the place of science in public life in the twentieth century: the authority of *scientists* to make unchallenged claims about nature and about their relationship to public political life, and to mediate the relationship between scientific knowledge and political values, decreased. At the same time, however, the authority of *scientific knowledge* itself increased. In this chapter, I provide an overview of the central arguments in the book and of the structure and content of the chapters to come.

THE MILITARY, SCIENCE, AND SOCIAL MOVEMENTS, 1945–1975

After World War II and increasingly through the 1960s, the idea that science and scientists were uniform forces for progress came under fire. Although criticism of science and scientists and doubts about the benefits of technology have a long history in America, the criticisms during this period were far more vociferous and diverse than ever before. Many were centered around the relationship between scientists and the military. As we will see, those who criticized science drew force from the social movements of the time. Although early challenges took the relatively genteel form of written and verbal debate, by the late 1960s radical critics of scientists and science were targeting the places where scientists worked and lived. They had gone beyond cool professional discourse and cerebral argument to personally identify, ridicule, and in some cases physically attack individual scientists. Some elite scientists fared the worst: jeered and heckled at meetings and forced to walk gauntlets of protestors in front of their homes and workplaces, scientists who considered their military-sponsored research a patriotic act were accused of being as responsible for the war in Vietnam as the generals who directed it. Other critics lambasted scientists for producing racist and sexist research under the guise of scientific objectivity.

That scientists attempted to reorganize their relationships with the public and the state in the period between 1945 and 1975 was not idiosyncratic. Other professionals have organized themselves to include public problems and concerns within their jurisdiction while still leaving a special set of tasks, skills, and responsibilities for themselves. For example, Kristen Luker showed that American physicians in the mid-nineteenth century removed women from decision making about abortion to establish their own professional jurisdiction.[6] Other professionals have taken on new research subjects as a result of their engagement with public political debates, as Lily Hoffman demonstrated in her study of the mobilization of physicians and urban planners in the 1970s, and Scott Frickel has shown in his study of the formation of the field of environmental toxicology.[7] Still other professions have expanded their jurisdictions to include service to new groups, as Christian Smith's study of the development of liberation theology among Central American Catholic priests and nuns showed.[8] More generally, in the late 1960s, professionals in most Western countries were rethinking their relationships with public political issues and considering how to better use their skills and authority to address immediate social and political concerns.

Before World War II, American scientists were no strangers to political engagement, of course. Engineers in the early twentieth century organized and advocated for "social responsibility" among their peers. In the late 1920s, some scientists who favored teaching evolution over creationism lectured and published to advocate using evolutionary theory and science as a basis for personal and public morality. Many eugenics scientists were active members of a broad movement to "purify" the American "race," working closely with politicians and citizen groups around the nation.[9] In the 1930s, scientists organized groups to fight fascism and racism and to seek ways to use science to end poverty and war.[10]

Yet the mid-twentieth century presents a special case. Some scientists wanted to continue with some of their prewar political activities and, more broadly, to develop a New Deal–style system for science funding that would be based on regional need rather than federal military priorities. Their hopes would not be fulfilled, however, because the promilitary sponsorship of science forces won out in the battles over who would control atomic power production and on what basis federal funding would be delivered.[11] The intensification of scientists' efforts to define the proper relationship among knowledge creation, war, and the public was not simply a philosophic or epistemological dispute, or a matter of intellectual positions. It was a response to the changing material conditions of science and to the political mobilization of Americans from many different political communities and walks of life.

The close association between scientific research and the military began after World War II. Government and military leaders, and some scientists, realized the importance of scientific talent and ideas in maintaining atomic and other forms of military supremacy. As a result, federal funding for scientific research and education swelled dramatically, from fifty million dollars in 1939 to nearly fifteen billion dollars in 1970.[12] Between 1947 and 1960, most federal funding for science came from the Department of Defense. Funding was distributed to a decentralized network of recipients that included universities and federally funded laboratories. New knowledge proliferated and more disciplines and subdisciplines formed, increasing the intellectual diversity of the field of science.

Scientists became important political advisors during the mid-twentieth century, too, providing recommendations on everything from which weapons to build to what students should learn in school.[13] In the 1940s and 1950s, scientists' contributions to defense were often lauded as contributions to democracy. Scientists were thought to provide the know-how to keep the nation safe, and to contribute to an "informed public," which was considered an important feature of a healthy democracy.[14] As Gerard Piel and Dennis Flanagan, the publisher and editor of *Scientific American* in the late 1940s, wrote, without information about science, "modern man has only the haziest idea of how to act in behalf of his own welfare, or that of his own family and community."[15]

Yet lavish funding, access to the highest levels of government, and association with national defense were not uniformly welcomed by all scientists, nor by the public, intellectuals, or political authorities. The new state-science relationship posed threats to scientists' ability to act on their political beliefs, and shifted funding toward a limited range of subjects. Above all, it raised questions about the extent to which science was an independent community and a force for the improvement rather than the destruction of society.

In the 1940s and 1950s, it was difficult for scientists to speak out on these issues. The national security system, which was intensified in 1947, swept up scientists in high-profile and routine investigations. As Jessica Wang has shown, scientists made up more than half of those investigated by the federal government between 1947 and 1954. The security frenzy included extensive surveillance of scientists who were peace activists, including Albert Einstein, and repeated public investigations of leading scientists such as Robert Oppenheimer and Edward U. Condon, whose reputations were damaged despite the failure of loyalty committees to find them guilty of security breaches. Restrictions on travel and security clearances for federal grants and contracts added to the atmosphere of suspicion and fear.[16] As a result, many scientists—but by no means all—were wary of taking explicit political positions that might be construed as contrary to the military goals of the United States or in any way "political." Part of the story I tell in *Disrupting Science* is of a small group of dedicated pacifist scientists who personally refused military funding and who urged their peers to find ways to use science for "productive" ends, even though they were at great risk for asserting their perspective.

By the late 1950s, as repression had eased, scientists began to raise new questions about the politics of science, this time about the extent to which democratic procedures were being subverted by the failure of scientists to provide the public with facts and information sufficient to allow their full participation in political debates about the wisdom of atomic testing. In the late 1960s, radical scientists went beyond calls to reform the behavior of scientists or democratic procedures; they called for the wholesale restructuring of society.

Yet scientists were not the only ones who questioned the new military-science relationship. In the early 1960s, members of Congress began to raise questions about the wisdom of using the majority of federal research funds for military purposes. They called instead for more spending on health and social problems. President Eisenhower's last presidential speech famously warned of the dangers of a "military-industrial complex," and of the dangers to the freedom of university research presented by massive federal funding. Politicians and presidents, however, played a relatively small role in generating the moral outrage that drove scientists to rearticulate their place in American public life. The social movements of the 1950s, 1960s, and 1970s provided some of the pressure and much of the moral argumentation and camaraderie that led to the creation of new means of organizing the relationship between knowledge production and power. The scientists who are featured in this book often considered themselves part of these broader political movements. The intersection of social movements and changes in the organizational, moral, political, and economic organization of science offers a window through which we can observe how scientists created new understandings of the place of science in American public life.

The critiques of science and scientists that scientists and other activists made in the three episodes I examine can best be understood as arguments stemming from two established and dominant American political traditions, liberalism and moral individualism, and one emergent perspective, that of a Marxian-inspired New Left. By political traditions, I do not mean static tools strategically identified and mechanically applied. Traditions are full of currents and countercurrents that people endlessly reconfigure as they creatively integrate them with real political problems. Even as the protagonists in this book drew on political traditions to formulate criticisms of science, they also transformed them in powerful ways. By the late 1960s, scientists' efforts to forge a new relationship with the public and the government were informed by the political analyses of the New Left and by Marxists. Both had roots in earlier American political thought, but compared to moral individualism and liberalism, they were more fertile ground for the development of novel ways of articulating how scientists could use their skills in the service of the public.

The least well known, but earliest, tradition on which critiques of science were based in the thirty years following World War II was moral individualism. In this tradition, transformation of the individual moral conscience is the source of broader social change.[17] Those who drew on this tradition argued that scientists had failed to take personal moral responsibility for the development and use of scientific ideas and products. Unlike liberal scientists and commentators, scientists drawing on this tradition did not turn to the government for solutions to what they saw as the moral corruption of science and scientists through association with the military. These scientists were inspired by religiously based activists and leaders such as Martin Luther King Jr. and the Fellowship of Reconciliation leaders Brad Lyttle and A. J. Muste. Those scientists who espoused this tradition had little confidence in either the government or organized political groups to effect real transformations in the science-military relationship. They believed that the relationship between science and the military could be decoupled only through the personal commitment and choice of individual scientists to refuse military work.

A second tradition from which ideas were drawn about the proper arrangements among science, the military, and citizens was liberalism. Scientists and other activists working in this tradition assumed that an educated and informed citizenry was the major means for making decisions about the proper role of science in public life. From scientists who called for scientific rather than government control of science after World War II to critics of "technocracy," those who argued from a liberal perspective believed that scientists' proper role was to inform the public of facts that citizens could use to rationally decide among alternatives.

These two traditions were the basis for the criticisms of the science-military connection through the middle of the 1960s. In the mid-1960s new voices were added. Marxists and New Left activists and intellectuals became critical of the relationship between capitalism and science, and feminists and antiracists associated science with the domination of women and blacks. College students were especially important in generating activism among scientists. In 1966, they began to gather information about how the facilities and faculties on their campuses contributed to weapons production. Some who were influenced by Marxism argued that science had been captured by the needs of the upper classes and by what they saw as imperialist goals of the United States bent on the material and military domination of its citizens and those of other nations. Many New Left activists, inspired in part by the ideas of Frankfurt School philosopher Herbert Marcuse, were critical of the ways in which capitalism and instrumental rationality left people bereft of the ability to imagine and create.[18]

In practice, few of the scientists whose activities I examine in this book called themselves "moral individualists," "liberals," or "radicals" when they engaged in challenges and defenses of science. In the episodes in which contradictions in the professed and actual relationships between science and politics were variously uprooted and exposed or vehemently defended, activists, intellectuals, and journalists often wove together elements of different traditions and perspectives. Moreover, the volatile intersections of science and politics were not abstract debates, but involved concrete political events. The arms race, the development of nuclear, chemical, and biological weapons, and the war in Vietnam were the substantive issues around which scientists struggled to make good on the promise of science to serve all people.

Scientists were ultimately trying to steer a course between two potentially contradictory positions. On the one hand, many asserted that the public political authority of science was based on the strong distinction between scientific knowledge and practices, and the messy world of values and moral concerns. On the other hand, some claimed that the value of science lay in its affinities with and beneficial effects on aspects of social life, including democracy, moral progress, and the economy. Both of these bases would be fundamentally shaken by the close ties between scientists and the military that characterized the mid-twentieth century and by the social movements that condemned that relationship.

After sixty years of building a professional field that increasingly centralized the power to make uncontested claims about nature in the hands of scientists, the two decades following World War II at first seemed to continue that pattern, given the lavish funding and the centrality of scientific ideas to the military and security projects of the cold war. Yet the cumulative pattern of scientists' political organizing against this new relationship helped to contribute, ironically, to a weakening of their political authority. The organizations they formed linked science to moral and political projects that called for more citizen and scientist involvement in technopolitical issues. In turn, this call led to a weakening of scientists' political authority, but also led to the greater importance of scientific claims and technologies in structuring and adjudicating political debate. Of course, this was not the only source of the disruption of scientific authority. The growth of the regulatory state beginning in the mid-1960s, especially the regulation of research on human and animal subjects, highly visible problems resulting from scientific technologies (thalidomide, atomic fallout, pollution), intellectual critiques of science, and the growing importance of the market also contributed to these shifts. *Disrupting Science* is an effort to demonstrate the role of scientists in contributing to the "unbounding" of scientific authority from scientists and its "binding" to other decision makers through social movement activity on the part of scientists.[19] In the next section, I turn to a theoretical elaboration of some of the main sociological claims of this book.

Although this book considers three key episodes separately, it will show that each group built on or responded to those that came before it, so that over time different visions of how scientists should act in a moral fashion proliferated and contended. This variation in itself helped to undermine the idea that science was a socially or morally unified field built on facts that could be used to constitute political or social life. Although concerned specifically with scientists, my analysis is situated in a longer tradition of analyses of contentious moral politics.[20]

I will seek to explain when and why scientists engaged in work that was simultaneously an attempt to redeem science from the moral pollution of its association with the state and a deliberate effort to encourage the values and moral meanings of other communities to affect scientists' behavior. Such action poses a puzzle for most analyses of scientific authority.

SCIENTIFIC AUTHORITY AND POLITICAL ACTION

The distinctiveness of science as a specific form of social action, and of scientists as a group, is a major theme in Western culture. Historians and sociologists have understood the emergence of science as a distinctive field in terms of scientists' efforts to limit access to the social and material means of producing credible scientific knowledge. Just as other forms of action were differentiated, so, too, did science come to be a distinctive practice.[21] Sociologists of science have also been concerned with the ways that scientists have established authority using material, and especially linguistic, tools to manage the "boundaries" of science. Such activity is normally considered to be *protective* of the areas in which scientists engage.

As Thomas Gieryn argued in his 1983 foundational article on this subject, "boundary work" is the process by which scientists "attribute certain qualities to scientific claims in order to draw a rhetorical boundary between science and some less authoritative, residual non-science."[22] Other writers have followed Gieryn, elaborating on the causes, processes, and consequences of how scientists contest epistemic claims among themselves and others in ways that reestablish the authority of science and scientists.[23] In contrast to earlier structural-functionalist analyses that assumed that scientific activity was autonomous from other forms of social action and that its authority derived from its objectivity and prima facie social value, studies of boundary work demonstrate that the political authority of science depends in part on scientists' active engagement in discursive, organizational, and material projects.

Although the concept "boundary work" is theoretically able to incorporate actions in which scientists engage in activities that are intended to lessen their control over decisions about science, it has not often been a subject of study. One of the assumptions common to studies of boundary work among scientists and other groups is that professional advantage and monopoly drive most decisions by scientists and others. As Gieryn argues in his analysis of "cultural cartography," or the making of cultural boundaries, "it makes little sense to argue that cultural cartographers are indifferent about how epistemic authority is allocated or that they would deliberately prefer tactics designed to lose it."[24] Whether the analysis draws on rational choice theory, Marxism, Bruno Latour's systems-based models of credibility, or Bourdieu's class-based analysis of science, the monopolization of expertise and authority is assumed to trump other motivations.[25]

This assumption is its own kind of boundary work: without allowing scientists to have a broader range of motivations, analysts reproduce the cultural idea of scientists as people who are not "like us." These analysts would grant that scientists may also work toward other goals, but they would not do so if it might cost them cultural credibility or other forms of power. Although this may be true most of the time, there are theoretical and empirical reasons to argue that it is not always the case. Sometimes scientists undertake actions that give others authority, but do not call such actions "political" or explicitly treat them as contentious. They may see them as continuous with routine professional behavior.[26] Scientists, like other people, must be seen as having a range of preferences, from advancing professional interests at all costs to sharing or limiting their own authority. They hold moral and religious beliefs; they are members of organizations outside of science, relatives of people with illnesses, residents of areas with particular kinds of problems, and a host of other identities. Some scientists decry any attempt to bring such identities to bear on scientific issues, of course, but others do not. Even without assuming steady, fixed interests derived from group membership, it is not difficult to imagine that in particular contexts, some scientists may be willing to act on political issues, even if it costs them, or their profession, social prestige. Thus, the flat, one-dimensional caricature of the scientist in the laboratory concerned only with professional advancement is as misleading as the image of the scientist seeking "truth" above all else. Simply put, like other people, scientists have multiple interests, all of which they may enact, or challenge and transform, depending on political, economic, social, and intellectual circumstances. The ways in which scientists experience and engage these circumstances, moreover, is not as individuals, but as members of overlapping organizations and networks to which they belong.[27]

In each of the cases of organizational formation I examine, founders were in organizational settings, such as universities, professional associations, political organizations, and religious groups, in which critical examination of professional roles was taking place. In such contexts, it is possible to see how scientists could conceivably limit their own professional power. Scientists' decisions about how to engage in political debates of the day were thus shaped not by a transhistorical idea of who they were, but by a dynamic, shifting political environment.

One of the results of these explorations of new science-politics relationships is that, at present, scientists act in a surprising variety of ways to give citizens the power to make better decisions about sociotechnical issues. These include the provision of information, and an understanding of how scientists work, to the public or to specialized groups such as patients or people living in contaminated areas. Consensus conferences, the use of amateurs to carry out research projects, "town meetings" on scientific issues, "participatory design" programs, scientist-citizen collaborations on health and environmental problems, and service on public information and communication committees in professional associations are other ways that scientists engage with the public, over and above the work they do in public interest groups. Although few, if any, scientists hope to give up all the benefits that expert knowledge can confer in debates over sociotechnical issues, there is clearly a range of preferences among scientists, shaped by historical events.[28]

Scientists who organized themselves into groups that took on issues that bridged the "purely professional" organizations and "purely political" organizations during the mid-twentieth century did so in the context of political and religious communities that were engaged in self- and group criticism that encouraged sacrifice for a greater good. In the 1940s and 1950s, pacifists in the United States, organized in the civil rights and the antinuclear movements, emphasized the need to take risks that might, in the short run, lead to personal or group harm but would ultimately lead to a greater social good. This same idea was adopted by New Left groups as well. In the late 1960s and the 1970s, political groups on the left, often inspired by the "new Marxism," especially Maoism, were critical of their own interests and agendas, and sought ways to bring them more in line with the interests of less powerful groups. In such contexts, limiting one's own or one's group's power is understandable.[29]

It need not be the case that willingness to give up professional power comes from immersion in political groups that hold values distant from mainstream politics. Scientists with more mainstream political views may also aspire to provide other groups (patients, the public, or people living in contaminated areas, for example) with access to scientific information and an understanding of how scientists do their work. These scientists use many of the methods noted above, such as establishing collaborative relationships with nonscientists or participating in public information sessions. They also offer individual-level testimony and assistance to advocacy groups. All of these are ways in which scientists have helped and continue to help empower nonscientists, and all bring to mind issues about the balance of authority between science and scientists.

Understanding the intersections of public political debates and the actions of scientists is part of the "new political sociology of science." This body of work is institutional in its approach and organized around questions of how interactions among scientists and other groups shape not only knowledge itself, but also the rules and resources that scientists and others have for creating such knowledge. Allowing for a wide range of situated motivations, it emphasizes the ways that existing formal and cultural rules, including those of states, economic relationships, and organizational forms, matter for understanding the power of science and scientists.[30]

The State as Catalyst and Constraint

The creation of and forms of action embodied in new kinds of scientist-based organizations in the mid-twentieth century were shaped by the state and by social movements. The state is examined here as a differentiated system, whose power is exercised in material, bureaucratic, and cultural forms.[31] Three key features of the American state are centrally important in this analysis. The first is its capacity for developing knowledge, equipment, and personnel for the creation of weapons and a war-ready military. This capacity developed rapidly after World War II. Scientists were not in a simple sense "captured" by the state; many scientists, especially physical scientists, were actively and willingly engaged in science advising and the development of weapons. As Patrick Carroll has demonstrated, scientists and engineers have long provided the material basis of state making, including maps, roads, bridges, public health systems, and policing techniques.[32] Yet some scientists in the mid-twentieth century were deeply unhappy about the entire relationship between science and the military, or some specific aspect of it. For some, it was the lack of autonomy scientists had in choosing projects; for others, it was secrecy constraints and the lack of public information; and for still others, it was the perception that science was now associated with destructive rather than constructive research. All asked whether the military-science relationship was serving broad social and scientific goals or narrow military ones.

In addition to generating dissent among those unhappy with the military-science relationship, state surveillance and repression shaped not only the extent to which scientists engaged in political action but, equally important, the content of their action. Historians of science and scholars of social movements have demonstrated the effects of surveillance and the restriction of civil liberties on the capacity of groups to dissent from the state.[33] Direct repression played a critical role in generating some of the opposition to the closer relationship between science and the military, between 1947 and 1955 in particular. But just as important, the *fear* of red-baiting in the 1950s and early 1960s shaped scientists' claims that their actions to reform the military-science relationship were "apolitical." As Christian Davenport, Gilda Zwerman, and Patricia Steinhoff have demonstrated, political repression may have little dampening effect on extremist groups and make them more committed to their cause.[34] In the analysis put forth in this book, repression played all three roles: it dampened the amount of dissent, shaped the "apoliticism" of scientists' actions, and spurred other groups to intensify their commitment.

The variation in forms that scientists' actions took were also shaped by the state. The decentralized system of scientific work and training in the United States meant that scientists in the mid-twentieth century were working and living in settings ranging from heavily policed federal research laboratories to research universities, to small liberal arts colleges, and to industry. Disciplines proliferated during the mid-twentieth century, and so too did the organizations to support them. This system was established after World War II, after a long series of debates about whether and how the federal government would support scientific research. Funding for research, while heavily focused on the military and a small number of research institutions, was never entirely centralized. This variation helped to create different communities of scientists who were exposed to and helped to create different visions of the relationship between scientists and citizens.

Social Movements

Scientists' ties to politically active groups of nonscientists were also critical in explaining scientists' desire to reform the science-military-public relationship, and the claims and actions that they created. The mid-twentieth century was one of the most active periods of social and political protest in American history. Peace, environmental, gay liberation, civil rights, feminist, and "power" movements—as well as their numerous branches—flourished. The critiques of American society that they posed went beyond economic demands or calls for personal moral reform. By 1970, there were few aspects of American social life, from its schools to its supposedly shared values, personal relationships, to the environment, that were untouched by political critique. What was perhaps unique in this period in American history was the totality of the targets of activism. Scientists were no different from other groups in organizing to transform different aspects of social life.

Scientists, like other contemporaneous political actors in the mid-twentieth century, took aim at a variety of targets. Some focused on the state, others on the hearts and minds of individuals or the reform of personal relationships, cultural values, or professions. The scientists who were collectively responding to the state-military relationship directed their attention to the public, to their peers, and to larger systems of scientific production that included the state and economic, race, and gender systems.

To understand the aspects of science that scientists thought were properly subject to moral and political scrutiny, it is necessary to attend to the process of organizing. Most of the analysis of social movement organizations has drawn attention to the important issues of how field-level resources are defined, competed for, and mobilized, and how broader political contexts shape the possibilities for the development of movements.[35] A new body of scholarship, some based on symbolic interaction theory and the sociology of religion, has focused attention on the interpretive activities of social movement organizations to understand how groups come to articulate grievances, create new forms of action, and project their futures.[36] By interpretive activities, I mean the processes by which groups come to identify who they are, what their interests are, and what kinds of actions and claims are viable. As Elisabeth Clemens has argued, the question of "how shall we organize" proves to be consequential for the development of actors as activists and the prospects for further organized collective action.[37] Organizing is the process of adapting traditions and cultures in ways that provide activists with ways of thinking about problems, notions of enemies and targets, probable strategies, and ideas of what the future might look like.[38] In the organizing process, groups come to *identify* interests and preferences as much as they *express* preexisting stable preferences. Opponents and potential allies are critical in shaping "who we are" in any organization, for their actual and expected responses shape what kinds of images and ideas are viable. The discussions that take place can be called "rim talk," in the words of Erving Goffman. "Rims" are the imagined edges of what kinds of things are possible.[39] Groups that are inventing new forms of action cannot know what might work ahead of time, so they engage in the work of identifying what might satisfy group members, allies, and targets. As James Jasper argues, "much of what protestors do can be understood as experiments aimed at working out new ways of living and feeling."[40] By examining these processes, it is possible to observe the creativity that is at the heart of vibrant democracies.

Organizational disruption and rebuilding is one important way that social movements undermine the economic and symbolic reproduction of their targets. As analysts of institutions have demonstrated, organizations are loci through which formal and informal rules, communication systems, images, material things, and social relationships are reproduced.[41] Although a law may have an effect because it is known to exist, its social durability is powerfully shaped by the collective, organized actions that are taken in response to it. To challenge such systems helps to symbolically disrupt the cultural order and values that underlie all organizations. The relationships among people, things, and ideas represent particular values, which are routinely reproduced through organizational action. When such routine work is disrupted, so, too, are the values and assumptions that undergird it. Because social movements are treated in this book as attempts to deconstruct and reconstruct an organized moral order as much as to redistribute resources, the disruption and rebuilding of organizations is a central concern here.

Most analysts of social movements are ultimately interested in the question of whether the movement made a difference in some way, often as examined through changes in laws, rules, and policies. Such a question is appropriate for movements that have such actions as their goal. But other movements, as noted earlier, are interested in transforming the actions of individuals, providing new norms, or changing institutions outside states or formal policy systems. The attention to policy alone, moreover, fails to account for the ways in which earlier "failures" are used as models and antimodels by later activists.

Scientists' collective attempts to respond to the military-science relationship have helped to culturally "un-bound" the making of moral claims about science from scientists, and "re-bound" it into networks of citizens, intellectuals, and government. During the period of scientists' professionalization in the United States between 1880 and the early 1930s, scientists consolidated their authority over claims about the natural world by building a professional field that allowed them considerable control over how they treated their subjects, and to claim approbation for their contributions to economic, social, and political life. In the twenty-five years following World War II, these claims to authority became "unbounded" in the sense that scientists consistently lost full control over their research subjects, and lost the capacity to claim that their work ultimately contributed to progress. Rather than seeing social movement in terms of beginnings and endings, here I examine the ways in which groups that were focused on a similar substantive concern cumulatively questioned ever-broader aspects of the scientific endeavor, resulting in a crisis of authority that has helped to shift the authority of scientists in political life.

Scientists made public debates that had previously occurred among scientists in private. The Committee for Nuclear Information, Science for the People, and the Society for Social Responsibility in Science aired their concerns about the moral and political status of science and scientists, with later organizations criticizing more aspects of science. Their actions gave nonscientists access to the political conflicts and debates over what constituted knowledge and how best to justify scientific claims and the credibility of those who made them. Access was sometimes formalized, as in the case of information provision, and in other cases it was made available through continued media exposure. Since scientists' own political activities were coupled with new rules about research on human and animal subjects, with a growing regulatory state that monitored the problems caused by industrial production and medicine, with feminist and community health movements that were skeptical of medical science's claims about the poor, blacks, and women, and with intellectual critiques by writers such as Herbert Marcuse and C. Wright Mills that condemned instrumental rationality itself, scientists' activities cannot, in some simple sense, be thought to "cause" the unbinding of moral claims-making about science from scientists. I aim, in this book, to tell a piece of this larger story.[42]

THE CASES

This preliminary introduction to the episodes of organization creation that are the central empirical focus of *Disrupting Science* introduces the key themes and protagonists in each. As scientists struggled to reconcile the promise of science to improve life with science's relationship to politics, they often began with goals that were quite vague. Scientists wanted to create something new, but were not entirely sure what new organizations might look like, nor were they entirely certain of the practices in which they would engage or the claims that they would make. The organizations I study in this book were created as means of exploring, and in some cases institutionalizing, ideas about the proper relationship between science and politics.

The first organized expression of scientists' opposition to an alliance between science and the military that had developed during World War II came shortly after the end of the war, and is not one of my focused cases here. Physicists and engineers who were participants in the Manhattan Project groups at independent laboratories and at universities in the United States created new organizations at and across their work sites. These organizations engaged in lobbying and public education to oppose the proposed military control of atomic energy and the rapidly developing arms race. Some organizers were motivated by an intense desire to ensure that scientists could freely exchange ideas that were necessary to develop their research. Others were motivated by humanitarian concerns about future use of atomic weapons. Underlying their claims and actions was a belief that scientists formed an international group whose devotion to truth and universalism could and should be the basis of international political relationships. Among them were scientists who had already been involved in antifascist, antiracist, and New Deal political campaigns in the 1930s. By 1947, national security programs and laws were used against scientists, quickly discouraging their political activities. Within two years, the organizations they had formed were either defunct or had become nominally apolitical groups devoted to promoting science and avoiding hints of partisanship for fear of investigation. This story has been told many times in a number of excellent accounts.[43]

Disrupting Science begins just as this story leaves off. My analysis begins with an examination of the 1949 formation and political views of the Society for Social Responsibility in Science (SSRS). Urged by the pacifist leader of the Fellowship of Reconciliation A. J. Muste and organized by the Columbia University engineer Victor Paschkis, scientists and engineers in the SSRS, some Quakers, refused to participate in professional work that they considered "destructive." The group urged scientists to consider themselves to be like any other individual faced with a moral decision about how to live one's life, and not to consider themselves unique because they were scientists. Of course, the SSRS was in fact a scientist group. To the extent that there *was* something special about scientists, they argued, it was that the new role of the military in fostering scientific research meant that the products of scientists' work were likely, they believed, to be used for destructive purposes. The SSRS rarely undertook collective actions in the conventional sense that social movement organizations or interest groups do. Instead, the group served as a source of support and a forum for discussion of the personal responsibilities that scientists had for their work. Individual members, by sharing their views and decisions with others, hoped to spark other scientists to undertake similar consideration of the consequences of their work.

In contrast, the second organization that I study, the Greater St. Louis Citizens' Committee for Nuclear Information (CNI), held that scientists formed a special community with a unique moral commitment and role to play in a democratic society. Formed in 1958 by scientists from Washington University and members of St. Louis religious, women's, and labor groups who were involved in the Ban the Bomb movement, the scientists of CNI prevailed on other members to focus the group around the provision of information to the public about the health effects of atomic fallout. The group's decision to serve as a conduit for factual information but not to advocate any specific position emerged from the political conditions under which the group was formed. For one, espousing any political position, aside from full support for the government's programs and policies, was politically risky for American scientists during the 1950s because they were heavily targeted by security investigators. Earlier attempts by founders to engage other scientists in taking stands on political issues of the day had failed; few could agree on a position, and others were worried about whether their views would be taken as evidence of disloyalty. Another reason for the turn toward the "information-provision" model was the controversy that raged among scientists about what, exactly, the health effects of testing were, and what kinds of decisions should be made based on the facts that *were* known. The public, founders of CNI thought, was largely cut out of this decision-making process. For democracy to function in its fullest form, the public would need information unfettered by the partisan interests of government and other scientists, so that ordinary citizens, too, could make well-informed political decisions.

The third major episode of organization formation, in 1968–1969, was characterized by far more dramatic actions by scientist and nonscientist activists, the involvement of scientists as targets of activism, and by the formation of "radical science" groups. Students and faculty, increasingly angry about the war in Vietnam and their own universities' involvement in weapons production, demanded that universities and individual faculty members sever financial and advisory ties with the military. Many participants were already deeply involved in the anti–Vietnam War movement, and as such, they drew on identities as both scientists and antiwar activists in their actions and claims. Activists made demands using a range of methods, from teach-ins to picketing buildings and faculty homes.

Like other professionals during this period, some scientists insisted, with widely varying success, that their professional associations oppose the war in Vietnam. Other science faculty formed organizations devoted to "social responsibility" through information provision and advocacy that did not fundamentally question the meaning of science or the benefits of science-military ties. The Union of Concerned Scientists, a faculty-based group formed at MIT in 1969, for example, revived the older information-provision and advocacy models that were pioneered by CNI.

The intersection of campus-based activism and professional commitments to science also led to the development of radical science groups, including Computer People for Peace, the Committee for Social Responsibility in Engineering, and the student-based Science Action Coordinating Committee. But by far the most important radical group to emerge during this period was the national, decentralized, and nonhierarchical group Scientists and Engineers for Social and Political Action, known more commonly as Science for the People (SftP). The group demanded that science be used to benefit the poor, women, and other disenfranchised groups, rather than for militarism and the benefit of wealthier classes. Rather than identifying themselves primarily as scientists, many members saw themselves as scientists who were also involved in radical political action around the war in Vietnam and, to some extent, the black power and women's movements. They rejected the liberal model of information provision that CNI had established, and wanted scientists to go beyond general expressions of responsibility that the SSRS had pioneered, even as some of their peers were rediscovering and building on those models. Instead, SftP wanted scientists to work on projects to help ordinary people, especially the poor, blacks, and women, who had suffered, members argued, at the hands of scientists.

SftP's challenge to how science was used to benefit the powerful might have been marginalized had the group not used dramatic means to convey their ideas early in the group's history. SftP disrupted many of the public rituals that had traditionally provided science with public demonstrations of unity around shared rules for social action, such as professional meetings and awards. They disrupted the meetings of professional associations by demanding that the speakers discuss the political implications of their work, and by using political theater to dramatize their point of view. They refused professional awards, and sometimes used the receipt of awards as occasions to forcefully argue that science had been corrupted by its associations with capitalism and militarism, and helped to perpetuate gender and race inequality. Meetings and honors are processes by which boundaries of social groups are reaffirmed. By undermining these processes, SftP members catapulted themselves into the midst of a growing debate over the proper role of science in public life.

If disruptions of ritual were one of the ways that SftP called into question the political relationships among scientists, the state, and citizens, SftP chapter activities explored ways of reorganizing these same relationships. These included carrying out public debates with sociobiologists, working with the Black Panther Party, sending scientific instruments and books to Vietnamese scientists, and assisting Central American farmers in designing new systems for growing crops.

ORGANIZATION OF THE BOOK

The remainder of *Disrupting Science* is structured around building an argument that shows how states and social movements shaped the timing and form of scientists' postwar antimilitary activity, demonstrating connections among the three main organizations that I have described. As a historical narrative, the book is structured around explaining how and why radical science emerged in the 1960s as a reaction against earlier attempts by liberals and moral individualists to end the relationship between science and military. It is not a history of each organization, but an analysis of how the intersections of scientists, social movements, and the state produced different forms of scientists' political action. The book moves from the efforts of scientists in the 1940s and 1950s to have their peers take responsibility for avoiding harmful work, to the development of the "public interest" liberal information-provision organization, and then to the emergence of radical science groups. Chapter 2 examines changes in the science-state regime between 1955 and 1970, and presents an overview of the major arguments that activists and intellectuals made in their critiques of science. The first part of the chapter examines the growing interdependence of science and the state, and the simultaneous increasing organizational, intellectual, and demographic diversity in the structure of science, between 1945 and 1970. The second part examines how liberal and moral individualist political traditions shaped science criticism.

Chapters 3, 4, 5, and 6 examine the three historical episodes in which the intersection of science and activism produced new organizations. Rather than treat them as radically distinct, I show the relationships among them as scientists understood them and at a more theoretical level. Chapter 3 studies the origins and development of the moral individualist group Society for Social Responsibility in Science, founded in 1949, in the context of a renewed pacifist movement and the dramatic increase in the surveillance and political control of scientists in the early cold war era. I show that the group was never able to move beyond its structure as a forum for discussion, and became increasingly less visible as other scientist groups took bolder steps to challenge the science-military relationship. In chapter 4, I offer an analysis of the development of the Greater St. Louis Citizens' Committee for Nuclear Information, founded in 1958, in the context of the Ban the Bomb movement and the lessening of political repression in the late 1950s. The emergence of the group was dependent on increased public and scientific concern over the health effects of atomic fallout.

In chapter 5, I examine a more contentious episode of scientists' public engagement with issues concerning the military, which took place in 1968 and 1969. The war in Vietnam drew scientists into activism, both as targets and as active participants in social movements. I chronicle the role of two MIT groups—the student-based Science Action Coordinating Committee (SACC), founded in 1968, and the faculty group Union of Concerned Scientists, founded in 1969—in mobilizing American scientists to engage in debates over the role of science in the war in Vietnam. I show how SACC's style of organizing and political arguments began to undermine the liberal model of science activism that CNI had helped to popularize. The second part of this chapter chronicles the development of radical science activism through an analysis of the formation of SftP.

Chapter 6 traces the activities of three local chapters of SftP. Here I show how the multiple intellectual and political commitments of SftP members, and their diverse locations, shaped the development of many different versions of "science for the people."

The final chapter of the book summarizes the main conclusions of this analysis and situates them in the debates over how and when scientists engage in public political action and the role of social movements in shaping scientists' actions. The book concludes with an elaboration of how scientists' mid-twentieth-century political activities "unbound" scientists' authority over public claims about the veracity of their work and the contributions of science to a broader public good.

CHAPTER 2

The Expansion and Critiques of Science-Military Ties, 1945–1970

In the decade and a half after World War II, the profession of science appeared to be at the height of its power. As a result of scientists' proven contributions to weapons production in the war, funding from industry and government reached levels unheard of for any profession in the history of the United States. Much of this funding was from the Department of Defense (DOD), and the major beneficiaries of this arrangement were physicists. An elite group of scientists, many of them physicists, had entered the halls of government after the war. They provided the federal government with scientific advice on subjects ranging from civil defense and armaments to vaccines. Government funding generated rapid development of scientific know-how, with applications that ranged from weapons systems to home appliances to pharmaceuticals. Intellectual developments were so rapid and innovative that new scientific disciplines sprang up almost overnight and were quickly populated by a growing phalanx of newly minted scientists. Many Americans admired and trusted the profession. Although scientists had long been lauded by Americans, their new importance to the nation's political, economic, and military life was unprecedented. It would be hard to imagine a moment of greater glory for the profession.

Yet the political, financial, and organizational arrangements that gave scientists new power and authority were filled with contradictions. They and their intellectual and journalist supporters saw scientists as an important bulwark against communism, generating economic benefits and modeling a way of thinking and working that was considered quintessentially democratic. Yet citizens were not encouraged to critically examine science in any way, but merely to "appreciate." Moreover, aside from voting for federal officials, citizens had little formal access to decision making about political issues that had scientific significance. Scientists contributed to national security, but by the late 1950s, citizens were beginning to raise questions about the wisdom and safety of atomic testing and civil defense and to protest them. "The bomb" continued to make Americans uneasy, even as many accepted the necessity of nuclear weapons. Scientists' capacity to respond to these concerns with a unified voice was compromised by growing demographic and intellectual diversity among scientists. They were working in diverse organizational settings and disciplines, and their ranks contained many young scientists. This diversity was a boon to innovation, but it also helped to shape a diversity of opinions about the proper relationship between scientists and public life in America.

In the early 1960s, members of Congress began to raise questions about the wisdom of spending so much federal money on military needs. Many members of Congress wished to redirect monies toward pressing social and economic issues of the day. The political access that scientists had to the Truman and Eisenhower administrations through committees and formal positions faded under presidents Johnson and Nixon. The new expectations for more socially "relevant" research worried scientists, and expensive new weapons systems were proposed and funded against their advice. To make matters worse, federal research funding began to level off by the end of the decade.

Not only was there unrest about the military-science relationship among the public and in government, but also groups of activists and intellectuals argued that scientists wielded too much political power. Others went beyond concern with elitism to charge that science had been misused and needed to be applied to nonmilitary purposes. By the mid-1960s, instrumental rationality itself was blamed for domination in politics and everyday life. Running across these critiques was a series of challenges to scientists: What was the relationship between scientific ideas and moral and political claims? Who was responsible for the negative consequences of scientific ideas and technologies? Should scientists have a special role in a democracy? Or was expertise antithetical to democracy? By the end of the 1960s, the questions became even more daunting: Was scientific objectivity even possible or desirable? Was military-sponsored scientific research corrupting universities?

Scientists were participants in debates over these questions, sometimes responding defensively to the challenges of other groups, and sometimes raising the questions themselves. To understand the questions that were asked and the responses that scientists came up with, it is helpful to situate these debates about the proper place of science in American life in the context of three American political frameworks: moral individualism, liberalism, and the New Left. These are not meant to be strict labels that shaped how people thought about what was wrong with the relationship between scientists and the military. These traditions provided the framework for the more concrete and specific debates that led to the formation of the organizations that I examine in chapters 3, 4, and 5.

This chapter is organized into two main sections. The first provides an overview of the science-military relationship between 1945 and 1970. The contours of this relationship will be familiar to many, but here I want to emphasize some of the main features of these close relationships between scientists and the military, and the links between scientists, the military, and universities, for these were relationships that critics took up. The second part of the chapter introduces the main charges that activists and intellectuals made about the science-military relationships. Few of the writers and speakers cited would have identified themselves as "members" of the New Left, liberals, or moral individualists. I use these labels to organize the logical basis and assumptions of the claims rather than to place individuals into political categories. Taken together, these two sections provide a broad overview of the contexts in which scientists' political organizing took place after the establishment of the close military-science tie in the late 1940s.

A BARGAIN WITH THE DEVIL? FINANCIAL DEPENDENCE AND POLITICAL ADVICE, 1940–1963

The Establishment of a Military-Based Federal Research and Development System

In 1940, Vannevar Bush, the entrepreneurial former director of the Carnegie Corporation and the former dean of MIT's Engineering School, was appointed as the director of the newly formed National Research Defense Committee, charged with organizing research and development (R&D) for defense purposes. In 1941, that committee was replaced with the Office for Scientific Research and Development, which oversaw most of the important weapons R&D during World War II. The most important military-scientist partnership of the war, however, was the development of the atomic bomb under the auspices of the new Manhattan District of the Army Engineers, commonly known as the Manhattan Project. Thousands of scientists carried out weapons-related research at universities and other research sites around the country. During the war, the federal government spent more than five hundred million dollars in research monies, most of it for military purposes, at universities and other research institutes.[1]

The scope of this mobilization of scientists in the service of state building was unprecedented. Scientists and engineers have long played important roles in state building, helping to constitute who counts as a citizen, to organize public health projects, to identify borders and resources, and to develop systems of surveillance and defense.[2] Yet the number of scientists involved—upwards of five thousand—and the engagement of so many university-based scientists in research were new. The results of the Manhattan Project and other federally funded wartime R&D impressed Bush and other political leaders, who saw a continuation of federally directed research as essential to a safe and strong nation.

Between 1945 and 1947, scientists, legislators, industry leaders, and bureaucrats hotly debated what role, if any, the federal government ought to have in funding R&D. At the center of the debate were Bush and the West Virginia senator Harley Kilgore, a New Deal Democrat. Kilgore pushed for a federal funding program that would award funds based on their potential social and economic benefits, and would distribute funds as needed across different regions of the country. In his vision, a federally appointed committee of labor representatives, business leaders, and scientists would control the research. Bush, by contrast, hoped for an agency that would be controlled by scientists and would be insulated from political influence. Much less concerned with broad social and economic needs, Bush's vision would continue the military-based R&D funding that developed during the war. Alongside the conflicting views of these two political leaders, debate raged among scientists, legislators, and members of the military and business communities over whether scientists or government should make decisions about the distribution of research funds; how research monies would be allocated to federal agencies; which regions of the country, industries, colleges, universities, and disciplines would be favored; and to what degree R&D was the business of government at all.[3]

Scientists played an important role in these debates. Between 1945 and 1947, scientists, organized as the Federation of American Scientists and the National Committee on Atomic Information, lobbied Congress and carried out public education campaigns in favor of civilian and scientist control over postwar research policy and international control of atomic energy research. Some scientists went beyond a focus on the United States: they played an active role in the world government movement, an attempt to end nation-based rivalries that might, in the future, lead to full-scale nuclear war.[4]

In the end, the federal system of R&D established between 1946 and 1950 was a far cry from Kilgore's vision and more in line with Bush's ideals.[5] The new system was composed of a variety of agencies, some of which were controlled by the military, some by civilians, and some jointly administered. Short-term contracts and grants were the major mechanisms for disbursing funds to universities, federal laboratories, and collaborative ventures that involved private firms. Atomic energy research was not organized internationally, as scientists had hoped, but rather through the United States Atomic Energy Commission (AEC), which at least, from the point of view of scientists, was a civilian agency.[6] One of the most striking features of this system was that the vast majority of R&D funds were disbursed by the DOD.

Between 1950 and 1963, the amount of money available from the DOD dwarfed the funds available from all other federal agencies combined. In 1958, for example, the DOD received 74 percent of federal R&D funds; the National Science Foundation (NSF), established in 1950 for the purpose of funding basic research, received a paltry 1.5 percent; the AEC received 17 percent; the National Institutes of Health received 2.5 percent; the Department of Agriculture received 1.8 percent; and the remainder was spread among all other agencies. Although the DOD funded both so-called basic and applied research, military needs clearly were given priority in the funding system. Although there were slight variations in the distributions of funds between 1950 and 1963, no agency came close to receiving the same levels of funding as the DOD. This distribution pattern remained relatively steady, but the sheer amount of money available increased dramatically. The Korean War in particular confirmed to many scientists, politicians, and military leaders that science and engineering could be put to great use in wartime, and as a result, R&D monies tripled, from a little over one billion to more than three billion dollars, between 1953 and 1956.[7]

That military needs were at the heart of the new funding system could not be misunderstood. In the words of the AEC historians J. Stefan Dupré and Sanford A. Lakoff, writing in 1962, "urgent requirements of military security are almost wholly responsible for the staggering amount of government-sponsored research ... [and] the contemporary competition with the Soviet Union continues to demand the accelerating development of new weapons."[8] A 1986 House of Representatives report on U.S. science policy drew a similar conclusion. During the 1950s, even so-called basic research funding at the National Science Foundation, the report stated, was linked to military needs. The arms race, the report concluded, became an important element in both the Cold War and science policy. "National security," the authors noted, "remained the principal rationale for the support of basic research throughout the period [between 1940 and 1985] even for the NSF."[9]

No discipline benefited from DOD funding more than physics. Between 1953 and 1959, the amount of money physicists received dwarfed that distributed to other fields. Physicists received almost ten dollars for every one dollar all other fields combined received from the DOD. The interdependence of physics and the military was clearly a boon to the discipline and the DOD. The funding system during this early period was heavily geared toward military needs; physical scientists, with their capacities to build guidance and radar systems and to apply subatomic physics to develop nuclear weapons, and with their host of other skills, were indispensable in building up the nation's weapons supply. Physicists were assured of lavish funding for fundamental and applied military research, while the military benefited not only from the tools that physicists helped build, but also from a pool of talent that was on tap for future research projects. By contrast, biology, the discipline whose star was beginning to rise after the discovery of the mechanism by which DNA replicated itself in 1954, received very little funding from the DOD. Most federal funding for biology R&D between 1953 and 1959 came from the Department of Health, Education, and Welfare, and more specifically, from the National Institutes of Health.[10] Yet here, again, the amount of money available for the biological scientists was many orders of magnitude smaller than it was for physicists.

Linking Universities, Scientists, and the Military

University-based scientists were among the major beneficiaries of the new funding system. In 1953, 54 percent of all research funding for colleges and universities came from the federal government; by 1965, it had risen to 73 percent.[11] No model of populism, the federal funding system before 1964 was organized around the distribution of competitive grants to individual scientists at large, established universities, such as the University of Chicago, Columbia University, Cal Tech, Princeton University, Johns Hopkins, and the University of California–Berkeley. MIT and Stanford were among the universities that built their postwar research programs around federal, but especially DOD, funds.[12]

Another way in which universities were connected to the military was through federal contract research centers (FCRCs). Modeled on the Manhattan Project, FCRCs were nonprofit laboratories where firms and universities carried out research. FCRCs were funded through the same competitive contract system used to distribute funds directly to firms and universities. By 1961, the government operated sixty-six such centers, including MIT's Lincoln Laboratories, where major research on radar took place.[13] In 1965, the FCRCs receiving the largest amounts of funding were the University of California–Berkeley's Lawrence Livermore National Laboratory and Los Alamos Scientific Laboratory, the University of Chicago's Argonne National Laboratory, MIT's Lincoln Laboratories, the Stanford Linear Accelerator, and Brookhaven National Laboratory.[14] Most FCRCs received the majority of their funds from the DOD and the AEC, the latter of which had both a military and a civilian mission. FCRCs on or near university campuses were officially separate from universities, but in practice FCRC monies and facilities were critical to university research and to graduate student training, especially in physics and engineering. This blurry relationship would come under fire from critics in the mid-1960s.

The intensity of the science-military-university tie grew after 1957, when the Soviets launched Sputnik, the world's first orbiting space satellite. Sputnik sparked a dramatic reassessment of U.S. science policy. The rocket engineer Wernher von Braun commented that Sputnik had "tripped a period of self-appraisal rarely equaled in modern times. Overnight, it became popular to question the bulwarks of our society; our public educational system, our industrial strength, international policy, defense strategy and forces, the capability of our science and technology. Even the moral fiber of our people came under searching examination."[15] American scientists had long believed that the Soviet Union did not have the capacity to develop advanced projects on its own, but relied on scientific ideas that were developed in the United States and elsewhere. That could no longer be assumed, as *Science* editor Joseph Turner wrote: "We have now had a rude awakening in a completely convincing demonstration of the excellence of Russian science and technology. One piece of evidence was up in the sky on 4 October 1957 for all to see and hear."[16]

When the Soviets launched another satellite a month later, this time with a live dog on board, it was evident that the Soviet Union had the capacity to land a person on the moon and to create intercontinental missiles. Adding to U.S. embarrassment and panic was the fact that the Sputnik crisis took place toward the end of the International Geophysical Year (IGY), an international program sponsored by the U.S. government in collaboration with sixty-seven other governments, the World Health Organization, NATO, the Organization of American States, and labor organizations, among others, to promote general knowledge of geology, oceans, and the atmosphere, with specific programs for Arctic and Antarctic studies.[17] The IGY was supposed to both highlight U.S. advances, many of which were funded through military sources, and provide an ideological image of a unified world, at least through its common dependence on and use of natural resources. The Soviets demonstrated that however collaborative they were on IGY projects, they were pursuing their own aeronautics agenda.[18]

One of the changes that resulted from Sputnik was the creation of new agencies and programs to increase the sophistication of weapons, and the speed at which they were produced. One of the most important new federal agencies was the Advanced Research Projects Agency (ARPA). ARPA was formed in January 1958 as a subunit of the DOD.[19] This shift to direct control of military research by the president underscores the importance accorded weapons research at this historical moment; it also made it more difficult for other government groups, such as Congress, to monitor the ARPA budget and programs. ARPA funding immediately skyrocketed, and it became one of the most important sources of federal R&D funding between 1959 and 1963.[20] Among the universities receiving the largest amounts of ARPA funds during this period were most of the same universities that received funds for FCRCs and research grants to individual universities. After 1965, as a result of congressional complaints about the inequality in the geographic distribution of federal research funds, some large public universities became recipients of DOD funding, including the State University of New York–Buffalo, the University of Wisconsin–Madison, the University of Arizona, the University of Michigan, and Pennsylvania State University.[21]

One of the FCRCs that ARPA funded was the Institute for Defense Analyses (IDA). IDA was formed in 1958 in response to Sputnik. A nonprofit organization, its sole purpose was to provide scientific advice and information to the Joint Chiefs of Staff and the DOD.[22] One of the projects that the IDA established was called Jason. Based on the model of summer study research groups that the federal government had been using to gather advice on military issues from scientists, Jason funded individual physicists across different universities to meet in the summer to propose solutions to problems of weapons development. The idea was to gather together "a freestanding, ongoing group of smart young physicists . . . to spend summers together working on defense problems."[23] Jason never had large numbers of physicists working on military-based problems at any given time; it usually had only thirty or so scientists on the payroll. But through Jason, physicists were now ensconced in a program whose sole purpose was to expedite weapons development by working outside the more cumbersome bureaucratic systems within the DOD. Because Jason physicists held academic appointments during the year, they were now an important link between universities, the military, and science.

Sputnik also spurred the creation of the 1958 National Defense Education Act (NDEA). It, too, was designed to more closely link scientific research and the military. Government leaders, physicists, and many journalists considered training enough scientists and engineers—but especially physicists—essential to the nation's security. The NDEA situated the demand for new scientists in explicitly military terms: "[The] security of the Nation requires the fullest development of mental resources and technical skills of its young men and women."[24] The act allocated $250 million for the improvement of education, mainly in the sciences, and established the Physical Science Study Committee plan for physics research, designed to make American students competitive with their Soviet counterparts.[25] It also provided a wide assortment of student loans, graduate fellowships, grants for equipment to teach science and languages, and funds for guidance counseling and vocational education.

More than 1,300 schools and more than 100,000 students benefited from NDEA funding in 1960 alone.[26]

One of the components of the program was the National Defense Graduate Fellowship Program. It awarded fellowships to 2,500 students for study at 139 colleges and universities.[27] The goal of the NDEA was more than technical training. "Not only is there a great opportunity to underwrite research for its direct contribution to the nation's welfare," argued one economist at the NDEA hearings, "but the opportunity exists to instill in [students] a consciousness of their responsibility to the nation's security."[28] The goal of the NDEA, then, was not simply the production of new scientists, but also the socialization of young scientists into treating military work as a routine job of the scientist. As the historian David Kaiser argues, although most physicists did not work on weapon's development in the 1950s, they "received money from defense-related bureaus to create an 'elite reserve labor force' of potential weapons-makers in the ranks."[29]

NDEA-generated programs supplemented other programs that linked military needs to science education. The AEC- and NASA-sponsored academic-year and summer fellowships, DOD student summer employment, and ROTC scholarships provided military-based educational opportunities.[30] The content of science teaching did not escape the influence of the military, either. Science textbooks reflected the newest ideas in science. In some disciplines, especially the physical sciences and engineering, what students learned was profoundly shaped by the military problems of the day.

Taken together, these varied systems of funding and education fundamentally changed the locus, content, and volume of scientific research in the United States. It is difficult to overestimate the effects that military funding had on American scientific research in the mid-twentieth century, particularly research in physics and engineering. Before World War II, most research was carried out by industry; after the war, industry still played a key role, but the federal government became the major funder of research in most physical and natural sciences. The monies for research in physics and engineering were free-flowing, as well: as Jack Ruina, the director of ARPA in 1961, recalled, "I had a budget of three hundred million dollars, which was like one and half billion in current dollars. I had no oversight—Congress never questioned giving us anything that we wanted."[31] Military funding had a particularly powerful effect on research on college campuses, for universities were an integral part of the research infrastructure set in place after World War II. Through FCRCs that were on or near university campuses, the Jason program, grants to individual scientists, and the influence of military-based research on the content of knowledge taught to students, military needs profoundly shaped the lives of American scientists. As the historian Stuart Leslie argues, "in many disciplines, the military set the paradigm for postwar American science. . . . [T]he military-driven technologies of the Cold War defined the critical problems for the postwar generation of American scientists and engineers. Indeed, those technologies virtually defined what it meant to be a scientist or an engineer." The challenges that they faced in solving these military-related problems, argues Leslie, "defined what scientists and engineers studied, what they designed and built, where they went to work, and what they did when they got there."[32] Scientists were hardly puppets of the state, however. They also played an important role in shaping how federal R&D funds were spent.

The Science Advising System, 1945–1963

In the decade after the end of World War II, scientists' most powerful advising role was serving as members of the AEC General Advisory Committee. The AEC was nominally a civilian agency, but most of the atomic energy produced was for weapons. Moreover, the agency had an extraordinarily broad mandate: its charge was to both regulate and promote atomic energy. The board played a key role in decisions ranging from weapons tests, to the building of nuclear reactors, to assessing health threats from atomic fallout.[33] Scientists also served as advisors to individual agencies within the DOD, and to the director of that agency. Through a series of changing committees and agencies during the 1950s, including the Research and Development Board (1946–1953), the office of the assistant secretary of defense at the Office of the Secretary of Defense (1953–1956), and the Defense Review Board (1956–present), scientists provided advice on how to develop new weapons and assess the effectiveness of existing weapons systems.[34]

Another early means by which scientists became part of the advisory system was through a new contract system with the National Academy of Sciences (NAS). The arrangement stipulated that the NAS would, when called on by a federal agency, carry out independent evaluations of government projects.[35] In 1951, President Truman created the Science Advisory Committee (SAC), the first permanent committee to directly advise the president. That committee was composed of fifteen scientists, mostly physicists, who were charged with providing scientific information and advice to the president. Meeting twice a year, the committee was primarily concerned with analyses of atomic weapons program decisions.[36]

Just as Sputnik spurred new patterns of funding, it also led to a vastly expanded science advisory system. The 1958 Defense Reorganization Act led to the creation of ARPA, and to the position of defense research and engineering; the purpose of these new systems was to more quickly and effectively develop weapons systems.[37] President Eisenhower believed that advice from scientists who could help direct science and technology policy and programs was a critical part of postwar science policy.[38] As a result of his interest and the general belief among military leaders that scientific advice was critical for weapons development, in 1958 and 1959, three major science advising positions and committees were created to help the federal government create a new set of policies and rules that would encourage the rapid growth of science personnel and weapons development. Moreover, SAC was reorganized as the President's Science Advisory Committee, and a new position, the special assistant to the president for science and technology, was established. The special assistant kept the names of all consultants secret.[39] Scientists' influence was growing in other parts of government, too. Between 1959 and 1963, scientists were appointed as permanent, full-time advisers to secretaries of seven major federal agencies.[40]

The total number of scientists who held advisory positions through the mid-1950s was probably not large. The historian Daniel J. Kevles estimates that somewhere between 200 and 1,000 scientists (mostly physicists) advised the government on a regular basis.[41] No matter what the numbers, when compared to earlier historical periods, in which scientists were rarely called on by Congress or the president, these new permanent advisory roles signaled a new faith in science advice as a necessary and crucial ingredient in postwar governance. Although scientific advice was dispersed throughout many committees and agencies, scientists had the ear of the president and of the director of the DOD.

The advising system was largely a closed loop, into which the public could not enter. Since most of the agencies to which scientists provided advice were in the executive branch, there was no formal way in which citizens could exert any influence on their decisions. Even if there had been, it is doubtful that scientists, who were largely skeptical of the public's capacity to understand how science worked, would have engaged their questions with any seriousness.

The new forms of political influence that were available to a small subset of scientists served to broaden the power of scientists, but they did little to provide ordinary people with access to decision making about science. There were simply few organizational or political means for citizens to directly influence the structure of the production and use of science. Many scientists considered the public ignorant of the most important aspects of science and therefore incompetent to judge scientific findings or participate in decisions about science. A 1957 *Science* editorial on the relationship among government funding, science, and the public, for example, lamented the supposed public desire for gadgets without an understanding of how science actually worked:

Unfortunately, a wish to manipulate the course of nature does not necessarily imply a wish to understand the natural laws upon which such control is based. To the consumer of scientific knowledge, that is to say, to the man who rubs the lamp and commands the jiini, the achievements of science are nothing more nor less than feats of magic. The various agencies devoted to science might just as well be given titles such as the National Academy of Magic, the National Magic Foundation, and the American Association for the Advancement of Magic, and one of the most pressing problems of the day might just as well be the shortage of magicians.[42]

The gee-whiz attitude that scientists accused Americans of holding was as much a product of how science was conveyed to the public by journalists and government as a sign of public ignorance. Like their nineteenth-century predecessors, such as the British science popularizer John Tyndall, those who wanted the public to appreciate science were campaigning for more social status and economic support.[43] As Bruce V. Lewenstein has argued, scientists and science journalists in the period between 1945 and 1965 worked hard to have the public "appreciate" science. This usually meant that the public recognize and be grateful for the benefits of science rather than critically examine scientific ideas, their uses, or the motivations behind particular scientific projects. Scientists and their government supporters maintained "a moral certainty about the social importance and efficacy of science," Lewenstein argues, that led scientists to meet a "'demand' for information about science, despite the lack of evidence that a demand existed."[44] The result was a steady supply of information about scientific findings and inventions provided through radio and television programs, books, and magazines, in the absence of the provision of tools needed for critical analysis of information and facts.[45]

When scientists did provide information to the public, it was often with a sense of noblesse oblige, as the political scientist James C. Wood argued in 1964. Caricaturing the attitude of disinterestedness and elitism that he saw many scientists espouse when they served on government committees or provided other public service, he wrote: "[Scientists] sacrifice professional careers in the interests of an informed debate on great public issues. Their function in the controversies over space, missile systems, or education is not to find new glory or power for their peer group, to promote new funds for a pet project, or to protect a vested interest. Compared to the self-seeking, the parochialism, and the limited knowledge of other participants, the scientists offer the welcome contrast of prescient men concerned only with explaining and using the powers of nature. Their higher duty is to their profession."[46]

Thus, as scientists, but especially physicists, became more deeply enmeshed in a system of funding and advising rooted in military needs, they were simultaneously disconnected from the perspectives of the public. The stage was now set. Scientists were in a high-risk situation. They were highly dependent on the state and especially the military, implicated in national security advising and decisions, and largely uninfluenced by citizen concerns. So long as the military focus for research did not come under fire, and the public was quiescent about the contradictions inherent in expert advising in a democratic system, scientists could probably count on continued levels of funding with few questions asked. Yet that was not what happened. Between 1962 and 1969, the science funding and advising system shifted considerably, in the context of broader critiques of science from intellectuals, activists, and politicians.

New Priorities: Social and Economic Benefits, 1963–1969

The early 1960s were the beginning of a sharply different period in the relationship between the federal government and scientists. Free-flowing federal funds for basic and applied research, the cozy science advising system, and the favoritism that a few major research institutions enjoyed in federal grants and contracts would all be undermined in the 1960s. As the historian Daniel J. Kevles argues, by the middle of the 1960s there was a broad consensus among the public and political leaders that the nation's scientific program had to move "from a politically elitist program to a politically responsive enterprise, from a best-science to a geographically more even distribution, from luxury to leanness in levels of funding, from pure to applied research."[47] These shifts were largely the result of pressure from Congress.

Between 1945 and the early 1960s, Congress played a relatively small role in deciding national funding priorities, nor did it concern itself with overseeing the payoffs (and boondoggles) of specific funding programs. In the early 1960s, Democratic and Republican members of Congress began to take a much greater interest in science policy and funding. Fiscal conservatives worried about the costs of R&D, and liberals expressed concern about whether R&D monies had actually provided much benefit for the majority of Americans. The highly publicized deformities of babies whose mothers had taken the drug thalidomide was one source of discontent, which resulted in the reorganization of the Federal Drug Administration. The civil rights movement had some effect too, swaying some leaders to push for more funds for jobs, health, and antipoverty programs. Other congressional leaders argued that the diversion of scientific talent from civilian to military purposes was a cause of slow economic growth.[48]

In 1963, the House Committee on Science and Astronautics convened a subcommittee headed by Emilio D'Addario (D-CT) to investigate how

R&D funds were being spent, to which institutions federal funds and fellowships were awarded, and how such programs might be coordinated. For D'Addario and his supporters, far too many funds were going to major research institutions concentrated in only a few geographic areas. The committee quickly produced a series of reports critical of the failure to expend funds on health, education, urban problems, and environmental issues.[49]

With the support of President Johnson, several key changes resulted from the D'Addario committee's reports. One was the formation of a new DOD program called Project Themis, which provided defense-related research grants to universities that had not previously received them. Suddenly, state universities such as Pennsylvania State were to be recipients of defense funding.[50] Military funding, once limited to a handful of campuses, was now spread across many universities. Another change that limited the military domination of science was the 1968 congressional authorization of the NSF, which required it to fund applied as well as basic research, made the social sciences eligible for funding, and required Congress to approve the NSF budget.[51]

The dispersal of funding and Congress's decision to place greater emphasis on nonmilitary applied research took place at the same time that federal funding for science was on the decline. By 1966, the expansion of federal funds for R&D had slowed, and in 1969 it dropped off for the first time since World War II. Industrial financing, by contrast, expanded, as did monies from universities, nonprofits, and colleges (although their contributions were tiny in comparison with those from industry and government).[52] The financial demands of President Johnson's Great Society programs were one of the reasons for this shift, but so too was the belief among many members of Congress that the investment in science had not paid off.

The social and biological sciences benefited from the turn toward social and economic research. Biological scientists were critical to many of the new health projects that were established in Great Society programs and President Nixon's War on Cancer. Physicists were still a highly favored group, but it was clear that by 1965 the government was considering a wider range of sciences as integral to national political priorities.[53]

By far the biggest shift, however, was to one of the most expensive scientific projects ever undertaken in the United States: the Apollo moon landing program. Proposed by President Kennedy in 1961, the program was organized under the aegis of the previously weak and poorly funded NASA. NASA was an external contracting agency that provided grants and fellowships to universities and businesses.[54] Although NASA clearly began as a response to cold war concerns, only one of its explicitly stated goals was military benefit.[55] NASA's funding became lavish, outstripping all other agencies except the DOD as the leading source of federal R&D

funds by the end of the 1960s.[56] As a result of the vast increase in funding for NASA, engineers had replaced physicists as the primary recipients of federal research funding by 1967.[57]

The new emphasis on non-weapons-related applied research meant that more scientists would be pulled closer and closer into the orbit of debates about the proper role of scientists in public life. Individual scientists still advised the government, of course, but new players were beginning to enter the debates in new ways. The National Academy of Sciences, for example, expanded its advisory role in 1962 by forming the Committee on Science and Public Policy to prepare reports for the federal government as a matter of course rather than of contract. The group's first project was a study of world population growth, placing the issue on the international public agenda.[58] Despite new relationships such as this one, the tremendous access to the machinery of government decision making—particularly in the executive branch—that scientists had enjoyed in the 1950s became spottier and weaker in the 1960s.

Scientists' power to influence government weakened even more as relationships between the executive branch and scientists cooled significantly after 1965. Few scientists were enthusiastic about the new research emphases that both Nixon and Johnson embraced. Relations were further strained after 1968 by a series of struggles between scientists and the Nixon administration over weapons systems and science advice to the president. Nixon was unhappy that scientists disagreed with him over two of his pet projects, the supersonic transport system and the antiballistic missile (ABM) program. In 1973, after five years of increasing hostility to science advisors, Nixon eliminated the President's Science Advisory Committee.[59]

As Congress and the president called for broader research applications, citizens were beginning to gain access to the information and political institutions that were critical for participation in decisions about science policy. The Freedom of Information Act (FOIA) of 1966, the result of a ten-year investigation into government secrecy spearheaded by the Democratic representative John Moss, enabled citizens to request any kind of government information that was not classified for security purposes. Although citizen groups did not immediately flood the government with FOIA requests, the law nevertheless provided a legal framework through which citizens could acquire information to which they would not have been privy a decade earlier.[60]

Two other legal shifts took place in the 1960s that provided citizens with the capacity to acquire and critically evaluate scientific knowledge and products in new ways. One was the rise of the consumer movement, spearheaded by Ralph Nader. Nader successfully used highly publicized class action suits and other forms of litigation against manufacturers of faulty products. If there ever was a time when Americans were unskeptical of technologies, that time was passing. The 1969 Environmental Policy Act (NEPA) marked another important change in the relationships among scientists, the government, and the public. It required citizen participation in the review of projects that might affect the environment, effectively inviting the public to become involved in challenges to the scientific arguments put forth on behalf of companies and government agencies engaged in work that affected nature.[61] Even if citizens did not immediately make use of FOIA and NEPA, these new laws signaled that citizens had more extensive rights to be informed about and to have input into government policy.

The shifts in funding priorities, the cooling of the close relationship between scientists and the executive branch, and the establishment of new means for citizens to gather and challenge government information meant that scientists, as a group, were in a much different political position than they had been ten years before. When these shifts took place, scientists as a group experienced greater and more varied demands, and there became fewer corners of the government where they might be insulated from these demands. Yet it would be a mistake to see scientists in monolithic terms. As I have argued, engineers and physicists, and then biologists, were the largest beneficiaries of government research funds. Demographically, some groups of scientists began to vary more as well. Although the median age of scientists remained 30–35 between 1950 and 1970, the median age for physicists and biologists, who were the most politically active in the 1950s, 1960s, and 1970s, dropped from 30–35 between 1950 and 1960, to 25–30 between 1960 and 1970.[62] The influx of young people was paralleled by the fragmentation of scientists into more disciplines. In 1950, one could earn a doctorate in 37 scientific fields. By 1960, it had jumped to 128. Some fields, especially chemistry and physics, differentiated more quickly than others, so that it was possible to receive a doctorate in narrower and narrower subfields.[63]

These demographic changes encouraged the development of new scientific ideas, but they also meant that scientists were more demographically heterogeneous (even though they remained overwhelmingly white and male). Divided by discipline, place of work, and age, their experiences of the shifting political and economic relationships between themselves and the state were varied. Some, such as chemists, worked in industry, where they were insulated from many of the intellectual debates about the role of science in public policymaking; whereas biologists, who were more likely than other scientists to work on college campuses, were likely to be exposed to them. The demographic arrangements, and the political traditions and communities on which scientists drew, would help to shape the ways in which scientists responded to the military-science relationship.

This brief overview of the growth of the military-science relationship in the 1950s, and the beginnings of its unraveling in the 1960s, provides the basic set of relationships that shaped the political actions of the scientists—mainly biologists and physicists—whom I study in subsequent chapters. These scientists responded to different aspects of the science-military relationship: some were dismayed that their peers did not personally refuse military funding, others that military funding was distributed to universities, and still others that scientific know-how was diverted from solving other pressing problems of the day. Yet scientists were not the only group that responded to the militarization of science.

Other activists and intellectuals articulated their critiques of science in writing, in public displays, and through personal action. Often the critiques were part of larger indictments of American society, and they portrayed science as part of what was wrong. Critics charged that science, long associated with "progress," was also a destructive force. The idea that scientists were value-neutral with respect to the outcome of their work, as Robert K. Merton had asserted in 1942, was steadily eroded. Pacifists argued that individual scientists were not pawns of the state or value-neutral interpreters of nature, but rather were individuals who were morally corrupt because they accepted funding from the military. Just as the American ideal of a nation of equals crumbled during the cold war in the face of the civil rights movement, so too the idea that the nation was a democracy came under fire from antiwar activists and intellectuals who saw scientific advising as an antidemocratic practice. Even scientific thinking did not escape critics' eyes: its spread throughout American society was blamed for alienation, and its use as a basis for political decision making was condemned as technocratic rule that paid no attention to moral values. More generally, these critics called into question whether science was really associated with progress and democracy, and whether scientific thinking was an antidote to ideology or a source of moral repression. In the next section of this chapter, I provide an overview of the arguments against some aspects of science that were based on three different political frameworks: moral individualism, liberalism, and the Marxist/New Left.

SCIENCE UNDER FIRE

Moral Individualism: Alliances with the Military as Failures of Personal Responsibility

Jubilant over the defeat of the Axis powers, in 1945 few Americans questioned the morality or wisdom of the government's decision to develop and use the atomic bomb. (The bomb itself was another story: Americans were fearful of its power and uncertain whether other nations would soon have its "secret.") Newspapers and radio stories lauded the scientists who helped create the bomb, and the president personally thanked them for their selfless efforts.

A small and determined group of pacifists were the exception to this rule. Among them was the peace activist A. J. Muste. Muste, along with other pacifists, was a strong critic of militarism in all its forms. They were among the earliest voices to dissent from the growing consensus that a powerful military made possible by scientific research was necessary.

A former Marxist, Muste had become tired of the violence that Marxist-Leninist groups were using in their struggles against capitalism, and in 1936 he embraced a philosophy of comradeship and nonviolence. A formidable and charismatic leader, Muste was a powerful figure in the war resisters' movement between 1941 and 1945. One of his most distinctive strategies grew out of his belief that social change had to emerge from changes in individuals in social institutions, rather than from states.

Writing in the Fellowship of Reconciliation (FOR) journal *Fellowship* in 1945, Muste had no sympathy for the scientists who, after participating in the Manhattan Project, were suddenly critical of military control over scientific research: "The scientist who supports or condones a nationalist foreign policy, a big military establishment, and peacetime conscription while professedly working to abolish the bomb and internationalize the control of atomic energy—the scientist who is not prepared to be a conscientious objector against any further work on atomic bombs or similar instruments of destruction and against any service to an administration which retains the bomb and prepares for atomic warfare—is, whatever his intentions, an enemy of mankind and a traitor to both science and religion."[64]

Muste's harsh words and his pacifist stance drew on a long history of "confessional" protest that developed in the United States during the abolition and temperance movements of the 1830s and 1840s. As Michael P. Young argues, these movements mixed personal and social transformation by calling on individuals to repent and reform as a means of saving the soul of the nation. Bearing witness to one's own sin through public testimony and personal action was critical to social transformation; without it, the nation would suffer ever greater moral corruption.[65] Underlying this form of social and political transformation was the idea that each individual had personal, moral responsibility for the well-being of the nation. Social transformation was not thought to occur through changes in law or policy alone, but through the person-by-person adoption of morally right ways of living, expressed in public forums. A form of Christian witnessing, such public displays worked through moral persuasion. One of the hallmarks of this style of social change is its radical egalitarianism. No matter what one's station in life, one is always susceptible to sin, and thus no person is exempt from the moral compulsion to transform self and society. This emphasis on personal responsibility was at the center of religiously based, especially pacifist, critiques of the relationship between scientists and the military.

Muste saw the relationship between science and the military as a moral corruption of science—and scientists. Working with other peace activists such as David Dellinger, Ralph DiGia, and the labor and civil rights organizer Bayard Rustin, Muste began a peace campaign organized through the War Resisters League, FOR, the American Friends Service Committee (AFSC), and newer groups such as the Committee for Nonviolent Revolution. Many of those who participated in the campaigns, including the physicist Don DeVault, were former conscientious objectors who had spent the war either in work camps for conscientious objectors, or in prison for their refusal to participate in the often humiliating and menial projects to which they were assigned in work camps.[66]

Pacifists were widely criticized by religious and nonreligious groups and individuals alike for what seemed to be a selfish commitment to moral purity that might leave millions of people dead. The liberal theologian Reinhold Niebuhr, for example, rejected the pacifists' individualism and especially "their equation of the political policy of states with those of individuals," and called on Christians to reject only "total war," not war itself.[67] Quakers, some of whom were pacifists, were divided over how to respond to the massive buildup of atomic weapons and the realization that the Soviet government was engaging in a militaristic foreign policy similar to that the United States was developing. Despite the fact that the AFSC won the 1949 Nobel Peace Prize, Quakers were divided over whether to educate the world about the need for peace, participate in overseas public service that paralleled other U.S. aid campaigns, or to rely on the practice of individual witness to convince others to withdraw from military activity.[68]

The pacifist critique of war had few adherents until 1955. That year, a committee of Quaker pacifists that included Muste published the seventy-page pamphlet *Speak Truth to Power: A Quaker Search for an Alternative to Violence.* Asserting that militarism was not only a prelude to war but also a danger to democracy, the authors asserted that society should be reorganized in a way that encouraged small-scale community and individual responsibility and that put an end to the "deification of the state," which had "perverted" science at the expense of the individual.[69] This document helped to catalyze new levels of peace action in the United States. The next year, Muste and some of the authors of *Speak Truth to Power* founded the radical pacifist journal *Liberation,* which would become the mouthpiece of the growing group of pacifists who believed that action was the key to creating a peaceful rather than militaristic world.

Pacifists used direct action, in the form of sit-ins, refusals to go inside during civil defense drills, and civil disobedience. In the late 1950s and early 1960s, American peace activists, most of them pacifists, sponsored a half dozen "walks for peace." Often taking months to complete, some of the largest of these were transnational walks thousands of miles long. Activists, carrying banners and singing as if on a religious pilgrimage, protested nuclear testing and the arms race. Their goal was to make peace by building direct relationships with ordinary people whom they met along the way or who participated in the march.[70]

Above all, pacifists insisted that all people cease to participate in any activity known to help the military. Typical of their actions was the distribution of flyers at the Polaris nuclear submarine manufacturing plants in New London, Connecticut. Yale Alumni against Polaris exhorted Polaris workers to take personal action to stop militarism. They urged workers to leave jobs in military industries, to insist on contracts that called for "useful" work instead of war work, and to refuse to pay federal income taxes or serve in the armed forces.[71]

Like many other critics of war, militarism, and the arms race from across the political spectrum, many pacifists saw technocratic thinking as a source of collective moral and political madness. Many of those who advocated a more action-oriented pacifism also believed that technocratic thinking and organization—the subversion of creativity to the discipline of rationality—contributed to the growth of mass society. Characterized by large-scale national structures, impersonal bureaucracies, and increasing rationalization spread by experts, mass society quashed the local, personal, and small-scale groupings and experiences that had prevailed in earlier eras. Rationalization, the very tool that was supposed to liberate humans from superstition and lead to progress, had limited the possibilities for individuals to live creative and liberated lives. The relentless pursuit of security through militarism did not lead to security, they argued, but rather to madness. On a flyer handed out in 1957 by the pacifist group Committee for Nonviolent Action, during a walk from Hanover, New Hampshire, to Washington, DC, pacifists wrote: "Who is secure? The children and young people who sense the madness of the adult world?"[72] Writing in *Liberation* in 1956, the peace activist Dick Bruner ridiculed the biologist Ralph Lapp, whose criticisms of the military were based on the calculation of potential deaths from variously sized bombs, rather than on the broader question of war itself. Bruner asked sarcastically of Lapp's calculations: "Do scientists use slide rules to arrive at figures like these?"[73]

The mix of the critique of instrumental rationality and the confessionalist call for individual responsibility for war merged in the speeches and writings of religious leaders during the anti–Vietnam War movement. One of the most energetic, visible, and active antiwar groups, Clergy and Laity Concerned (CALC), an interdenominational group of religious leaders and activists, was one of the most outspoken about this position. At the 1968 Mobilization against the War demonstration in Washington, DC, scientists and instrumental rationality were the targets of CALC leader John C. Bennett, the president of Union Theological Seminary. He asked, "When will we take seriously our moral responsibility for the effects of our actions upon the hundreds of millions of people who have no part in the decisions and who do not even share our view of the issues at stake? . . . [T]he same thing can be said about the decision to engage in nuclear tests that have consequences not foreseen by the scientists who plan them and affect distant nations that have no part in the decisions."[74] CALC leader Rabbi Abraham Heschel made a similar point. "Pestilence that stalks in the darkness, infecting the minds," he said, "is politics of an isolated, autonomous science following its own rules." The isolation would end, Heschel argued, when "we take responsibility for the pestilence."[75] Heschel's and Bennett's sentiments were echoed by Martin Luther King Jr., also a member of CALC. He reminded his audience that individual Americans—not just the government—were responsible for subverting the creative to the technocratic: "Each of us lives in two realms of life—the within and the without. The within of our lives is that realm of spiritual ends expressed in our literature, morals, and religion. The without of our lives is that complex of devices, techniques, mechanisms and instrumentalities by means of which we live. The problem we face today is that we have allowed the within of our lives to become absorbed in the without."[76]

Running through the moral individualist critique of science and militarism, thus, is a key theme: that anyone who does not explicitly refuse to cooperate with the military is complicit in its destructiveness. Pacifism and moral individualism as a means of transforming society were not popular political perspectives at the height of the cold war. Yet during the late 1940s, when they first appeared, and as they grew in intensity in the 1950s, they offered a striking contrast to the idea that scientists were the objective interpreters of nature or apolitical purveyors of facts to the government. Pacifism and moral individualism asserted that scientists were, like other people, morally responsible for their actions. Advising the government against the development of a particular weapon was not a moral action, according to the moral individualist tradition. Only the refusal to participate in any kind of military research or advising would constitute "right" action. In the 1950s and early 1960s, champions of this perspective demanded that individual scientists take personal responsibility for the uses made of their work. Their arguments were in marked contrast to the liberal critique that emerged in the mid-1950s.

A Liberal Critique: Technical Elites Unconstrained by Values and Government Scrutiny

By far the most visible critique of scientists in public life in the postwar era was drawn from a liberal perspective. As the historian Alan Brinkley argues, liberalism had contradictory meanings and expressions in the mid-twentieth century; indeed, defining liberalism was a major struggle during the postwar era.[77] Whatever the substantive disputes, of paramount concern in classical liberalism is how to ensure the effectiveness of democratic procedures. Democracy, in the liberal view, requires an open "public" forum, properly regulated by the state, in which participants with different values and preferences debate and act on their interests. In theory, through open debate, the right decisions will be made.[78] For this system to work, no one group should dominate, and certainly the range of material discussed should not be limited.

Based on these assumptions, one major concern about the scientist-military relationship in the mid-twentieth century was that scientists had become an elite out of touch with and unbeholden to the public and Congress. As a result, scientific ideas had been "misused" because of failure to balance the cold logic of scientific thinking with explicit attention to human values. Few in the liberal tradition were concerned with reforming the actions of individual scientists, or with questioning the relationship between science, capitalism, and the corruption of human creativity, as other critics were. The debates about science, from a liberal perspective, were centered around how to subjugate scientific thinking to values in public decision making without allowing ideology to overtake reason.

To understand the tensions surrounding values as a basis of decision making, particularly in the 1950s, it is useful to consider that in the early postwar period, the fear that demagogues and tyrants could sway Americans, as Hitler and Mussolini had their citizens, led a generation of intellectuals to emphasize reason as an alternative to ideologies such as communism and totalitarianism that would, through passion, transform entire ways of life. Daniel Bell's famous remark that the historical moment marked the "end of ideology" reflects the feeling among some that reason had triumphed over passion and revolution.[79] Viewed as a politically untainted form of thought, scientific thinking was considered, in this view, an important basis for social life. Scientific claims could be separated from scientists themselves, since science was thought to be a form of thinking based on universalistic rules. If reason and instrumental rationality were the antidote to emotion-based decision, then a modified form of Hamiltonianism, in which elected representatives controlled political decision making, was a way to avoid mass-based politics at a historical moment when life-and-death matters of atomic weapons were the major political concern of the day.[80] Still, striking the right balance between Hamiltonianism and Jeffersonianism, facts and values, was not easy, and some in the liberal tradition expressed concerns about whether scientists had become an illegitimate "elite."

This tension is not new, of course; from the establishment of the American Constitution, tensions between the value of elite versus mass rule were at the center of political debate. Yet another problem, too, was arising: some intellectuals began to see scientific problem solving as antithetical to democratic practice. As David A. Hollinger articulates the problem in its general form, "In an ideal science, there is no power; in an ideal polity, there is no truth."[81] In the ideal, liberalism is characterized by agreements arrived at through the consent of the group. Yet in science, facts supposedly trump the collective will; in liberalism, the collective will trumps all else. Thus, the goal of science is to reveal truths unfettered by human interests, whereas in democracy, it is to bring parties to a consensus that was not necessarily "truthful." Thus, for many liberal critics of the relationship between scientists and the military, the major problem was that the instrumental logic of scientific thinking had come to dominate other forms of thought, pushing values aside in the quest for answers and decisions that were unquestionable and universalistic. The solution was to restructure formal decision making.

These problems were present from the beginning of the 1945 debates over scientists' role in government. When the war ended and debates about control over atomic energy were taking place, some alarms were raised about whether it was reasonable for scientists to participate in these debates. One worry was that scientists' worldview was incompatible with democracy. David Hawkins, a Los Alamos Laboratories historian, wrote in 1946 that there was much concern over whether the "scientific point of view is incompatible with the spirit of democracy." Reminding his readers that Winston Churchill had worried earlier that year that, just as the world had once suffered from a theocratic government, now there was a danger of a "scientistic" government, Hawkins nevertheless assured his audience that scientists were not autocrats but sincere people who wished only to inform the public of how science would now shape international political action.[82]

Despite such reassurances, not all observers were convinced. The philosopher Sidney Hook, an anticommunist social democrat, argued that although scientists were well-intentioned, they were overstepping their professional bounds by offering political opinions about foreign policy. Hook argued that scientific logic was fundamentally different from political democratic logic. In science, he wrote, "no one is free to think as he pleases about the objective evidence and compulsions of fact. . . . [I]n human affairs, however, freedom requires a pluralism of values and tastes."[83] This same refrain could be found in debates about nuclear policies in particular. The policy analyst and physicist Herman Kahn, in his widely read *On Thermonuclear War* (1961), argued that nuclear war was winnable, and that even after a war, human life would go on.[84] Although Kahn himself was far from a wholehearted supporter of the cold war, Walter Goldstein and S. M. Miller, writing in *Dissent*, still took him to task for his technocratic thinking. Kahn's calculations of the probabilities of a Soviet attack, and his projections of fatalities based on different sized weapons, were arrogant and misguided, they argued. His argument, in "seeking flawless syllogisms . . . disdains the subtle nuances of political action and human aspiration."[85] Hans Morgenthau, a political scientist and close advisor to President Kennedy, made a similar point. He argued that Kahn's emphasis on instrumental rationality as the basis of political action was reductio ad absurdum, since "the possibility of universal destruction obliterates [a] means-ends relationship . . . by threatening the nations and their ends with total destruction."[86]

The two most well-known criticisms of the domination of scientific values and elite, secretive decision making came from Dwight Eisenhower, in his farewell address in 1961, and from the chemist-turned-writer C. P. Snow. Eisenhower argued that a "military-industrial" complex had taken over American life, intruding into the economy, mental life, and politics. He warned that "in holding scientific research and discovery in respect, as we should, we must also be alert to the equal and opposite danger that public policy could itself become the captive of a scientific-technological elite. . . . We must never let the weight of this combination endanger our liberties or democratic processes. We should take nothing for granted. Only an alert and knowledgeable citizenry can compel the proper meshing of the huge industrial and military machinery of defense with our peaceful methods and goals, so that security and liberty may prosper together."[87] Snow argued that the humanities and the sciences were farther apart than ever before, and to the detriment of both. Although Snow was more enthusiastic about the capacity of scientific thinking to generate solutions to political problems, he was critical of its independence from the broader moral concerns of humanists.[88]

The concern over this imbalance was also visible in an emerging critique among liberals: scientific knowledge had been "misused" because it had not been subjected to proper political oversight. Like congressional critics in the early 1960s who deplored the misdirection of most research funding toward military projects to the neglect of urban issues, the environment, and health problems, intellectual critics were determined to see that scientific knowledge be harnessed to solve nonmilitary problems. The most vivid expression of the misuse of science theme came from the emerging environmental movement. Although conservation had long been the domain of elite "preservationists," and hunters and fishers who focused on wilderness issues, the 1962 publication of Rachel Carson's *Silent Spring* was the opening salvo in the environmental campaign to redirect the use of science. Rather than seeing "better living through chemistry," Carson argued that chemicals, particularly DDT, were the cause of the systematic pollution of the environment and, by extension and ultimately, of humans: "The most alarming of all man's assaults upon the environment is the contamination of air, earth, rivers, and sea with dangerous and even lethal materials," Carson wrote. "In this now universal contamination of the environment, chemicals are the sinister and little recognized partners of radiation in changing the very nature of the world—the very nature of life."[89] Without a dramatic halt to contamination of this sort, spring, she argued, would be silent, because there would be no birds left to sing. The procedural solution was to redirect the ways that scientific knowledge was used.

By the mid-1960s, liberal critics of the relationship among scientists, the military, and elite forms of governance proliferated. Political scientists charged that the elite advising system was distorting democratic politics. As James C. Wood argued in 1964, scientists put forth the impression of disinterestedness when they served on government committees or provided other public service. Wood charged that scientists were not in fact the independent, high-minded advisors that they pretended to be—or that governments wanted them to be—but another interest group with their own careers to protect.[90] Don K. Price, another political scientist, argued in his 1965 book *The Scientific Estate* that scientists had become an "estate" acting outside the boundaries of legitimate democratic practice.[91] In 1963, the political journalist Meg Greenfield wrote a scathing critique of the role of scientists as political advisors. She argued that scientists held an unfathomable number and variety of advisory positions, but that their advising had not been properly constrained by the rules of governance. "The factors that put the scientists beyond accountability and their work beyond review—the maze at the working level, the fuzziness of authority at the middle and the top [of government], and the unorthodoxy of present operations . . . have undeniably encouraged on more than one occasion a quick, casual, and even sloppy approach to problems."[92]

The solution to these problems, from a liberal perspective, was to reinscribe instrumental rationality and scientist advising within the proper channels of democratic control and discourse. In contrast to the moral individualist position, which hoped to reform individuals, the liberal position urged reforms of political institutions. By the mid-1960s, however, another critique, this time much broader than the criticisms of government and immoral individuals, was developing among New Left thinkers.

The New Left: Instrumental Rationality and the Corruption of Social Institutions

Before the 1950s, liberals and conservatives shared the belief that economic prosperity would lead to a harmonious society. Without material want, conflicts would be reduced, and Americans would be able to pursue leisure or other activities that satisfied human needs. Ironically, the postwar economic boom helped to challenge this idea. The lessening of material want had not, it seemed to some critics, created satisfying relationships among people or generated feelings of contentment. In the 1950s, critics including C. Wright Mills, William Foote Whyte, and David Riesman, among others, laid out the ways in which the rational bureaucratic systems whose efficiency had led to high productivity had also produced malaise and alienation, and had dissolved old bonds of community. For these critics, the root problems of American society were to be located not in distortions of democratic systems or failures to act in a moral fashion, but rather in broader economic, organizational, and knowledge structures.

This critique of the alienating effects of bureaucratic systems was an important intellectual element of the "New Left," which emerged in the early 1960s.[93] In contrast to the "old left," the largely communist groups who pressed for changes in class relations to the exclusion of other concerns, the New Left saw material deprivation as only one of many problems that would need to be solved to transform the United States into a humane, economically egalitarian, community-based society in which individual expression was valued. For some members of the old left, technical experts and instrumental rationality were tools for political transformation. New Left intellectuals and activists by contrast, saw overreliance on them as sources of oppression.[94] To create more "authentic" human experiences, social structures from the economic system to personal relationships would have to be transformed. One major problem, however, would face all who wished to do so: the real sources of power were often hidden in everyday life, for they were no longer located in formal politics, but in an array of institutions, styles of thought, and in personal relationships.

The sociologist C. Wright Mills was among the earliest intellectuals to influence the development of New Left politics, particularly among young people. In 1956, Mills published *The Power Elite*, in which he argued that in modern democratic societies, power increasingly took the form of manipulation by economic, military, and political leaders. Rather than acting separately, they worked together, Mills argued, creating a society dominated by enormous bureaucracies that provided little possibility for the expression of human creativity.[95] By 1958, Mills had homed in on the way that the military had taken over political decision making. In *The Causes of World War Three*, he described the ways in which close associations between the military and governance had corrupted politics. Here, again, one of Mills's main messages was the invisibility of power and ways in which it was infused into all aspects of social life. In both the Soviet Union and the United States, he wrote, "the Science Machine is made a cultural and a social fetish, rather than an instrument under continual public appraisal and control; and to the Machine's economic as well as military aspects, the organization of all life is increasingly adapted."[96] Mills argued that one way to prevent World War Three was to remove from the private economy all scientific research and development contracts. Since profits were tied to the endless preparations for war, Mills argued, removing the profit motive might put an end to one of the sources of the sense that a war-ready state was important and necessary.[97]

If Mills's analysis described the ways that institutions were corrupted, Herbert Marcuse's pointed to the role of instrumental rationality in creating an alienated society in which it was difficult to dissent. Like other Frankfurt School theorists, such as Hannah Arendt and Eric Fromm, Marcuse argued that political oppression was rooted in ways of thinking.[98] In *One-Dimensional Man*, published in English in 1965, Marcuse argued that scientific ways of thought had limited the possibilities for expression and imagination. He wrote that "operationalism," the reduction of what is real to that which can be turned into an "operation," or system of manipulation, pervaded the modern world. Because operationalism and its cousin in the behavioral sciences, behaviorism, prevented the articulation of viewpoints that might not be measurable, society had become "one-dimensional." Far from a politically autonomous and disinterested system of knowledge and enlightenment, scientific thinking was an essential component of political oppression. "The government of advanced and advancing industrial societies can maintain and secure itself," he wrote, "only when it succeeds in mobilizing, organizing, and exploiting technical, scientific, and mechanical productivity."[99] Unlike liberals, whose solution to the problem of a society whose decisions were shaped more by technical rationality than values was to reform the political system, for Marcuse, a major solution was for individuals to gain "effective social control over the production and distribution of the necessities."[100] Equally important was that scientific thought itself had to be transformed. Science, he said, had to become political, "recognizing scientific consciousness as political consciousness, and the scientific enterprise as a political enterprise."[101]

Like Marcuse, the psychologist Erich Fromm criticized instrumental rationality because it denied the centrality of emotional and psychological experience. Like the religiously based pacifists, who warned of the psychological effects of militarism and instrumental rationality, Fromm and the philosopher Hannah Arendt worried about the effects of instrumental rationality. For Fromm, this way of thinking was a source of the "madness" that had taken over society. Anyone who could present "a 'budget' of from five to a hundred and sixty million fatalities without shrinking," wrote Fromm in response to the commonplace calculations of deaths in hypothetical nuclear war, was afflicted with "the pathology of alienation."[102] Fromm's and other critics' notion that society had gone mad was not invented in the middle of the 1950s, but its roots were in the collective responses to the invention and use of the atomic bomb, an abomination and horror so great it was almost inconceivable. When the great mushroom cloud had evaporated, writers were at a loss for words. Religious people grappled with questions about the existence of God and ordinary people wondered if the end of the world itself was near. When the Ban the Bomb movement emerged after 1955, the idea that advocates of the atomic bomb could talk about its use in the same terms as more trivial, routine political decisions seemed utterly insane. Political discussions that might have once appeared rational—that is, about the best choice given particular calculations of means and outcomes—seemed out of touch with a new reality.

Film and other mass media condemnations of instrumental rationality as irrational reached far larger audiences. In popular films, scientists were very often portrayed as heroes or saviors, but science fiction films in the 1950s and in the 1960s also articulated Americans' anxieties about the lack of control over the very technologies humans were inventing. Technology was frequently depicted as out of control and as having consequences that humans were unable to anticipate. The dark seriousness of science fiction films in the 1950s contrasts starkly with Stanley Kubrick's 1964 film *Dr. Strangelove, or How I Learned to Stop Worrying and Love the Bomb*. In both content and form, this surreal comedy captured the insanity and absurdity of the cold war. As they try to avoid an accidental nuclear war, none of the major characters, including the mad scientist Dr. Strangelove, appears to be aware of the devastation that could be wrought by the bomb. Cheerfully discussing the means of destroying the planet, the characters use rational, instrumental language, with seemingly no understanding that real people's lives are at stake. Kubrick's film portrayed political leaders and scientists as deeply psychologically disturbed, teetering on the brink of insanity, and who enthusiastically and playfully make decisions about the destruction of the planet.[103]

The ideas of intellectuals were one source of the challenges to scientific thinking and its domination in a wide range of social institutions. Equally important were the actions to change those institutions undertaken by students and others, who rebelled against participating in systems dominated by instrumental rationality. A student at the conspiracy trial against activists who had participated in the disruption of the 1968 Democratic National Convention in Chicago explained that her actions were motivated by a rejection of a faceless system in which the logic of instrumental rationality was unquestioned: "The more I see a system that tells middle class whites like me that we are supposed to be technological brains to continue producing CBW [chemical and biological warfare], to continue working on computers and things like that to learn how to kill people better to learn to control people better, yes, the more I want to see that system torn down."[104]

In an often-quoted statement, Mario Savio, the student activist leader of the University of California–Berkeley free speech movement, said, "There is a time when the operation of the machine becomes so odious, makes you so sick at heart that you can't take part even tacitly, and you've got to put your bodies upon the levers, upon all the apparatus and you've got to make it stop."[105] For these activists, the "system" was the problem, and by that they meant a complex system that involved not just formal government, but ways of thinking and a range of social institutions.

In the late 1960s, the most visible expression of hostility toward science—not only scientific thinking—were the protests against war research on college campuses that grew in frequency and ferocity beginning in the middle to late 1960s. In the Port Huron Statement, the manifesto written by Students for a Democratic Society in 1962, the authors argued that the university had to be one of the central bases of activism of the New Left. The university, they wrote, was a "permanent position of political influence," the "central institution for organizing, evaluating, and transmitting knowledge," and was open to many viewpoints. But most important for the critique of science, the university helped perpetuate "immoral social practice," exemplified, they wrote, by the "extent to which defense contracts make the universities engineers of the arms race." Thus, campus-based scientists were one of many cogs in the war machine. Whether motivated by a desire to purify the university or not, some of the most violent and dramatic anti–Vietnam War activism in America was directed against university-based and military-funded science research projects.[106]

Between 1965 and 1970, on at least ten major college campuses and on dozens of smaller campuses, military-supported research buildings and laboratories were the sites of antiwar protest. In each case, protestors directed their actions against the physical representations of the alliance among universities, scientists, and the military, usually the campus laboratories, associated FCRCs, and other institutes where military-sponsored science research took place.[107] At Stanford University, antiwar activists targeted the FCRC Stanford Research Institute (SRI). Nominally separate from the university, SRI had been created to attract defense contracts and it employed scientists from Stanford and other locations to carry out research projects. The historian Stuart Leslie recounts that in 1965 "radical student and community groups (apparently with inside help) began leaking details of SRI's chemical weapons and counterinsurgency research through the underground press."[108] Similarly, demonstrations against the FCRC Lawrence Livermore National Laboratory, which developed and tested weapons and was managed by the University of California, were routine by 1970. Livermore Laboratory was the site of major research to develop the multiple independently targetable reentry vehicle (MIRV), the ABM, and nuclear weapons development. Activists mobilized both workers with antiwar sympathies within the laboratories and activists from the Bay Area to picket outside the labs.[109]

Soon after the Stanford leak, students and workers ferreted out similar information on other campuses. In 1965, students at Pennsylvania State University discovered information about war work on their campus, which showed that the DOD was funding the university's Institute for Cooperative Research. The institute's brochure, procured by students, revealed that the university was being funded for research on "the feasibility of the use of temporarily incapacitating chemical and biological agents, including psycho-chemicals, in specified military situations. Both actual and potential uses have been examined. In addition, present and anticipated accomplishments in the biological and chemical research and development program of the U.S. Army are being studied."[110]

The May 1968 student occupation of Columbia University arose in part because of Columbia's association with the IDA. Columbia University was one of the universities whose science faculty received IDA monies. In a letter to President Grayson Kirk, dated one-and-a-half months before the student takeover, representatives of the SDS wrote, "We make the following demands: 1. that Columbia University publicly and formally resign from the Institute for Defense Analysis. 2. that all Columbia professors currently employed by the IDA be obliged to resign their posts as IDA military intellectuals." They concluded that "until Columbia University ends all connections with the IDA we might disrupt the functioning of those involved in the daily disruption of people's lives around the world."[111] The campus takeover was catalyzed by the administration's punishment of anti-IDA protestors for their demonstration in front of Low Library. Columbia SDS activists were outraged by the university's punishment, and again demanded that Columbia "sever all ties with the Institute for Defense Analysis so that Columbia's resources are not used to develop the techniques that prevent oppressed people at home and abroad from gaining control over their own lives."[112] Activists held a small demonstration at Low Library, joined by neighborhood and campus activists who were protesting against Columbia's proposal to build a gym in Morningside Park. They then took over Low Library and Fayerweather Hall.[113] Just ten years earlier, the IDA was drawing the best and the brightest to its summer retreats under the auspices of the Jason project. Now, it was under attack for its role in political oppression. As I will discuss in chapter 5, the 1968 events were not the end of Columbia activists' anti-IDA actions; in 1971 and 1972, they made Jason scientists the direct targets of protests.

The frustration of antiwar and neighborhood activists at Columbia University can be seen on other campuses as well, where the physical destruction of property was used to protest the war in Vietnam and universities' roles in it. One of the most dramatic events of the campus-based anti–Vietnam War movement was the 1970 bombing of the University of Wisconsin–Madison's Sterling Hall. Sterling Hall housed the physics department and the Army Mathematics Research Center. The attack followed years of protests, often violent, against campus recruitment by Dow Chemical and by ROTC on the Madison campus, during which activists smashed windows, brandished clubs, and firebombed the office of the associate dean for student academic affairs.[114]

The campaign against Dow Chemical, the manufacturer of napalm, was the largest and best organized anticorporate campaign in the movement against the war in Vietnam. On dozens of campuses, students organized anticomplicity campaigns that sought to rid their colleges of association with Dow. Activists protested at Dow plants, asked faculty to sign noncooperation pledges, and disrupted Dow's campus recruitment visits, often carrying "Dow Burns Babies" signs.[115] Activists made the point that science no longer served the social good, but had become an agent of death.

The Dow campaign was part of another dramatic moment in the antiwar movement. In 1968, student protest at Kent State University in Ohio was directed against the campus's Liquid Crystals Institute (LCI), which developed motion detectors used in Vietnam, and against weapons makers' recruitment on campus. A later demonstration at Kent State gained more national attention when, on May 4, 1970, Ohio National Guard troops fired into a crowd and killed four people. Although part of the demonstration was motivated by President Nixon's decision to bomb enemy targets in Cambodia, which expanded the war into another country, student demonstrators were specifically targeting the LCI and Dow Chemical on campus.[116]

Student protests against military research centers and projects on college campuses were in direct opposition to the hopes of those who developed and administered the NDEA. Rather than accepting a seamless marriage of higher education and the military, some students refused to allow any military research on campus at all. But student protest against military-based research on college campuses was more than an effort to return the campus to a pure state. It was an effort to protest against a war that some of them might be drawn into and that they believed was unjust.

The Convergence of Opportunity and Critique

This chapter has provided a sketch of the paradox that scientists confronted in mid-twentieth-century America: flush with funding and responsible for many of the innovations that led to American economic and military superiority, but by the mid-1950s, increasingly criticized for their ways of thinking, the uses made of their work, and for their failure to accept responsibility for the products they helped to make. In the next four chapters, I demonstrate that scientists did not simply react to the critiques that were made: they helped to make them as well. Not all scientists' lives intersected with political activism; in fact, very few scientists participated in debates about the politics of science between 1945 and 1975. And for some scientists whose lives did intersect, the encounter was confusing, an Alice-in-Wonderland world in which they were told that down was up and up was down: the old rules of political action and meaning that had been so helpful in a less contentious era no longer applied. But other scientists were eager to become involved in movements that were critical of science and to find ways of transforming science to make it more commensurate with their political values. Whether scientists were enthusiastic or reluctant, they found that joining science and politics in new ways would not be easy, for many solutions were fraught with problems that threatened to compromise the bases on which scientific authority rested.

Scientists as Moral Individuals: Quakerism and the Society for Social Responsibility in Science

> What Gandhi did in his life, every man can do. He can refuse every command, every claim that he considers immoral. . . . If every man resolves, with utter courage, and without bitterness against his ruthless enemy, to stand fast at all costs by the truth that he has seen for himself, to follow the gleams of light that he has, with faith that the God of truth is mightier than the devil of deceit and despair, if he can hold to this faith, though he may perish, others will be inspired by his example, and reason and goodwill and love will be triumphant on earth.
>
> —Horace Alexander, *Everyman's Struggle for Peace*, 1947

> It would be sad if the SSRS ever imagined that it had all the answers. With everchanging circumstances the way forward will also change. We must become known for our approach to problems rather than for our answers to them.
>
> —O. Theodore Benfey, "The Task Ahead of Us," 1953

Founded by scientists and engineers who came from a pacifist and moral individualist political tradition, the Society for Social Responsibility in Science (SSRS) eschewed professional work that was morally unconscionable to individual members. The main organizer of the group, Victor Paschkis, was a Quaker and an engineer at Columbia University. Thirty-five scientists and engineers joined him in forming the group in 1949. Many were members of historic peace churches or worked at colleges traditionally associated with those churches. A handful were conscientious objectors during World War II. The SSRS was formed to serve as a source of fellowship for individuals who refused to work on military research projects, as a visible exemplar of personal responsibility among scientists, and as a forum in which discussions about the proper role of scientists and engineers in public political life could take place. All members had a strong commitment to taking personal responsibility for their role in war and in peace, whatever the consequences.

The SSRS was founded at a moment in history that might have seemed auspicious to anyone connected to organized religious pacifism. In 1947, the American Friends Service Committee (AFSC) shared the Nobel Peace Prize with the British Friends Council for their relief work during World War II. The Society of Friends, one of the historic "peace churches" in the United States with which some members of the SSRS were affiliated, was undergoing a renewal that strongly encouraged peace testimony. And only two years earlier, Mohandas Gandhi had led a nonviolent revolution based on the principle of peaceful noncooperation. Signaling an openness to the kind of activities the SSRS advocated, these changes pointed toward a supportive environment for a group of scientists who refused to participate in research or teaching that they found morally objectionable.

Other contemporaneous political changes offered more dismal prospects for moral individualist activism, however. In 1947, the American federal and state governments initiated an extensive network of loyalty programs and laws that disproportionately affected scientists because of their involvement in weapons production. Fears of investigation ran high between 1947 and 1955, as hundreds of public figures were humiliated and lost jobs and careers as a result of security investigations. One immediate consequence for scientists was a quick ending of the "scientists' movement" begun at the end of World War II to maximize scientists' influence on federal funding for science and to ensure international control over atomic energy. These two conditions—a turn toward peace action among Quakers and other peace churches, and the rise of political surveillance and investigations of dissidents—shaped the SSRS's struggle in the early 1950s to find a way to refuse participation in military activities without drawing the attention of the security state.

The SSRS's heyday was between 1959 and 1965. The SSRS never had a large group of members who were involved in the day-to-day business of the group—a dozen or so scientists took care of it—but it was supported by eminent American and international scientists, including Kathleen Lonsdale, Max Born, Hideki Yukawa, Konrad Lorenz, and Albert Einstein. Later in the 1950s, when membership in the group expanded, the SSRS attracted scientific luminaries such as Salvador E. Luria, Linus Pauling, Albert Szent-Györgyi, and Herbert Jehle. Despite its small membership, over its twenty-five-year history the SSRS played an important role in sustaining debates among scientists in the United States and around the world about the proper role of scientists in public life. Through the SSRS newsletter (whose readership of more than several thousand peaked in 1959), letters to the editor, and participation and organization of conferences, members of the SSRS engaged their peers in discussions about whether individual scientists or a scientific "community" were responsible for the uses made of scientific knowledge. From talks to high school students, to conferences on disarmament sponsored by Nobel Prize winners, to pamphlets, activities of the SSRS's membership were a fixture in the political debates about science and public life in the United States. As the group expanded in the 1960s, however, its members became divided over the group's mission and who qualified for participation. They debated whether to place emphasis on moral acts by individuals or on engagement in collective actions.

Two major tensions led to the inability of the organization to remain cohesive and to spread its perspective. The first tension involved the moral basis of the group. A founding principle of the SSRS maintained that social transformations would take place through individual revelations of individual moral choice. In keeping with Quaker principles, founders of the SSRS, including Victor Paschkis, Franklin Miller, Nelson Fuson, and Otto Theodore (Ted) Benfey, hoped to offer moral support or fellowship to those who refused to work on military research projects, and to use moral persuasion through individual witnessing to convince other individual scientists to withdraw from military and other research they found objectionable. The principles have their roots in Quaker theology, which teaches that each person has an obligation—and an opportunity—to seek God's "leading," or moral direction, through the expression of and response to one's "inner light." This spiritual light is the manifestation of an individual's personal relationship with God, an experience so profound to the founders of the Society of Friends that it was said to cause members to "quake." This "light" was subject to continual interrogation by individual Quakers and by the communities in which they lived. Simple living, the everyday symbolic practice of the sacraments rather than their ritual enactment, the accordance of equality to all people, and the belief that peace will come from love rather than fear are among the central tenets of Quakerism; they were also central to the SSRS's early mission.

The SSRS's founders believed that scientists did not have a special obligation to avoid immoral action. All humans shared the obligation to make moral choices to act in ways that were helpful rather than harmful. Because scientists and the military had drawn closer together, many SSRS founders believed that scientists were obligated to resist.

This idea, however, was contrary to another that was developing among scientists and intellectuals: that scientists were a unique moral community. Still, many members of the SSRS *did* see themselves as unique because science was the only profession that had such close ties to the military. The tension between specificity and universalism eventually became a wedge that irreparably divided the group. This tension between the universal and the particular self made it difficult for the SSRS to incorporate some new members and to work with other scientist groups.

As the peace movement reemerged in the late 1950s after a period of quiescence between 1950 and 1956, SSRS's pacifist branch became much more involved in direct actions such as vigils and marches. Some SSRS members did not view the Quaker principles of individualism and persuasion through personal witness as this kind of collective action. Over the next decade, scientists and science students motivated by the urgency of solving problems such as atomic testing, civil rights, and the war in Vietnam would call on the SSRS to take more "action." This tension was compounded by many founding members' idea that the purpose of organization was to support others morally, much as a Quaker meeting offered participants a sense of membership. Ultimately, the two impulses—toward individualism and fellowship on the one hand, and toward proactive, organized collective action on the other—would come into strong tension.

By 1969, the SSRS's original perspective was marginal in both the scientific and the antiwar communities. It had become a catchall organization for people who used the group to organize around specific issues such as overpopulation, computerization, and the war in Vietnam, rather than around the general principle of "right livelihood," which had been the organization's original mission. In this later phase, the SSRS competed with other scientist groups that were more effective at undertaking collective actions.

Nonetheless, during the 1950s, the SSRS played a critical role in sustaining debate about the responsibility of scientists for their products and for the way that scientific rationality was applied to political and moral questions. Through speeches, a newsletter, and letters to the editors of science and nonscience periodicals, and at conferences, in classrooms, and in meetings with individual scientists, SSRS members refused to let their peers believe that the only viable and moral form of engagement with public political issues was simply to produce ideas and knowledge, and let others make decisions about their use. The SSRS's approach never sat well with many other scientists, but the SSRS's persistence made its perspective hard to ignore.

This chapter traces the origins of the individualist model of personal responsibility as scientists in the immediate postwar period struggled with questions about their new role as advisors and providers of ideas to the military. It then moves on to examine how the central features of the SSRS's philosophy continued to be shaped by its founding principles, by red-baiting, and above all by the rise of liberal and direct action approaches to scientists' close ties to the military.

Contradictory Contexts: The Rise of Anticommunism and the Quaker Renewal Movement

Anticommunism and the Collapse of Liberal Science Activism

As discussed in the previous chapter, immediately after the end of World War II, military leaders, members of Congress, government bureaucrats and scientists debated how—and whether—to institutionalize a system of federal research and development (R&D) funding for scientific research and whether atomic energy should be subject to international control. For scientists, what was at stake went beyond policy questions. Scientists struggled with how to reconcile the tension between expertise and popular democracy, the meaning and extent of the "autonomy" of science, and the obligations of scientists in the face of the growing threat of nuclear war. The scientists who were involved in these debates included luminaries such as the nuclear scientist Harold C. Urey, who had long been involved in leftist politics, Hans Bethe, Albert Einstein, and the physicist Edward U. Condon. Younger scientists were involved, too, including the physicists William Higgenbotham, Katherine Way, H. H. Goldsmith, Irving Kaplan, and Lyle B. Borst. Scientists expressed their views in the hallways of Washington, on the pages of journals such as the *Bulletin of the Atomic Scientists*, through lobbying campaigns in Congress, and through a novel public education campaign to promote international control of atomic power. In these ways, scientists developed and expressed their own views about the ideal relationships among scientists, the state, and the public.[1]

The scientists' public education campaign was led by the National Committee on Atomic Information (NCAI), which was loosely associated with the Federation of American Scientists (FAS). The NCAI was a coalition of labor, scientist, education, and religious groups, including the National Council of Churches, the National Council of Negro Women, the United Steel Workers, the National League of Jewish Women, the National Farmers Union, and the Catholic Association for International Peace. In one remarkable year, the NCAI sponsored hundreds of meetings with churches, synagogues, schools, unions, fraternal organizations, and other civic groups, not only providing "the facts" about atomic energy, but also discussing their implications. The NCAI distributed film strips, study kits, and discussion outlines, copies of the *Bulletin of the Atomic Scientists* and the federal Acheson-Lilienthal report on atomic energy, and thirty thousand copies of the pamphlet *Education for Survival*, which suggested to citizens what they could do to prevent nuclear war. Although it would be easy to treat these materials as nonpartisan in the strict sense, the materials clearly indicated that human survival was at issue if weapons proliferated.[2]

Engagement with political issues was no novelty for American scientists. Their closest experience with it was in the 1930s, when American scientists assisted European Jewish scientists, combated fascism, and supported American civil liberties through groups such as the Emergency Committee for Displaced German Scholars, the American Committee for Democracy and Intellectual Freedom, and the American Association of Scientific Workers (AASW).[3] Many scientists involved in these and other groups were committed to an ideal of scientists as an international community based on a commitment to truth through reason. In the aftermath of the war, many American scientists continued to espouse internationalism as a scientific and political ideal. Typical of this perspective was what the physicists Philip Morrison and Robert R. Wilson expressed in the *Bulletin of the Atomic Scientists* in 1946: "It ought to be the task of science to lead the badly divided world to unity of tolerance, to understanding through cooperation, and not to close doors, to hole in, and to start work on bigger bombs."[4]

By 1947, faith in liberal internationalism began to fade among politicians and scientists. But more important in limiting scientists' public political activities were the laws, executive orders, and Supreme Court decisions between 1940 and 1953 that limited citizens' capacity to associate freely, made federal employment conditional on its consistency with national security, and, for a time, limited citizens' right to refuse self-incrimination. Prewar committees and laws such as the House Committee on Un-American Activities (HCUA, formed in 1938), the Hatch Act (1939), which barred communists from federal employment, and the Smith Act (1940), which made it a crime to belong to an organization that advocated the overthrow of the American government, had given the government new powers to suppress dissent. But it was not until 1947 that the purges and investigations had a significant effect on the political activities of American scientists. That year, President Truman signed Executive Order 9835, which called for loyalty investigations of all government employees. In February 1950, Senator Joseph R. McCarthy charged that the Department of State knowingly harbored communists. Hearings on McCarthy's accusations gripped the nation and damaged the careers and lives of hundreds of Americans. As a result of McCarthy's charges, anxiety about domestic communism intensified. In 1950, the Supreme Court upheld the conviction of eleven top leaders of the Communist Party under the Smith Act. The following year, the United States Senate created the Senate Internal Security Committee, designed to rival the HCUA. At the same time that the federal government was establishing these restrictions on liberties, colleges and universities were enacting their own loyalty policies, as did individual states.[5]

As a result, scientists' capacity to dissent—and to carry out their normal work and personal lives—was sharply curtailed. By 1948, about three out of four papers produced by Atomic Energy Commission laboratories were classified, and all Atomic Energy Commission fellowship recipients were required to have security clearance. Foreign scientists were frequently denied visas to the United States, and American scientists were frequently denied passports.[6] The result was a web of restrictions that limited scientists' capacities to freely exchange information (particularly, but not only, if it was military-related), to travel, and to speak freely about political issues of the day. The web was both visible and insidious: scientists could not know if they were being investigated and, until 1953, if dismissed from a position on suspicion of disloyalty, they did not have a right to hear the charges. Although many Americans suffered from these rules, scientists were hit especially hard because of their access to scientific ideas that were, or could be, used in national defense.

Among scientists under FBI surveillance were Albert Einstein; Harlow Shapley, the president of the American Association for the Advancement of Science (AAAS); Edward U. Condon, a physicist and former president of the American Physical Society; the physicists J. Robert Oppenheimer and Philip Morrison; and Linus Pauling, the chemist and future Nobel Peace Prize winner. Many other lesser-known scientists would pay a high price for the government's anticommunist zeal, enduring embarrassing and time-consuming investigations into their personal, political, and professional lives.[7] Professional organizations, too, were targeted, and sometimes it was members of these organizations who served as informants.[8] By the mid-1950s, the frenzy had slowed with the discrediting of Senator Joseph McCarthy, but not before it had ruined the careers of dozens of scientists and instilled fear and paranoia into thousands of others. As a result, although they were involved in the debates over the control of atomic energy between 1945 and 1947, many scientists began to lose their taste for public political debates, given that persecution was a possibility even when one was not involved in political activities, subversive or not. The alliance with the military in this early period proved to come with heavy costs. Reporting on the discussions about the mobilization of scientists for the Korean War, the science journalist A. G. Mezerik captured the scientists' dilemma: "And, year by year, it seems to many scientists that they are more and more becoming prisoners of their own creations. The more powerful those creations are, the more the scientist seems to lose his independence and with it his will to speak out."[9]

Between 1945 and 1947, key pieces of legislation passed that settled the pragmatic questions of control over atomic energy, federal sponsorship of scientific research, and the roles that scientists would play in decisions about how federal funds should be spent on R&D. These decisions, and the development of more intensive security programs, played an important role in the demobilization of the "scientists' movement" that the Federation of American Scientists had organized to support international control of atomic energy and to lobby for an important role for scientists in decisions about federal funding for R&D. As a result, the flurry of lobbying, education, and other liberal actions among scientists slowed and then died out. By 1948, the NCAI was collapsing, and the FAS was becoming mired in concerns about communist infiltration. Scientists' attempts to place themselves at the forefront of a new, liberal international political and intellectual community came to a halt as almost as quickly as they had begun.

Quaker Renewal

Although the political climate was becoming much chillier for those who wished to dissent from cold war politics and policies, not all dissenters withdrew from political engagement. Quakers and pacifists were among those who remained engaged. After a century of decline and organizational disarray, a new movement within Quakerism during World War I called for a renewal of peace testimony and activity, a practice that had fallen out of favor in the Society of Friends during the nineteenth century. During the 1920s and 1930s, more people became Friends largely because of the new emphasis on peace. More educated, more elite, and younger than other Quakers, some of these "convinced" Friends, as well as those who were born into the Quaker religion ("birthright" Friends), served in conscientious objector camps or were imprisoned during World War II. Rather than halting the renewal project, camps and prisons served as meeting places for like-minded Quakers, who frequently organized interracial groups to fight against injustice in the camp and prison systems. Among the organizers were David Dellinger, A. J. Muste, and Ralph DiGia, who would go on to become leaders of antiwar activism in the 1950s and 1960s through groups such as the War Resisters League and the Committee for Nonviolent Action. Some of the founders and most active members of the SSRS, including Don DeVault, Malvern Benjamin, and Edward Ramberg, were also in camps or prisons, where they were participants in actions for social justice as well.

Immediately after the war, some conscientious objectors and other pacifists formed the short-lived peace groups Peacemakers and the Committee for Nonviolent Revolution, which were intended to serve as organizational vehicles for peace projects. One of the main debates in these groups and in the AFSC was about what would constitute peace activity. For many participants, it meant any effort that would change society rather than change individuals, as Quakers had traditionally done. The distinction between these two practices is a fine one, but worth explaining because it proved to be critical in the direction that the SSRS took in the 1950s.

Traditionally, Quakers believed that their major religious mission was to save individual souls. Pioneers in the formation of prisons in the United States, they believed that conversion to moral action and thought would come from individual contemplation and attention to one's "inner light," or the manifestation of God within each individual. Living an exemplary moral life was both evidence of attention to inner light and an example for others to follow. The peace renewal movement was organized around this same process of recognition, but it aimed to persuade larger numbers of people to engage in peaceful activities. During the 1950s, the peace renewal movement split over just what sorts of actions ought to be undertaken to promote peace. Some, like Robert Scott, the main author of the 1955 AFSC publication *Speak Truth to Power*, advocated that peace activists take "direct action" using tactics such as peace vigils, statements of conscience that advocated a specific position on an issue, peace walks, and civil disobedience. Other renewalists, such as Norman Whitney, rejected collective action and statements of conscience. Whitney advocated a return to the traditional practice of public, personal testimonies to promote conversions of other individuals. Whitney also thought that weekend conferences, which had been sponsored by the AFSC since the 1930s, were a critical means of both educating people about peace and taking action. Action, for Whitney, was peace testimony. By the late 1950s, it was clear that the peace renewal movement was moving toward direct action and statements of conscience rather than testimony and weekend retreats as means of social change.

When the SSRS was founded in 1949, the AFSC had just won the Nobel Peace Prize, and the peace renewal movement, though small, was gaining adherents and promoting new means of changing society. Despite the increasingly repressive political climate, scientists and engineers attuned to this movement would have reason to believe that there were others who would be supportive of their turn away from military applications of their work.

THE FORMATION OF THE SSRS, 1947–1949

The idea for the formation of the SSRS originated with Victor Paschkis, a Quaker who arrived in the United States from Germany in 1938. Deeply influenced by the German Fellowship of Reconciliation (FOR) member Sigmund Schultze, Paschkis was a member of the American FOR and the Society of Friends. Paschkis worked as a heat flow analyst on the faculty of Columbia University, and made his home in Hidden Springs, New Jersey, near a cooperative Society of Friends community. A "convinced" rather than a birthright Friend, Paschkis was an active participant in Quaker meeting in New Jersey, and was friendly with other Quaker scientists and engineers at Earlham, Haverford, and other Quaker colleges. The impetus for the formation of the SSRS came from the peace activist and pacifist A. J. Muste.

In 1946, Muste wrote an impassioned letter to Albert Einstein, a member of the Emergency Committee of Atomic Scientists, a small group whose main activities were fundraising for the NCAI. He implored Einstein to make a personal statement renouncing any involvement in weapons production.[10] In his letter, Muste was especially critical of the scientists who were working through the Federation of American Scientists to lobby for international and civilian control of atomic energy, many of whom continued to engage in research on atomic energy. "How can scientists who are building a stockpile of atomic bombs," Muste asked, "teach men that they must abandon nuclear weapons to preserve civilization?"[11] Arguing against the theologian Reinhold Niebuhr's claim that there was such a thing as a just war, Muste fervently believed that all war was evil. For him, radical pacifism was the only morally justifiable position vis-à-vis war. As the most important voice of radical pacifism in the United States, Muste held that the horror of nuclear war was so powerful that humans should—and would—renounce war forever.[12] Few scientists were willing to sign on to Muste's call for conscientious objection to work involving the production of weapons.[13]

In 1947, Paschkis wrote an article in the *Friends Intelligencer* that echoed Muste's view. Paschkis argued that all individuals, regardless of occupation, were personally responsible for the uses made of their work, and must choose livelihoods that were in line with their moral values.[14] His colleague at Columbia University, the educational psychologist George W. Hartmann, made a similar critique in *Fellowship* that same year: "How much faith can we place in the depth of conviction of those technicians who two years ago were literally 'evaluating' the worth of their well-paid contributions to the 'war effort' in terms of the number of dead Japanese for which they could claim credit?"[15] After reading Paschkis's article in the *Friends Intelligencer*, Muste telephoned Paschkis and suggested that he "get together a group of conscientious engineers and scientists," although he warned Paschkis that the task would not be easy, since "most scientists and engineers rarely [join] pacifistic societies because most of them strictly separate their work from their conscience."[16]

After an encouraging conversation with James Vail, the past president of the American Institute of Chemical Engineers, Paschkis used the FOR membership list as well as his own network of Quaker scientists and medical doctors in the Philadelphia–New York area to contact 114 scientists and engineers who he thought might be interested in forming a new organization. Seventy-two of those contacted expressed interest. With the enthusiastic support and participation of Haverford College president Gilbert F. White, thirty-five scientists and engineers met at Haverford in June 1948 to lay the foundation for what they would eventually call the Society for Social Responsibility in Science[17]

These scientists and engineers were not the only ones who were looking for ways to opt out of the new military-science relationship. Shortly after the publication of Muste's views, the MIT mathematician Norbert Wiener refused to accept military funding or share his knowledge in the service of war.[18] The physicist Leo Szilard, with whom Albert Einstein had first proposed the idea of the atomic bomb to President Roosevelt, had become an active opponent of war and the arms race. In a 1947 essay, Szilard criticized scientists who advocated building a hydrogen bomb: "It is remarkable," he wrote, "that . . . these scientists . . . should be listened to. But mass murderers have always commanded the attention of the public, and atomic scientists are no exception to this rule."[19] Founders of the SSRS were inspired by the actions of these eminent scientists.

Participants in the June conference determined that the AAAS, FAS, and the National Committee on Atomic Information, all concerned with the politics of atomic energy, were not venues for the expression of the group's views. They decided that "none of these organizations would wish to tie up with the group of people taking the minority view . . . and certainly would not want to help spread this view."[20] There was a certain irony, in the founders' minds, in the attention that the atomic scientists who participated in making the bomb but now opposed it were getting, while those who had refused to participate in the military at all were either ignored or disdained. Conference members decided that their organization would be a support group and referral service for those who refused to participate in military projects. Certain research projects, such as chemical warfare, submarines, missiles, and large-scale atomic development were unequivocally to be avoided by members. But this still left the more general problem: Where were they to draw the line for other projects? Should the group define what was unacceptable? Should individuals have to request approval from others for work they would undertake? What if, in ten to fifteen years, the research could be expected to have military applications? The group decided that it would be up to each individual to decide, on the basis of conscience, from which kinds of work they would abstain. Yet this decision did not resolve the problem of future military uses of one's research, a concern that would arise for new, younger members of the group in the late 1950s and the 1960s. These scientists would be less content with only avoiding direct participation in military research as a means of ensuring that scientific knowledge was used for nonviolent projects.[21]

The SSRS was formally organized at a meeting in September 1949 at Swarthmore College and elected Paschkis as its first president. William F. Hewitt Jr., a physiologist from Howard University, was elected vice-president, and Vincent Cochrane, a Wesleyan University biologist, was elected secretary-treasurer.[22] The name for the new group was the subject of much discussion, in part because the founders wanted to ensure that they were not a political organization, but an organization of moral individuals engaged in mutual support and witness.[23] Organizers rejected any name with the word *peace* in it, including Scientists and Engineers Committee for Peace and Justice; Scientists and Engineers for Peace; Scientists' and Engineers' Peace Association; and Scientists' and Engineers' Committee on Social Responsibility and Peace.[24] They settled on the Society for Social Responsibility in Science because it conveyed the idea that each scientist had to choose what sorts of work would satisfy the criterion of being responsible to the larger community. Once the group was formed, Paschkis sent out more letters to colleagues urging them to join. His letters were critical of scientists who wanted to educate the public but who continued to advise the government on atomic energy research. He reminded his letter recipients that "in order to be really effective in such [public] education, they themselves [scientists] have to set an example by absolutely refraining from what they see as destructive work."[25]

The initial response from the scientific community and the press to the organization was generally favorable. In a letter to the editor of *Science*, Albert Einstein urged others to join the group, as he had, because membership "will make it easier for the individual to clarify his mind and arrive at a clear position as to his own stand; moreover, mutual help is essential for those who face difficulties because they follow their conscience."[26] The *New York Times*, too, initially held a favorable view of the goals of the SSRS, stating that the group's "heterodoxy" could be traced back to Leonardo da Vinci's refusal publicly to describe his design for a submarine for fear it would be used for evil purposes.[27] Clearly, the SSRS position was understood as a bold political move for scientists on the cusp of the cold war.

The SSRS quickly made it clear to those who shared their commitment to refusing to participate in morally objectionable practices that their actions would be professionally and personally costly. In the first of a series of occasional pamphlets, titled *Social Responsibility in Science*, SSRS member Arthur E. Morgan, a former president of Antioch College and later chair of the Tennessee Valley Authority, wrote:

One element of insurance to enable a man to live by his standards in case of emergency is the maintenance of a very moderate standard of living, and the accumulation of a financial cushion. He should be able at any time to tell an unethical employer to go to hell without thereby losing immediate food and shelter for his family. Disciplined, simple living and reasonable savings are a great help toward professional freedom. . . . It is necessary to give up definitely any idea that scientific reputation is the primary aim. One may find it necessary to live his life and to do his work in relative obscurity in order to live by his convictions.[28]

It is difficult to imagine that any but those American scientists most deeply committed to noncooperation, or very secure in their careers, would be willing to make the sacrifices that Morgan suggests were necessary to survive the lean times and even professional obscurity that he believed SSRS members would likely encounter. Those willing to make sacrifices would face an additional hurdle that Morgan did not mention: the political risk of joining what amounted to a conscientious objector group for scientists and engineers at a time when security investigations were growing in number and intensity. Scientists facing such investigations would not find many supporters among the public or in government, either, because the growing consensus was that the Soviet Union posed such a great threat that there were few real options other than participating in the arms race. The historian Paul Boyer succinctly described the change in the perception of nuclear weapons in the five years after the war: "The dread destroyer of 1945 had become the shield of the Republic by 1950."[29]

The political risks that SSRS members faced were exacerbated by the decline of the religiously based national and international peace movement by 1950. Among the traditional Protestant peace churches, only the tiny Brethren and Mennonite denominations were uniformly committed to pacifism. Although there were a small number of Jewish pacifists, their numbers, and those of the small pacifist Catholic Worker community were in decline by 1950 as well.[30] The Quaker renewal movement, energetic and lively between 1945 and 1948, was in decline by 1950; most of the organized peace activities carried out by Quakers were done through the AFSC in overseas projects. Thus, at the time the SSRS was founded, ties to the peace community could not be relied on as a source of social support.[31]

ISOLATION AND SOLIDARITY IN AN ERA OF ANTICOMMUNISM AND COMMUNITARIANISM, 1949–1956

The first seven years of the SSRS's existence were lonely ones for the core members of the group, a handful of devoted members who held SSRS offices, wrote for the newsletter, and went to meetings. Many were scientists and engineers in the Philadelphia–New Jersey area, including Paschkis, Edward G. Ramberg, Malvern Benjamin, Ted Benfey, Nelson Fuson, and Leonard S. Dart, or had lived there, such as the newsletter editor and publisher Franklin Miller, who lived in Gambier, Ohio. The SSRS could count among its members prominent international scientists, including Kathleen Lonsdale and Max Born, and the American émigré scientists Albert Einstein and Hans Thirring (a Princeton physicist), but these individuals were largely unconnected to the day-to-day workings of the SSRS. During these years, the relatively intimate Philadelphia-area group explored ideas and practices at the intersection of individual responsibility and science. The SSRS served as a source of moral support and a forum for discussion among a relatively small group of scientists. Anticommunist, avoiding statements that might be construed as "political," and committed to the practice of individual witness rather than group action, the SSRS struggled to hold itself together until peace activity and concerns about atomic fallout after 1956 provided a more favorable climate for action. It is difficult to see how the SSRS could have been sustained without members who had preexisting religious commitments.

Anticommunism and the Avoidance of "Political" Positions in the SSRS

Anticommunist loyalty investigations, which were commonplace during the early life of the SSRS, affected the group in two major ways. The first was that the group took great care in ensuring that the government, citizens, and other scientists did not identify them as communists or as a group that communists could infiltrate. The preamble to the SSRS constitution stated that the group "stood against any war trend, whether in the United States, Russia or any other country,"[32] a phrasing that the founders included lest others believe that they might support war against the United States. The preamble, though, was not enough to dispel some members' concerns. In a letter to his colleagues in June 1949, Paschkis wrote: "One major difficulty . . . is that of 'Communists.' The executive committee believes that the Communist doctrine is definitely not antiwar, but rather only against certain types of war, or against war waged against certain countries. A person believing in this Communist doctrine, could therefore not, in sincerity, subscribe to the aims of our organization. But there is no way to detect people who, while not in agreement with our statement of aims, might be willing to sign it [the SSRS membership list]."[33]

To ensure that they were not inadvertently associated with communists, the SSRS scrutinized groups and individuals who invited them to participate in collective events. Invited to sign a scroll for a 1955 dinner honoring Harold C. Urey, a prominent chemist who was staunchly anticommunist but who was nonetheless the subject of a loyalty investigation in 1949, the SSRS decided not to attend because they were "suspicious of a Communist front activity in this dinner."[34] Some members feared that identifying themselves as members would hinder their ability to work with the state department on developing a small tools program for developing countries, because "the State Department might take our organization as tainted by Communism."[35]

Efforts by the SSRS to avoid association with communism may have been persuasive, but the claim that the SSRS was not political and posed no threat to the country did not convince everyone, including the FBI. One former SSRS member recalled his encounter with FBI agents:

> My wife was at home, it was during the school year, and I was off on campus teaching a lab. There were two young gentleman, in blue serge suits, hats, and ties, who came to the door. And my wife said, "FBI?"— she thought that's just what they looked like—and they said yes, they were from the FBI. They looked like FBI men to her, though we'd never seen one personally. They called me over from the lab. I took them over to my office, my spare bedroom where my files were and my addressograph machine. They asked me "Well, do you have any members in Cuba?" I said, "I don't know, maybe we do. We don't consider it to be of importance." I thought we might have a Cuban member. I went over to my file of stencils, and I randomly pulled out one of my 1,500 stencils. It was the stencil for the only Cuban member! I showed them the address, and told them that the group was entirely for humanitarian purposes. That seemed to satisfy them.[36]

Ted Benfey, one of the founders of the group and its second president, said in response to a query about whether the group members were afraid of the security investigations of scientists during the 1950s, "In retrospect I am amazed that I was willing to be president."[37]

The SSRS's claim to be a supportive fellowship rather than a group with a political position—or one that could be infiltrated—was unconvincing to the AAAS in the early 1950s. After years of operating as a generalist professional association, in 1951 the AAAS changed its mission. Under the leadership of Warren Weaver, the AAAS turned its attention toward public understanding of science. The AAAS leadership and membership did not agree on exactly what that meant in practice, but as the historian Bruce V. Lewenstein argues, members of the scientific community were "morally certain of the effectiveness and certainness of their worldview. . . . [They] hoped to see their values deeply influence the culture around them."[38] The new focus of the AAAS, then, would seem compatible with the SSRS's purposes. And indeed, the SSRS was able to organize a panel discussion on "Individual Responsibility of the Scientist" in December 1951, with speakers Charles Price of the University of Notre Dame, Carroll C. Pratt of Princeton, and the SSRS member William F. Hewitt.[39] Yet in 1954, when the SSRS requested formal affiliation with the AAAS, the executive committee turned down the request. They cited the SSRS's failure to furnish them "with sufficient information about its membership and its objectives to warrant favorable action on the request," and the "limitations" of the group's objectives, which left questions about the appropriateness of affiliation with the AAAS.[40] The association's decision appeared in the *New York Times*, the *New York Herald Tribune*, and other newspapers; in each case, the SSRS was characterized as an organization that desired to find constructive applications for science, not as a politically threatening group.[41]

The SSRS's firm and unwavering commitment to individual-level witness clearly made government officials and the AAAS uneasy. It is difficult to know how much of this uneasiness was because some participants in the group had been conscientious objectors, because of their assertion— difficult for many Americans to understand—that they were a fellowship group, not an "action" group, or because some members refused to work on military research. In an era in which the flimsiest reasons, or no reasons at all, could cause a person or group to be suspected of disloyalty, it is likely that uneasiness with the SSRS was based in large part on the simple fact that members did not conform to existing political expectations for scientists, namely, that they should willingly serve the government, or avoid any statement about politics at all. Unfortunately for the SSRS, the group's individualist stance was also out of sync with the emerging idea that scientists formed a special moral and political community.

Religiously Based Individualism in an Era of Scientific Communitarianism

The SSRS never called itself a religious group, nor did it formally espouse Christian principles or insist that its members were religious people.[42] Yet publications sponsored by the SSRS and by its members between 1949 and 1956 frequently evinced Christian sentiment. For example, in SSRS pamphlet no. 1, *Scientists and Social Responsibility* (1949), the group grounded moral obligation in Christianity. A year later, SSRS member Halbert Dunn's "A Prayer to be Said by the Men of Science" appeared in the *SSRS Newsletter*. He urged scientists and engineers to place fidelity to God above all else: "And if I work in science, but I remember not the Spirit of the Lord, and I fail to do the creative work of God, and I serve the dead god of theories; then take me from science and theory, for I would work for You in all I think and do."[43] A similar point was made in a quotation from the wall of the Stanford University chapel, published in the newsletter in 1952: "There is no narrowing so deadly as the narrowing of an individual's horizon of spiritual things. No worse evil could befall him in this course on earth than to lose sight of heaven. . . . No widening of science . . . can indemnify for an enfeebled hold on the highest and central truths of humanity."[44]

The SSRS member Kathleen Lonsdale was a British Quaker X-ray crystallographer jailed during World War II for her refusal to serve in the British military.[45] In a defense of conscientious objection for scientists written for the *Bulletin of the Atomic Scientists* in 1952, she argued that "the Christian in particular owes allegiance to God before his government. True democracy recognizes this, and permits that man who has a genuinely conscientious scruple to refuse work he believes contrary to the purpose of God and therefore injurious to mankind as a whole."[46] These examples of religiously based calls for individual action suggest why observers might see the SSRS as religiously based, even though all people were welcome in the group and some prominent members were not Christians.

The public espousal of Christian values in science was inconsistent with key intellectual and demographic developments in the postwar period. Before 1938, Protestants dominated American science. Protestantism served more as a source of credibility in scientific circles, a marker of social status rather than a specific religious practice. After World War II, intellectual culture in the United States became more secularized. As David A. Hollinger has argued, this occurred in part because of a widespread revulsion against the Nazi's extermination of millions of people singled out because they did not fit a particular vision of a Christian culture. Christianity thus became less powerful as a means of conveying legitimacy and worthiness in elite intellectual circles, especially in science, where many of the best scientists in the country were Jewish.[47]

Equally important was that the SSRS's Christian individualism was incongruent with the idea that scientists formed a distinctive moral community that shared unique norms and traditions. Intellectuals, including Robert K. Merton, Michael Polanyi, and Edward Shils, were among those who articulated this position. Merton's famous argument, published in 1942, was that science, as a social system, was characterized by an ethos of universalism, collective scrutiny of claims, group ownership of ideas, and disinterestedness.[48] For Shils, the community of science was based in part on traditions, ongoing practices, and values that were sustained through interactions among scientists. In his 1955 essay "The Autonomy of Science," Shils argued that traditions were not frozen in time, but underwent "continual and gradual revision, clarification and improvement in accordance with the judgment of scientists."[49] Other influential writers, such as Michael Polanyi, though not arguing that scientists were completely autonomous, saw them as a community that generated knowledge using shared social knowledge beyond objective fact.[50]

Intellectual arguments about scientists as a community were complemented by the collective actions that scientists undertook during and after the war. Members of the Manhattan Project acted, and were recognized by others, as a socially distinctive group that worked, on the basis of moral imperatives, on the development of the bomb and, after the war, worked almost as an interest group through their lobbying and education campaigns. These activities also helped to shape the idea that scientists were a distinctive group of political actors, rather than simply a collection of individuals.[51] The political pressures on scientists as a result of the anticommunist fervor during the first decade after the war also helped to constitute scientists as a socially recognized group. Indeed, to the extent that scientists were effective in responding to the tightening security noose, it was through the work of the FAS and the AAAS. These groups issued statements against secrecy requirements and opposed—somewhat pallidly, given the risks of dong so—the attacks on scientists for their supposed lack of loyalty.[52] To be sure, individual scientists, including Albert Einstein, Leo Szilard, and Norbert Wiener, routinely acted in the political arena on the basis of individual conscience, but the main tendency was to see scientists as a political and moral group.

Thus, the apparently religious, and individualist, approach that the SSRS took moved against the tide of collectivism that was articulated by intellectuals, the state, and scientists themselves. In this climate, the SSRS struggled to gain new members. This struggle, however, did not prevent a core group of U.S.-based and international members from making the question of scientists' moral responsibilities an important part of the discourse of scientists and other intellectuals.

Navigating the Political and Cultural Waters: The SSRS as an Exemplary Action and Fellowship Group

The SSRS was structured as a bureaucracy, as many American advocacy groups and mutual benefit associations are. Yet unlike these groups and associations, the SSRS undertook neither to serve only its members, nor to advocate only on behalf of other people. The SSRS thus does not fit neatly into standard categories of American civic action. During the first half of the 1950s, much of what SSRS members did can be characterized by what Max Weber called "exemplary" action.[53] As SSRS president William Scott reminded participants in the 1956 annual meeting, joining the SSRS was not akin to joining a church, a political party, or a mass movement. SSRS members "witness to our refusal to engage in activity that appears individually to us to be destructive"[54]

One kind of exemplary action was through the jobs that SSRS members held. Many taught at liberal arts colleges or worked for private firms or universities researching subjects such as television, artificial fibers, industrial machine design, heat flows, and theoretical physics that were understood by those engaged in them to meet their own moral standards. Some members coupled this with public speaking at conferences and other public forums and publishing articles and letters to the editor. Perhaps the most indefatigable speaker from the SSRS was Victor Paschkis. Evangelical about the need for scientists to accept personal responsibility, he spoke at national and international meetings of scientific societies, at colleges and universities, and in all manner of public forums in the United States and Europe.

Other SSRS members were also participants in meetings of other groups, often pushing speakers and other participants to consider what was meant by the assertion that scientists had moral responsibilities. Ted Benfey, for example, at a meeting about science and society at Harvard University pressed the speakers I. I. Rabi, a Columbia University physicist, and Harry Gideonse, the president of Brooklyn College, about whether they thought that scientists' responsibility was personal or collective.[55] Paschkis, Cuthbert Daniel, and two other SSRS members attended the Nation Associates meeting in May 1952, again pressing participants to consider the personal responsibilities of scientists.[56] The SSRS's arguments were frequently mentioned in venues such as *Chemical and Engineering News*, the *Bulletin of the Atomic Scientists*, and UNESCO publications, but the SSRS had little luck in attracting new members. Paschkis, returning from a trip to California, reported that members of chapters in Los Angeles and elsewhere were isolated and lonely, having few people with whom to meet for fellowship and support.[57]

Never actively persuading others so much as hoping to convert them through testimonials about their own choices, members of the SSRS took care to ensure that audiences understood that they spoke about their own views, not those of the SSRS. Another form of exemplary action was publishing articles that justified and explained exemplary moral action. SSRS members Ted Benfey of the Haverford engineering department, Franklin Miller of the Kenyon College physics department, and William F. Hewitt Jr. of the Howard University physiology department were among the most prolific authors of articles on the subject of personal responsibility. They wrote articles about the difference between individual and collective responsibility, what scientists were responsible for, and to whom they were ultimately responsible, which appeared in publications for scientist and nonscientist audiences such as the SSRS newsletter and the *Bulletin of the Atomic Scientists*.[58]

In a pattern that would continue for the duration of the organization's life, members of the SSRS sometimes attempted to participate in activities typically associated with "emissary" organizations, but they were largely unsuccessful. For example, in 1950, after a year of correspondence, interviews, and meetings with one another and possible sponsors, a group of SSRS members lit upon the idea of providing technical assistance to needy individuals and groups overseas and in the United States. Organized by SSRS member Norman Polster, an engineer at Leeds & Northrup, and Leonard S. Dart, a physicist with American Viscose Corporation, the SSRS Small Tools Committee project drew on circumstances found in Pennsylvania during the Great Depression. During this time, thousands of urban families acquired small plots of land and became part-time farmers, sometimes out of current need and sometimes to protect against future shortages. These families relied heavily on advice from government extension agents. Polster and Dart specifically sought to develop a small international agricultural tools program to assist family farms.[59] Initial contact with the Food and Agricultural Organization of the United Nations (FAO) to propose the "small tools" idea met with a favorable response, but the FAO informed Polster and Dart that in order for such a program to receive funding, they would need to find a laboratory where small tools could be developed and tested. Polster and Dart continued to pursue the project, but were not able to secure laboratory space. The Small Tools Committee itself, however, continued to provide advice on an ad hoc basis to different groups, including the interracial Koinonia Christian community in the state of Georgia.[60]

The SSRS attempted another type of emissary action through their job placement service for scientists and engineers who sought jobs that were compatible with their consciences. The grandly named Occupational Division of the SSRS was charged with this task. Those who were looking for suitable work published announcements in the SSRS newsletter describing their qualifications and the sort of work that they wished to find, or they contacted the SSRS directly. Yet the SSRS rarely found jobs for these scientists, mainly because job seekers seem to have found work without the use of the placement service. On occasion an agency or individual contacted the SSRS for assistance in finding appropriate staff. For example, in 1952 the United Nations Technical Assistance Administration asked whether it could consider the SSRS as a source for personnel recruitment. There is no record of the SSRS providing any assistance to the UN program.[61]

Although the SSRS's annual meetings did not help get the organization's message out, they did provide members with moral support for their unpopular position and with a venue for exchanging ideas with other members. This sense of fellowship was captured by Herbert Jehle, a Princeton physicist who had been deeply involved in the worldwide leftist World Christian Movement in Germany in the 1930s. Jehle turned down a position at a German university and refused to participate in research that would help rearm Germany. He was interred in a Vichy concentration camp in 1940 for refusing to serve in the German military, and then rescued by members of the world peace movement, who provided aid to people in concentration camps. He arrived in the United States in 1941.[62] In 1951, Jehle wrote that the SSRS conferences provided more than a means of conducting business and sharing ideas. There was a "more subtle and at the same time more profound meaning to a conference. . . . [It served the] need to discuss it [a moral position] with some friends who have gone through and thought about similar situations . . . and we need to exchange friendship and love so that we can stay away from resentment and bitter feeling"[63]

At least for the American rank-and-file members of the SSRS, being a member of the organization involved individual witnessing to other members and nonmembers, and, for those who could attend, participation in annual meetings for purposes of fellowship, SSRS business, and engaged discussion about the meanings of personal responsibility. The large number of elected positions and committees in the SSRS, though not used to structure advocacy or other such actions, did seem to have a useful effect nonetheless. This structure forced members to take responsibility for the continuity of the organization—essential if the group was to survive during a time when reasons for dropping out of the group were abundant. This core group of SSRS members helped to sustain an organization whose international members were to bring the group more public attention in the United States than the American members themselves had been able to muster.

The SSRS always attracted overseas members, who tended to be drawn from Western Europe (especially Germany), Japan, and India. Many early overseas members, such as the Japanese chemist Shingero Oae, were known personally by American members of the SSRS through professional or pacifist circles. Thus, Paschkis knew Oae through professional ties and the small pacifist movement that grew in Tokyo after World War II. That movement had the sponsorship of pacifist and religious groups in America, including FOR. Since overseas members did not often attend meetings of the SSRS or have other SSRS members in their vicinity, they maintained relationships with the SSRS through the newsletter and correspondence with individual SSRS members.

Association with some of these overseas members gave the SSRS a level of legitimacy and prestige. The SSRS counted among its members the

1954 Nobel Prize winner in physics, Max Born, who was active in international anti–nuclear proliferation activities. Hideki Yukawa, the Japanese 1949 Nobel Prize winner in physics, was also a member of the SSRS. He joined in 1953 after the thermonuclear bomb tests in the South Pacific in the early 1950s killed a fisher on the *Lucky Dragon* and poisoned the other crew members. The Austrian physicist Hans Thirring and the English chemist Charles J. Coulson, the author of *Science, Technology, and the Christian*, were also members.[64] Overseas members were not always eminent, of course. In 1955, the SSRS had a small core of thirty-nine members in Germany, Norway, and ten other countries.[65] Like those of their American counterparts, the actions of overseas members often gave rank-and-file SSRS members a sense of hope about their work. For example, in December 1954, the British mathematician Bertrand Russell, in a speech on BBC radio titled "Man's Peril," called for an end to the arms race. He asked, had humanity dropped to such a level that it would "carry out the extermination of life on our planet?"[66] Albert Einstein and Max Born joined Russell in his efforts. They worked together on a statement, called the "Russell-Einstein Manifesto," signed by eight other eminent scientists, some of them Nobel Prize winners, from around the world.[67] Before the manifesto was released, Franklin J. Miller and Paschkis met with Einstein at Princeton to discuss what role, if any, the SSRS might play in publicizing or circulating the manifesto. Although Einstein suggested that they allow Russell to take the lead, they were heartened by Einstein's willingness to discuss the manifesto with them, particularly because Einstein had publicly advocated taking personal responsibility for one's scientific work since the end of World War II.[68]

Efforts to Increase Membership

In 1955, the SSRS had perhaps only thirty-five to forty core members, and a subscription list of two or three hundred. Its funds were limited. Attempts to form chapters in Los Angeles, New York, and Chicago quickly faded, often because participants were "talked out" and because the SSRS did not have a program of organized group activity. Once local SSRS chapters discussed their views on participation in military or other kinds of research, or had committed to avoiding a specific kind of work, local members felt that there was little reason for the groups to continue to meet in the absence of concrete actions that often characterize voluntary association membership in the United States. Undaunted, the SSRS consistently sought ways to increase its numbers.

In 1953, the group asked the Harvard astronomer Harlow Shapley (who was himself a subject of loyalty investigations) and A. G. Mezerik, the author of *The Pursuit of Plenty: The Story of Expanding Man's Domain*, to provide an external review of the organization.[69]

Shapley noted that the group was "not communist" and "loyal to American traditions," and more functional as a fellowship between scientists than a "wide association." Yet he thought it "too passive and good mannered to be effective." Despite the publicity that famous scientist members provided, Shapley felt that there was little to attract most scientists, especially when many people did not want to join organizations due to "fear of smear." He contrasted SSRS with other groups that he deemed successful, including the AASW. Other groups Shapley believed the SSRS should emulate included the Conference on Science, Philosophy, and Religion in New York and the Foundation for World Government. Each had money and leadership, said Shapley, whereas the SSRS did not. Shapley was pessimistic about SSRS's future. He reported that the SSRS was considered by some to be communist, yet it did not openly espouse a political position or course of action that might have attracted a broader range of scientists. He recommended that if the group were to continue, it would need more funds (he did not say for what) and it would need to change its name because "many people associated the S.S.R.S. with the U.S.S.R." The group's careful consideration of what to call itself when it organized in 1948 evidently cost it, rather than drew new members.[70]

Mezerik was more optimistic than Shapley. Although he saw the group as too remote and its newsletter too ethereal, he observed that because "[Senator Joseph] McCarthy had recently suffered some defeats while Gandhi and anti-segregationists seemed to be gaining power," the opportunity for the SSRS to expand was available. Both observers recommended that the group take much bolder action in the future.[71]

Despite the continued lack of success in attracting the attention of the AAAS, by 1956, the SSRS was acquiring new members. Heartened by having the 1954 chemistry Nobel Prize winner Linus Pauling join the organization and by a rethinking of "the entire SSRS program" at their March council meeting, the SSRS core members clearly were becoming more energized about the possibilities for attracting more members.[72] There were new candidates for SSRS offices, too, after six years of having the same small core serve. Some of the candidates proposed new projects. The English biologist Alex Comfort, for example, thought that the group should continue to witness, but should work to resolve specific problems, such as increasing the life expectancy of the poor, and investigate problems such as aggressive behavior in humans. The computer scientist Kenneth Knowlton, also running for office, proposed that the group become more involved in expressing its political viewpoints.[73] Founders of the SSRS who were running for office, by contrast, advocated a limited expansion of the existing practices of witness and fellowship.

New members and new visions are always a double-edged sword for existing organizations, at once energizing them and potentially threatening the status quo. New members who came from outside the close circle of Quaker and other Christian scientists who formed the core of the SSRS were calling on the group to espouse political viewpoints—much as some eminent scientists were doing—and to develop activities that the group could undertake. New members posed great opportunities to expand and exposed the group's strengths and weaknesses. The cozy, intimate Quaker-based group in the United States, and the far-flung network of elite and rank-and-file scientists that characterized the SSRS between its founding and the late 1950s would become a larger-scale forum for debates among scientists and engineers about their proper role in public life, paralleling a similar expansion in debates about public life among leftists, who were also looking for new forms of action. By the end of 1963, the SSRS membership could be found engaging in a wider variety of activities and providing a range of answers to the question of how scientists ought to be engaged in public political life, but few would agree on any one idea.

COMPETITION AND COMPANIONSHIP: LIBERAL ACTIVISM AND DIRECT ACTION PEACE GROUPS, 1957–1963

In 1957, there was an unmistakable renewal in leftist political organizing and debate in the United States. The weakened credibility of anticommunist witch hunts following Senator Joseph McCarthy's fall from public favor in 1954 made it possible for open dissent concerning civil rights and above-ground atomic testing to develop. It is difficult to overestimate the excitement, however cautious, that many leftist activists began to feel after a decade of political repression that had been preceded by a war. From intellectuals such as C. Wright Mills, Herbert Marcuse, and Erich Fromm to activists such as Al Haber, Tom Hayden, and Ella Baker, and in new magazines such as *Dissent* (founded 1954), *Liberation* (1956), *The Monthly Review* (1955), and *I. F. Stone's Weekly* (1954) came new ideas about how to organize a society based on personal liberation and collective responsibility. In the first issue of the peace journal *Liberation*, its cofounder David Dellinger captured some of the main features of an emergent political philosophy based on a "here and now" revolution that began with change in one's own life that united politics and religion.[74] These new ideas were implicitly and sometimes explicitly critical of the technocratic rationality and impersonalism that seemed to members of the New Left to characterize public and private life. Liberals, too, were calling for a reimagining of public life, in part in response to the civil rights movement. President Kennedy's famous 1961 exhortation of Americans to ask what they could do for their country, rather than what the country could do for them, exemplified the new liberal spirit of engagement and responsibility.

Scientists were part of this revival. The AAAS Interim Committee on the Social Aspects of Science, founded in 1955, released a report in 1958 stating that as a group, scientists ought to consider the social consequences of their work. The SSRS leadership had a variety of responses to the AAAS statement. Paschkis was elated and wrote that the report was "of signal significance" for the scientific community, remarking that such a momentous decision was always the result of many forces, but "we may be grateful that we dare believe that in however small a measure, our voice was one, helping and urging this move." Edward Ramberg was more cautious, for he noted that the AAAS report had called for more scientist participation in public life, but that this was not necessarily a good thing. "It is only so," he wrote, "if the scientists have high ethical motivation."[75] Yet the engagement of scientists in public political debates continued, with little consideration for the personal morality of individual scientists. Also in 1957, Eugene Rabinowitch, the editor of the *Bulletin of the Atomic Scientists* and a proponent of atomic testing, and Joseph Rotblat, the head of the British Atomic Scientists' Association, spearheaded the formation of a conference of scientists to discuss and advocate international cooperation in science, and more internationalist political alliances among nations. The first meeting took place in Pugwash, Nova Scotia, in July 1957, with attendees from ten countries, including SSRS members Hans Thirring and Hideki Yukawa. They issued a statement advocating that the rest of the world follow the historical example of "cooperation" among scientists, and extend that tradition into other aspects of human endeavor. The ideological basis of the conference was almost identical to that which motivated the atomic scientists at the end of World War II when they argued for international and civilian control over atomic energy. In both cases, scientists (and physicians, in the case of Pugwash) thought that the inherent internationalism of science and its basis in reason could triumph over the nationalist ideologies that led to war and undermined scientific progress.[76] The special role of the scientist was to model political relationships that were based on the free exchange of ideas in a context of reason.

Another new group on the political scene was the Greater St. Louis Citizens' Committee for Nuclear Information (CNI), formed in 1958. This group is the subject of the next chapter, but its presence was a challenge to the SSRS, so a brief description is in order. Scientists, physicians, and members of a women's group in St. Louis, alarmed by reports of high levels of radioactive fallout falling on American cities and suspicious of government reassurances that the fallout was harmless, formed CNI to address the problem. CNI emphasized that it had no political position on whether atomic testing should stop, and saw its activities not as acts of personal volition on the part of individuals but as a manifestation of scientists' *duty* to provide the public with information that derived from their role as experts.

CNI, Pugwash, the AAAS Committee on Science in the Promotion of Human Welfare, and other science groups stood apart from the SSRS in advocating the social responsibility of *science*, not of *scientists*. This distinction, though seemingly small, is important for understanding why the original goal of the SSRS—to encourage scientists to take personal responsibility for their work—remained outside the mainstream of new political currents in science between 1957 and 1963. The concept of "science" comprises a range of ideas and activities, including knowledge, the methods by which it is made, the people who make it, the ideals associated with it, and the materials that are used and produced by it. The new liberal thinking situated the responsibility of "science" not in individual choice but rather in the group itself. That does not in any way imply that individual scientists ought to behave in particular ways, only that the institution itself has a specific role in or duty to society. This means that responsibility for some aspect of science—what that meant was not clear in the calls for "social responsibility"—could be given to representatives of the collective, or as the Columbia University philosophy professor Charles Frankel remarked in a 1959 debate, to no one at all. "When you talk about collective responsibility," he said, "you are saying that no one in particular is responsible."[77] The notion of collective or "social" responsibility also reinforced the idea that scientists were morally unified. The SSRS position, by contrast, left no doubt that scientists differed in their perspectives on the proper relationship between scientists and the military.

CAUTIOUS OPTIMISM AND ATTEMPTS AT EMISSARY ACTIVITIES

The new president of the SSRS, Edward G. Ramberg, was eager to take advantage of scientists' new political momentum. A prominent physicist at RCA Laboratories in Princeton, a Quaker active in Pennsylvania Friends meetings, and a conscientious objector who served in the Civilian Public Service during World War II, Ramberg remained committed to using exemplary action and witness. At the 1957 annual meeting, the membership voted to formally incorporate as a nonprofit in order to more effectively raise money for a conference on the "Constructive Uses of Science." Still concerned about associations with communism, the membership also voted to protect the SSRS name from unauthorized use.[78]

Yet again, anticommunism played a role in restricting the group's capacity to gain members and have their views heard without prejudice. On the day they filed their application in Buck's County, Pennsylvania, an attorney for the American Legion asked the presiding judge not to issue the charter of incorporation. The attorney said that the Legion could not understand the purposes of the organization, and also noted that several members of the group had a record of refusing military service. The Legion believed that the SSRS activities "might not necessarily [be] related to the protection of the strongest offensive military interests of this country."[79] The American Legion was one of a number of organizations that enthusiastically supported HCUA and other security committees during the 1950s by recording and circulating the names of those who were accused of being communist. The Legion's blacklist and those of other groups were used in Hollywood and by corporations and other employers to rid their organizations of suspected communists.[80] The judge sided with the Legion, and told the SSRS that it had to reapply with a clearer statement of purpose.

The SSRS reapplied for nonprofit status after rewriting the preamble to its constitution to include a set of practices in which members might engage, including abjuring nonmilitary work that they found objectionable (such as working in the liquor or tobacco industries) and working with other groups on positive applications of science. They were granted nonprofit status after their second application. Red-baiting had forced the group into articulating emissary programs that looked more like those of an interest group or an advocacy organization than one based on types of fellowship and witness that the SSRS had undertaken in the past. The experience did not embolden the SSRS. Rather, it signified to the group that it was an "inopportune time to reapply [for membership] to the AAAS."[81]

As is common in social movements, an opponent and legal requirements helped to shape the identity and strategy of the group.[82] The experience of the American Legion challenge helped to orient the SSRS, at least formally, toward collective action projects that had never been its strength, rather than toward the "right livelihood" focus that had characterized the group earlier. It is also true, however, that even if the application for nonprofit status had gone smoothly and had ratified the SSRS's perspective on individual responsibility and choice, the group's commitment to allowing moral scrutiny and encouraging a spurt of humility suggested that it would be slow to take advantage of some of the opportunities presented to it.

Despite the red-baiting, the SSRS pressed on with its mission. One of the activities that became more frequent was the sponsorship of forums for the discussion of science and peace. At meetings of the American Physical Society, the AAAS, and the American Chemical Society and at the Pugwash conference, SSRS members generated discussion and debate at well-attended and lively sessions. In 1960, they sponsored a meeting on chemical, biological, and radiological warfare, cosponsored by the Friends Medical Society. It drew more than four hundred audience members. Rather than work through the major professional association in the country, the AAAS, members began to organize sessions and informal meetings in other professional associations, such as the American Societies for Experimental Biology and Medicine. Other members spoke on the radio and gave talks to students and service groups about individual responsibility.

Victor Paschkis continued to travel internationally to meet with groups about the importance of personal responsibility and to make contact with existing or potential members of the SSRS. In 1959, he visited twelve countries, including Italy, Yugoslavia, East Germany, France, Czechoslovakia, and England, sharing his message with members of a worker-owned cooperative, a school to train young men to be mechanics, engineers at a Daimler-Benz factory, and individual SSRS members such as A. J. Coulson in England. His reports do not suggest that he was proselytizing or seeking organizational connections, but rather that he was "witnessing" and observing the ways that other people tried to act on their consciences.[83] Paschkis believed that "one of the finest accomplishments of the year is the remarkable campaign carried on by our member Linus Pauling to acquaint men everywhere with the dangers of the nuclear arms race and the need for international agreements."[84] Pauling's campaign included co-organizing a petition against the arms race that was signed by more than nine thousand scientists worldwide.

Finally, the SSRS continued to publicly praise individuals whose perspectives they felt were sympathetic to those of the SSRS, including the Harvard biologist John Edsall, an outspoken opponent of chemical and biological warfare, and *Scientific American* editor Gerard Piel, whose essay in the *Bulletin of the Atomic Scientists* urged individuals to refrain from destructive work.[85]

The SSRS perspective, however, was still not in the mainstream of the debates about scientists and the public political sphere. The new liberal ideal of scientist as expert in a democracy was becoming the favored articulation of the relationship between scientists and public political life. In March 1959, CBS televised a debate about whether scientists were socially responsible for the applications of science. It concluded that scientists themselves ought not to make political decisions about science, but advise the government and leave decisions to politicians. Witness and personal perspectives did little to impress agencies such as the Congressional Joint Committee on Atomic Energy, at whose hearings on nuclear testing Herbert Jehle spoke in 1960. Jehle spoke as an individual with moral concerns about testing. He was told by the Committee to "keep to the objective facts" and to present his organization's perspective, not his own.[86]

Deepening Divisions

The 1959 annual meeting marked a turning point for the SSRS. In his tenth anniversary address to the group, Paschkis asserted that the SSRS was in crisis. "The association is at a crossroads of so basic a nature that I want to speak in mundane terms about the future, if any," he said. Paschkis noted that three changes faced the group. The first was that membership was spreading around the world but not as much in the United States. Second, he noted that there was a growing awareness of the "secondary effects" of scientific research. No longer could a scientist be sure that avoiding certain kinds of work guaranteed noncomplicity in morally objectionable applications. "A project with a morally impeccable goal can become unacceptable," he argued. Finally, rather than seeing the explosion of groups of scientists engaging in the political debates of the day as an opportunity, he saw it as a challenge for the SSRS.

Paschkis believed that the SSRS should return to its roots and emphasize personal responsibility, and leave the business of influencing power structures to the FAS and the Pugwash Conferences, and leave public education to groups such as CNI. "The SSRS stand, whatever that may be, is so distinctive and so unorthodox that those who made the plunge need fellowship and need it so desperately that this is enough for the existence," he said. Yet Paschkis also acknowledged the need to make more "converts." To that end, he reminded his audience that the other purpose of the group was for what he called outreach. "Fellowship is not necessary for this. Members are driven by ideas. We must replace fellowship or outreach with fellowship through outreach."[87] What this meant in practice was not clear. Despite advocating a return to the roots of the group, Paschkis suggested that the information movement offered some possibilities for reaching other scientists. "Scientists understand technical reports better than laymen and have a duty to utilize this ability for the instruction, and therefore the benefit of our fellow citizens."[88]

Paschkis was responding to the influx of new members who brought new ideas about what the group should do, rather than to the SSRS's imminent collapse—for between 1954 and 1963 the SSRS grew. In addition to adding new members, some of whom ran for SSRS offices, newsletter subscriptions increased dramatically, from twelve or so in 1954 to more than twenty-five hundred in 1963. Membership went from almost three hundred to more than eight hundred during the same time period. International membership also increased, and a vibrant new chapter was formed in Boston.[89] Yet many of the newer members were drawn from outside the physical sciences and engineering, the fields in which founders and core older members worked, and fewer were Quakers.

Many new members were taking up subjects that had more in common with the scientific interests and specializations of biological scientists. Biologists were visible in the rolls of officers and newsletter writers in the SSRS between 1957 and 1963, and many biologists were involved in "direct action" and work that was more like interest group activity than witnessing. E. W. Pfeiffer, a University of Montana biologist and a member of both SSRS and CNI, for example, was active in research on the effects of atomic testing on the air and soil of North Dakota. He delivered talks and wrote articles in journals in which he presented information about how fallout affected genetics. The SSRS membership was clearly interested in this subject, because they invited CNI member and Quaker Walter Bauer, a pathologist at Washington University's School of Medicine, to serve as the keynote speaker for the 1959 annual meeting. Bauer reported on CNI's decision about whether to become an "action" group. In contrast to the SSRS strategy, Bauer said that CNI's success was based on the confidence inspired by its strategy of avoiding moral or political statements.[90]

Other members of the SSRS were participating in direct action activities to stop the spread of chemical and biological weapons (CBWs), in response to a new government campaign to recruit scientists for work on these weapons systems as individuals. These actions were prompted by the government's major CBWs initiative begun at Ft. Detrick, Maryland, in 1958. For the program to be successful, military leaders felt that considerable work had to be done to convince the public that CBWs were in the best interest of the country. General J. H. Rothschild, interviewed by *Harper's* magazine, argued, "We must make it clear that we consider these weapons among the most normal, usable means of war. . . . We must confront the existence and possibilities of these weapons and learn to live with them, with their potentials for human warfare as well as for destruction of life. . . . We cannot start too soon."[91] Another problem, military and science leaders argued, was that the program would be stalled if they could not recruit more scientists. The Army's chief chemical officer, Marshall Stubbs, saw the reluctance of scientists to work on such projects as a major obstacle to the program, and advocated that the military "spend time finding ways to overcome such reluctance." Conservative science publications also advocated for more CBW research. *Chemical and Engineering News* editor Walter J. Murphy, for example, supported the CBW program, and argued that "further recrimination [against scientists who worked on CBW] should cease. The professional societies have a solemn obligation to get the facts [about CBW] to the public; more scientists and engineers will be willing to work once the public is behind it."[92]

In response to what the SSRS called this "notorious public relations campaign" to recruit scientists to work on CBWs, in September 1959 the

SSRS circulated a resolution deploring the campaign to popularize germ warfare and rejecting official calls for scientists and individuals to "withdraw moral scruples and revulsion" in order to participate. They offered to help scientists find non-CBW jobs.[93] In 1959, FOR and other peace activists began a peace vigil at Ft. Detrick.[94] As a group, the SSRS joined "in spirit the men and women in vigil."[95]

By 1961, more SSRS members wanted the organization itself to promote engagement in vigils and other actions. The new push toward direct action was encouraged at the national level by William Davidon, a pacifist Quaker physicist from Haverford who was elected vice-president of the SSRS for 1962–1964. Davidon's public political activities as a scientist bridged the liberal and pacifist worlds. One of the original members of the Pugwash Conference, he was serving as the secretary of FAS when he was elected to office in the SSRS. Davidon also was an active member of FOR, and a participant in direct action peace activities, including the Ft. Detrick vigil. Davidon argued for the liberal vision of enjoining scientists to act as mediators between the Soviet Union and the United States.[96] Yet he also urged SSRS members to use direct action to pressure government leaders.

Paschkis and other longstanding members of the SSRS did not advocate either the direct action or the liberal information methods of engagement in public political issues, despite the growing interest in these activities among leaders and rank-and-file group members. Tension also existed over whether to allow social scientists to join the group as full members. From the beginning, the SSRS had welcomed social scientists as "associates" without voting privileges. When their full inclusion was considered in September 1951, Ted Benfey argued that it would be strategically ineffective to include social scientists because "there is such mistrust of social scientists on the part of natural scientists."[97] At that juncture, the SSRS decided not to allow them as full members.

In 1962, the debate about inclusion arose again, when social scientists expressed interest in joining the group. The debate exposed an underlying tension in the organization: Who counted as a scientist? Should all scientists wishing to avoid harmful applications of their work be full members? Or was there something special about natural and physical scientists? At the 1962 annual meeting, Don DeVault argued vehemently against the inclusion of social scientists because "social scientists don't have the same kinds of minds as scientists." DeVault also echoed Benfey's earlier argument. "If there are too many social scientists," said DeVault, "then the natural scientist will be in the minority, and the group won't appeal to scientists whose technical success threatens humankind. The natural scientist must walk his own path and can do so best in the company of those who must walk the same way."[98]

DeVault's arguments were in tension with two of the key ideas upon which the SSRS was founded: that there was nothing in particular about scientists that separated them from ethical responsibilities that other people faced, and the idea that the SSRS was specifically organized to assist those scientists and engineers who wished for fellowship as a result of moral choices they made about their work. The important point is not whether social scientists were scientific or not, but that some group members were worried that natural scientists would lose influence in the SSRS, and that their special relationship to militarism would become only one of many different bases for participating in the SSRS. Despite DeVault's impassioned pleas, in 1963, SSRS members voted by a margin of 3 to 1 to include social scientists and students. This decision was opposed by the old guard of the SSRS, including DeVault, Paschkis, Norman Polster, and Edward Ramberg, who continued to argue that social scientists ought not to be included.[99] The splintering of the SSRS would continue over the next decade, as the tension between the old guard and newer members grew even greater. The development and spread of both the information-provision model and a "direct action" model among scientists also gave scientists who advocated peace more organizational options.

ANTI–VIETNAM WAR ACTIVISM, YOUNG MEMBERS, AND THE COLLAPSE OF THE PERSONAL RESPONSIBILITY MODEL, 1964–1969

The Development of "Outreach" Witnessing

The decline in the SSRS model of exemplary action and the rise of "emissary" action can be seen in the Boston SSRS chapter, under the guidance of Herbert J. Meyer, an engineer employed with Physical Metallurgy of Arlington, Massachusetts. Meyer had long hoped that the SSRS would vastly enlarge by expanding student membership. In late 1963, Meyer helped to form campus groups in the Boston area. At Brandeis, the students Jim Dow, James Funston, Elaine Tarmy, and Charles Bolthrunis were the core of the SSRS group. At MIT, Michael Sobel, Stephen M. Kaiser, Sherman Rigby, and Peter Ralph were the major organizers. The SSRS Boston chapter held intimate discussions about the personal responsibility of scientists, considering issues such as whether the institution of an ethical pledge for scientists comparable to physicians' Hippocratic Oath would be meaningful and powerful. Unlike the earlier SSRS, the student groups in Boston quickly decided to engage in emissary activities that went beyond witnessing.[100]

Expanding the SSRS was important for these younger members. To that end, Meyer and Kaiser sent out a questionnaire to 2,700 graduate students at MIT, Brandeis, and Boston University, asking about their opinions on nuclear weapons and research funding, and their interest in taking action on these issues. Six hundred responded, and of those 150 said they were interested in participating in some kind of action group. Yet few people showed up for the student meetings. Kaiser felt that this was because "many young people, attracted to the militancy and relevance of the civil rights and student peace movements, will find the SSRS too philosophical or not sufficiently involved in political action."[101] Meyer worried that political conditions made students and workers fearful of speaking out. In an article in the SSRS newsletter, he argued that "trustees and alumni are heavily weighted on the side of industry from which the big donors come, that atmosphere already pervades the MIT campus. Therefore the freshman, mindful of the future paycheck, becomes swiftly inclined not to rock the boat, not to question. Everyone wants to play it safe. Because of the lingering memory of Fort Monmouth and J. Robert Oppenheimer he [the student] is painfully aware that he may have to pass through security and loyalty mills in the course of his career."[102]

MIT was simultaneously an obvious place to try to organize students around war-related issues and the last place one would expect to find people who wished to investigate the meaning of their work. MIT was one of the largest university recipients of R&D funds in the United States. Projects ranging from missile guidance systems to basic research in biology were funded by the military, and secrecy around such projects was common.[103] Yet students at MIT, like their peers at other top science and engineering schools, were not known for their involvement in extracurricular activities or political engagement. Those who wished to organize students at MIT would have their work cut out for them.

They soon found an issue that would draw the attention of the students. The Boston SSRS chapter assisted two assistant professors of history at Harvard University who were challenging the constitutionality of a national loyalty oath required for all high school and college teachers in the state. The SSRS chapter members concurred that the law was discriminatory and "ineffective in inducing loyalty in employees," wrongly required "support of governmental interpretations of the Constitution," and "intimidat[ed] the political non-conformist."[104] The case was highly publicized on the Harvard campus, and the professors in question received considerable support from faculty and students.

Emboldened by the interest that faculty and students showed at Harvard, in 1965 and 1966 Kaiser, Rigby, and Barry Blesser organized a series of actions and campus talks at MIT. They pressed MIT officials to tell the students which kinds of organizations students and faculty ought to avoid if they wished to keep their security clearances. MIT would not (and probably could not) provide an answer. The students decided to publicize the issue by organizing a talk on "Federal Policies in Denying Security Clearances." Joseph A. Fanelli, a former first assistant at the Justice Department and the defense counsel for many scientists who were prosecuted during the McCarthy era, and John F. Doherty of the Internal Security Division of the Justice Department were invited to debate the security clearance issue. Fanelli decried the association criterion—that scientists could be denied funding if they were associated with suspicious groups—because it stifled free expression. Doherty was surprised by students' concerns about the security clearances. He said that he had "never met any students concerned about what effects present groups might have on their future."[105] The Boston SSRS chapter reported that the "question period revealed a great deal of fear that security procedures might jeopardize their careers, unless they took great care in public statements and in joining organizations." Although the meeting was sparsely attended, those who did participate were highly engaged. As a result, SSRS decided to sponsor more campus discussions, but on topics with a broader appeal than the narrow issue of security clearances.[106]

A subsequent public discussion of the effects of a technical education on the social consciences of scientists and engineers was well attended, lively, and contentious. The SSRS followed up in January 1966 with another meeting on social responsibility in science. The group used BBC films about the relationship between MIT and the military as springboards for debate about the political culture—or lack thereof—on the MIT campus. A faculty panel composed of Jerome Lettvin (physics), David Schalk (humanities), Dean Brown (engineering), and Cyrus Levinthal (molecular biology) and the overflow audience of 250 held a free-flowing discussion about the role of scientists in public political life and about MIT-specific issues such as whether MIT's Lincoln Laboratories, funded by the military to engage in weapons research, ought to be allowed to continue.[107]

Blesser and Rigby concluded, "Our experience seems to indicate that mass exposure to new ideas stands a better chance of inducing responsible attitudes than prolonged discussion of the exact nature of responsible behavior." Few people had responded to the old-style SSRS discussion and fellowship meeting, but clearly many people were interested in discussing broader issues. The Boston SSRS thus continued to sponsor debates and meetings throughout the year, on topics ranging from the role of government in science to the amount of military funding on the MIT campus. The SSRS talks were complemented by other MIT public talks, including one by the media scholar Marshall McLuhan and the historian Lewis Mumford, who cautioned audiences about what was lost, as well as what was gained, with the growing influence of science and technology on human life. These and the SSRS meetings were well attended and full of lively discussion.[108]

The Boston chapter also used the materials developed by CNI, the scientist group whose model of political action was based on information provision. The chapter distributed CNI's newsletter outside movie theaters showing *On the Beach*, Ernest Gold's film about the last survivors of a nuclear war who wait on an Australian beach for a radioactive cloud to engulf them. A year later, the Boston chapter developed a Boston version of CNI's best-selling *One Year After*, a fictional account of the effects of a nuclear attack on St. Louis.[109] Several members of the chapter helped to collect teeth for CNI's Baby Tooth Survey, which collected baby teeth from parents around the country and tested them for the presence of Strontium-90, a by-product of fission that was present in atomic fallout.[110] Other chapters were picking up on the CNI model, too. At the Chicago regional meeting of the SSRS in 1963, for example, most of the annual meeting was a report on the formation of the Scientists' Institute for Public Information, an umbrella group that included local fallout information groups from around the country.[111]

For many young people on college campuses, these were not merely philosophical debates. In 1965, President Johnson sent the first of what would eventually be hundreds of thousands of American troops to fight against a communist-led insurgency in Vietnam. Students at Cornell, Michigan, Harvard, and Princeton had already begun to protest the war in 1963; by 1966, thousands of students across the country, as well as long-standing and newer peace groups, were holding demonstrations, burning draft cards and political figures in effigy, and protesting against ROTC recruitment on college campuses.

Irreconcilable Differences? Politics vs. Morality; Exemplary vs. Emissary Action

Despite this new interest in antiwar activities among young people, the momentum that the Boston chapter generated would not last. The chapter, and the SSRS as a whole, became involved in heated debates over the forms of action that were ethical and useful. No longer was the SSRS's problem, as it had been the 1950s, that few scientists were interested in engaging in political action; the problem now was that there were many competing means of doing so. The arguments within the chapter and in the larger organization are instructive for understanding how the SSRS slipped away from its original purpose.

Herbert Meyer espoused the old SSRS view. He continued to believe that "to assume responsibility for one's own acts would be enough" to carry out the moral obligations of scientists. To find out how best to engage other scientists, Meyer suggested that the SSRS study the views of scientists and engineers by reviewing research and carrying out a large survey. He hoped that the decentralized, autonomous groups that characterized New Left and pacifist groups could be adapted for scientists. This form, he believed, could be translated into laboratory training groups that would involve "the young and the old in spontaneous, autochthonous concerns."[112] Meyer's perspective was shared by the student SSRS member Stephen Kaiser, who argued that the SSRS should never become a partisan group because the moral purity of science had to be protected at all costs. "Our predecessors have fought for centuries to keep Science protected from the corrosive influence and control of the political sphere," he wrote. "Certainly, we should never make the mistake . . . of accelerating the process of the political infiltration of science."[113]

Other student members in Boston thought that the SSRS old guard didn't recognize how important it was to take action against the war in Vietnam. Peter Ralph, for example, wrote, "Our society must make unequivocal statements on political events. . . . U.S. action in VN [Vietnam] is a most blatant and violent negation of SSRS principles, a naked, ugly show of U.S. imperialism." To those who wished to "hide truths so that the SSRS might reach more potential members," Ralph asked, "may I suggest a front organization?"[114] Blesser expressed a similar view. Declaring the need to recruit students "with the same success as the SDS and SNCC," Blesser criticized the majority of the SSRS membership for their arrogance and sense of moral purity. The SSRS membership, he said, comprised "a select few scientifically trained . . . citizens who by virtue of their personal commitments are using the prestige of modern technology to gain respect for their morality." Far from the humble individuals they considered themselves to be, Blesser argued that the behavior of SSRS scientists was "not very different from the way that actors and astronauts use their public image as a lever to try to influence public and government alike."[115]

The activities of the Boston chapter didn't go far enough, argued William Davidon. In April 1966, A. J. Muste, Davidon, and four other radical pacifists made a trip to Vietnam to contact Buddhist leader Thich Nhat Hanh and other monks who were leading a peace initiative to stop the war. Davidon was inspired by his trip, and pushed the SSRS to take more direct, collective action against the war.[116] Davidon argued that the collision of an American B-52 bomber with an oil tanker off the coast of Spain, the use of herbicides on Vietnamese rice crops, and the dramatic increase in U.S. bombing of Vietnam meant that SSRS members had to be "concerned with these and other specific events, and not with creating a sense of euphoria by repeating pious generalities." Further, wrote Davidon, the SSRS ought to espouse pacifism and reject all forms of warfare rather than allowing individual scientists in the group to make up their own minds. Davidon thought that this would not be a difficult task, "since collective action was already necessary for scientists to understand elementary physical processes." He hoped that the SSRS membership would become "participants, not bystanders" in the struggle for peace.[117] Davidon had the support of many newer SSRS members. For example, Klaus Arons, an SSRS member in Germany, considered "vague discussions, without any stand on concrete situations" to "lead to an abstract idealism having no relevance to the problems facing society."[118]

Already stumbling under the debates over the membership of social scientists, the questions about what constituted useful and righteous action weakened the SSRS. To be sure, individual scientists from the SSRS continued to participate in the same round of conferences, to write letters, and to give speeches, but now more of them were also participating in vigils, speaking out against the war in Vietnam, and generally calling for something other than the slower-paced conversion process that the SSRS had originally been based on.

A Renewal of Liberal Science Activism in the United States

The formal apoliticism of the SSRS was decidedly at odds with developments in the scientific community. Taking explicit political positions was becoming more popular among American scientists. In 1964, scientists organized Scientists and Engineers for Johnson-Humphrey. Fearful that if the Republican candidate, Barry M. Goldwater, won the presidential election the chances of atomic war would dramatically increase, the group pulled no punches in its campaign, harshly criticizing Goldwater for his plans to develop more kinds of nuclear weapons.[119]

In 1967, the biologist Salvador Luria and the chemist Albert Szent-Györgyi called the war a "moral blight on our country" in a letter to the editor of *Science*.[120] Most of the scientists who wrote letters to the editor responding to Luria and Szent-Györgyi disagreed with them. "Not helping our country will not shorten the war. It will merely increase the number of casualties," wrote one scientist.[121] Others used a straightforward anticommunist rationale for refusing to oppose the war. "Fighting Communism in Vietnam is a dirty, complicated, and exasperating task," John M. Pfeiffer wrote, "but unless we win, we shall surely have to fight again closer to home."[122] For some scientists who agreed with Luria and Szent-Györgyi's opposition to the war, the issue was not the relative effectiveness of the war, but whether the alliance between the military and science was a Faustian bargain that corrupted science. John M. Reiner, of the Department of Microbiology at Emory, wrote, "There is an old saying: 'He that sups with the devil must have a long spoon.' When the devil is an institutional monster that threatens to destroy our honor even before it claims our lives, there is no spoon long enough to tempt a free man to such a feast."[123]

The biologists who opposed the war in Vietnam based their claims on appeals to general standards of morality, not on scientific study or personal responsibility. Given that biologists did not have the moral stigma that physicists had because of participation in the creation of the atomic bomb, their appeal to moral rather than instrumental logic helped to distance them from the kinds of criticisms that activists and intellectuals were making about the growing dominance of instrumental rather than moral logic in political life. Furthermore, biologists were beginning to overtake physicists in the status hierarchy of science. Geneticists claimed nothing less than to be able to read the "Book of Life" that contained all of the "information" (genetic sequences) to which life could be reduced.[124] In claiming to have discovered the very basis of life, biologists espoused a cultural position that was the antithesis of the cold, mechanistic model of physics as the study of dead matter and the development of weapons.[125] The optimism and cold calculations of the physicist were undermined, too, by ecologists, who sounded the alarm about the destructive capacities of science and called for more thoughtful consideration of the interconnectedness of life on the planet.

Not only were individual scientists taking strong, public political positions, so too were science organizations. In 1965, the AASW submitted a resolution to the AAAS urging an end to "hostilities in Vietnam. Like all scholarship, the sciences cannot fully flourish," the AASW submitted, "and may be badly damaged, in a society which gives an increasing share of its resources to military purposes."[126] The AAAS council approved a version of the resolution that called for "efforts toward negotiation and speedy settlement" of the war.[127] The next year, the AAAS debated a resolution submitted by E. W. Pfeiffer urging that the AAAS oppose the war entirely.[128]

The Fragmentation of the SSRS, 1965–1976

The SSRS never drew in those scientists and engineers who were opposed to the use of their ideas and products for the war in Vietnam. Internal conflict channeled much of the group's energies away from the development of new programs of action. Despite having relatively similar views about the direction of the organization, Paschkis and Meyer were always at odds. Each desired that the group focus on a slightly different aspect of the SSRS mission, with Paschkis calling for a continuation of witnessing, and Meyer calling for use of empirical methods to discover what would interest scientists. Paschkis felt that witnessing was a way to motivate others to act; Meyer, mindful that many young people were drawn to civil rights and antiwar activity, wanted the group to engage in activities in addition to sponsoring talks and publishing, but not to participate in direct action. Compounding this division was that Meyer wanted the group to include social scientists as well as natural and physical scientists, while Paschkis did not.

The election of Edward U. Condon, the former director of the National Bureau of Standards and a well-respected physicist, to the presidency of the SSRS in 1967 gave the SSRS new visibility, but it did not lead to a reinvigorated organization. Condon was well connected to other science associations that were active in politics, including FAS, the AAAS, and the American Physical Society, as well as the Committee for a Sane Nuclear Policy, the University of Colorado, and many government agencies. The struggle between Paschkis and Meyer overshadowed the group's efforts to accomplish very much. When Condon was elected president in 1967, Victor Paschkis wrote to him, "You may have seen that there was considerable questioning on the part of some people regarding the goals of the Society. Inevitably, as time goes by, some people may have lost sight of the purpose, namely to relate scientists and engineers who are willing to be guided in their choice of work by their conscience."[129] Franklin Miller, hoping to warn Condon about what he would face, said that there was "severe friction and mistrust" between the Philadelphia group and the Boston group that was led by Meyer. "He (Meyer) is perfectly sincere," said Miller, "and feels that the 'in group' [led by Paschkis] don't trust him (they don't) or his active group of young scientists in the Boston area SSRS. I have tried my hand at reconciliation, without much success, I fear."[130]

In 1968, Condon was reelected as president and Alice M. Hilton as vice-president. Together they gave the SSRS an even higher public profile. Hilton was a mathematician who had cofounded the Cybercultural Institute, a forum for the exchange of scientific ideas and ideas about the effects of computing and automation on culture and jobs. One of the authors of the influential 1964 article "The Triple Revolution," which was critical of those who believed that a highly automated world would bring ease and pleasure to everyone, Hilton was integrated into a wide network of engineers and intellectuals that went beyond those of the old guard who made up the majority of the SSRS's core membership.

The SSRS membership was optimistic about the group's chances for renewal with the election of Hilton and Condon. Edward Ramberg wrote that "the fact that [Condon] is a past president of the AAAS and has excellent standing in the scientific community as well as his straightforward logical suggestions, such as [to nominate] Alice Mary Hilton, [and] that the SSRS undertake an interdisciplinary research study on a subject such as the interrelation of the economy and war, with SSRS members in different fields serving as resource persons, may well prove fruitful."[131]

His optimism was not well founded. The SSRS had never been good at organizing research projects of the sort that Condon suggested, and the old guard's advocacy of witness and personal responsibility was out of step with new ideas about how scientists should be involved in public political life. New organizations, including the British Association for Social Responsibility in Science[132] and Science for the People, advocated that scientists become politically involved in undermining social institutions from universities, corporations, and professional associations to capitalism. These groups were generating energy and interest among young and old scientists. Still, although the SSRS never attracted large numbers of members, those who were in the group continued to be active. In 1970, the Boston chapter sponsored a meeting on Human Values in a Technological Society that included Condon, as well as members of other peace groups.[133] Other members continued to be engaged in individually-based projects, but the SSRS never returned to its role as one of the few forums for debates about what the proper role of scientists in public life should be.

The SSRS might easily be consigned to the dustbin of history because its effect cannot be easily measured according to the standards used for most studies of organizational effectiveness, such as membership numbers, coherence, stability, fund-raising, or policy enactment. Its diverse membership over time, the many varied activities in which those members were involved, and the relative autonomy of various chapters keeps it from having a neat organizational history or developmental trajectory. There was a consistent desire to increase membership, but the SSRS was never organized to serve purposes such as influencing policy. Over its twenty-year life span, the SSRS served, in the words of one member, more as a forum than a closed society. The organization attracted Nobel Prize winners, working scientists, students, science teachers, and doctors, each of whom took part in an extended conversation about the proper place of scientists in political, moral, public, and private life. But these groups of people came from varied circumstances, with varied interests and different understandings about how moral responsibility ought to be enacted. From a small group that weathered the political repression of the early 1950s to a contentious network of people with wildly different views about who scientists were and what they should do to improve the world, those who participated in the SSRS explored perhaps the most wide-ranging possibilities of any group of scientists who tried to disentangle scientists from the military in the twentieth century.

That said, the SSRS did not have the political impact on other scientists that it might have had. As the narrative here has shown, a deep commitment to individual witness and several fundamental ideological tensions meant that the SSRS was more effective at morally sustaining small groups of scientists than at mobilizing large numbers of scientists, and more effective at raising questions and issues than at influencing institutional policies. SSRS was more like a big tent under which people with ties to many different groups and activities could gather for intellectual and moral debate—a forum—than a typical social movement organization. The individualism so prized in Quaker theology and so integral to the SSRS's theme of personal choice and responsibility made concerted collective action consistently a problem for the organization.

This prime tension—between individualized means and a desire for social ends—was abetted by a second tension: whether scientists were unique in their situations, their expert knowledge, or their moral responsibility. After all, the group first organized as a collection of natural scientists (especially physicists and chemists) and engineers, most of whom worked either for universities or in industry. They were tied to an institutional sector increasingly connected to federal military funding, and were presented with a professional and social situation that was fairly new in American society. Those who founded the SSRS in 1949 believed that this relationship to the military gave them the authority to testify about what was wrong with this relationship.

And yet, the basic ideological claim, by the Quaker Paschkis and others, was that each person has an individual, personal responsibility to follow a moral calling and do only that work that can be morally justified. Scientists were not given a pass on this responsibility just because science was a neutral body of knowledge, and they were not exempt from personal moral choice simply because of their important role in helping the United States defeat fascism and hold communism at bay. The SSRS's values and ideology rejected pragmatic realism and realpolitik. And far from treating scientists as heroes above reproach, if anything the SSRS put more pressure on them to work only on morally justifiable projects.

The individual responsibility position relied on each person to make a judgment about expected effects and uses. Perhaps this position was possible before the era of Big Science, but as the relationships between the military and science became more difficult to trace, some SSRS members wondered how one could avoid entanglement with all objectionable activities. Some SSRS members dismissed this problem, arguing that many scientists knew perfectly well what they were working on. But two thornier problems would not go away: the first was that all scientific research could in theory be used for military purposes, calling into question how one could be a scientist and yet avoid moral responsibility for the uses made of science; the second, related issue was that to reject all funding from the military meant that scientists were asked to put their research projects and careers on the line in order to be true to their convictions. This model of critique and change thus had self-limiting properties, since the structure of scientific research funding in the United States through the early 1960s offered vastly more opportunities for military-based research than it did for other kinds of work.

Information and Political Neutrality: Liberal Science Activism and the St. Louis Committee for Nuclear Information

> The Greater St. Louis Citizens' Committee for Nuclear Information stated its objective as the collecting and distributing "in the widest possible manner" information concerning potential use of nuclear weapons in war, the testing of such weapons, and non-military uses of atomic energy. The group will undertake to stimulate public discussion of nuclear information and will, after a preliminary period collecting and distributing data, take a position on such issues as nuclear weapons testing and nuclear warfare.
>
> —*St. Louis Post-Dispatch*, April 22, 1958

The Greater St. Louis Citizens' Committee for Nuclear Information (CNI) was formed in 1958 by a coalition of scientists, lawyers, community members, doctors, dentists, and women activists in response to what they perceived to be the failure of the government to provide citizens with accurate information about the health effects of atomic fallout. Over the next seven years, CNI developed a new model of collective political action for scientists: the provision of information to the public. It is now commonplace for scientists, individually or in groups, to provide information to the public about a variety of issues, but in the late 1950s it was not. The reigning liberal view at the time was that public opinion and mass-based politics were potentially dangerous; the best forms of rule were enlightened executive leadership and the taming power of interest group politics. CNI challenged that idea by encouraging more public political engagement with sociotechnical issues by providing citizens with information that they could use to make political decisions. Unlike the Society for Social Responsibility in Science (SSRS), CNI's founders believed that scientists had a special collective responsibility—not a personal responsibility—to promote democracy by providing information that would foster public participation.

"Informing" the public was not the same as educating the public. The scientists' movement at the end of World War II had encouraged citizens to oppose the unilateral and military control of atomic power. Basing their philosophy on the Deweyian idea of the importance of using scientific methods to engage all aspects of a political debate, these scientists unapologetically advocated specific political positions with regard to the control of atomic energy, and encouraged the public to do the same. In contrast, the "information" model that CNI used was based on information theory. During the 1950s, information theory was rapidly developing in mathematics and the computing sciences. Bits of "information" (e.g., an electrical impulse, an equation) were coming to be understood as transmittable entities that stood outside any context. Information was thus theorized as something detached from origin or destination but which could be combined and used in different ways.[1] As Bazerman points out, information has had many different meanings historically. During World War II, it was associated with "gaining information about the enemy, transferring information among allied forces secretly and securely, and disrupting the communication of enemy forces."[2] CNI worked to do the opposite: to circulate facts, divorced from explicit frameworks of interpretation, to a wide public audience.[3]

CNI was founded because of the rising concern among scientists and the public about the health effects of atomic testing. Although scientists had worried about this earlier, it was not until 1954 that it became a public issue as a result of an incident in which Japanese fishers who had been showered with fallout from the U.S. atomic test BRAVO in the South Pacific became ill. Although the Atomic Energy Commission (AEC) consistently provided information stating that atomic fallout had a minimal effect, scientists equally consistently challenged that claim. The public, having read newspaper accounts of the Japanese fishers and of new controversies among scientists about the biological effects of radiation, clamored for facts. It was becoming increasingly clear, however, owing to the efforts of scientists and grassroots groups, that political considerations lay behind the decisions on what aspects of radiation and health ought to be studied. Was it best to consider radiation's effects on future generations? On children's health? Should all atomic fallout be considered? Or just the types known to lodge in human bones? Both opponents and proponents of atomic testing accused each other of distorting facts for political purposes. All of this complicated information provision to the public.

Thus, the idea that scientists could present "the" facts about radiation and human health was becoming difficult. Scientists began to worry that the public exposure of their scientific disagreements, normally safely hidden in laboratories and in conference meeting rooms, was undermining their scientific and political legitimacy. Compounding the legitimacy

problems caused by scientists' disagreements over testing was the parallel disagreement over what role, if any, scientists ought to play in public
life. CNI's espousal of "neutrality" was in part a response to these legitimacy threats.

CNI did not begin as an information-provision organization. The
group's decision to treat information as "neutral" and nonpartisan was
based on the political context in which it worked, not on a transcendent
notion of science. Formed by members of the St. Louis labor, women's,
religious, and university communities, CNI was a scientist-citizen group
built on the success of local St. Louis women's anti–atomic testing campaigns and the popularity of the speaking engagements physicians and
scientists from Washington University had undertaken since 1956 to explain the health and environmental effects of radiation to the public. Some
early members of CNI wanted the group to take specific political positions
based on scientific research on the health effects of fallout. One of the
group's founders, Barry Commoner, urged the group not to do so, but
rather to wait a year and reconsider the issue. In part because of the attention the group had attracted at the national and local levels during the
previous year, CNI adopted a stance of political neutrality, focused on
information provision.

Even though red-baiting had lessened by the late 1950s, when CNI was
founded, the group, like the SSRS, was still dogged by accusations that it
was a communist front group and that its members were associated with
communists. One reason that CNI drew attention was that some of its
highest-profile members personally opposed atomic testing or were associated with scientists who did. To defend themselves against charges of
being communist sympathizers, CNI members more fervently asserted
that the facts they provided were untainted by partisanship. Thus, rather
than seeing the claim of nonpartisanship as a natural mode of action dic-
tated by a transhistorical idea about the neutrality of science, it can be
seen here as emerging from a specific political context of attacks on the
political loyalties of scientists.

Yet espousing political neutrality did not work quite as well as CNI
had hoped. During the cold war, the idea of neutrality was often associated with internationalism, which some conservatives believed was little
more than communism in disguise. Neutrality could also be taken to mean
a lack of patriotism. Failure to take a political position that clearly favored U.S. interests—and this often meant promilitaristic, pro–weapons
development positions—might be viewed as evidence of a lack of loyalty.
After all, the lesson that scientists learned between 1946 and 1956 was
that even scientists who appeared to be demonstrably loyal to the country
and in favor of its policies could be charged with disloyalty. Just as the
SSRS had difficulty convincing others that its stance of personal responsi-

bility was nonpartisan, CNI had difficulty convincing others that its information was politically neutral.

Ultimately, however, CNI prevailed. Debates among scientists, intellectuals, and politicians gave way to the clear public desire for the information that CNI provided. Supported by foundations and public donations, CNI, along with other information-provision groups, was able to institutionalize this model during the 1960s. How and why the information-provision model developed and what made the model spread are the subjects of this chapter.

LEGITIMACY PROBLEMS FOR SCIENTISTS: THE CONTROVERSY OVER THE HEALTH EFFECTS OF ATOMIC TESTING

CNI was formed in direct response to concerns among citizens, scientists, and politicians about the wisdom of atomic testing. Concern grew dramatically after the United States tested its first hydrogen bomb on March 1, 1954, over the Bikini Atoll in the South Pacific. One thousand times more powerful than the bomb that had been dropped on Hiroshima, this bomb's explosion created an enormous fireball that could be seen more than one hundred miles away and blew a mile-wide hole in the ocean floor. The test was not a complete success from a public relations perspective. The winds had shifted unexpectedly during the test, carrying radioactive fallout much farther than the military had anticipated, showering the inhabited Marshall Islands and other areas outside the fifty-thousand-square-mile "danger zone." It fell on the Japanese fishing boat *Lucky Dragon*, which had traveled near the test zone. The twenty-seven fishers on board felt the blast but were not concerned about the ash that drifted lightly over the boat. When the boat reached Japan two weeks later, however, twenty-three of the crew were in advanced stages of radiation sickness. Japanese citizens were outraged. They boycotted fish markets and marched in the streets against Americans, ultimately helping to spawn a mass-based movement in Japan against atomic testing.[4]

The fate of the *Lucky Dragon* fishers placed radiological effects at the forefront of new debates about atomic testing. Shortly after the March 1 test, Japanese officials announced that the bomb's fallout contained the element strontium 90 (S-90). Not found in nature, S-90 was known to mimic calcium in the body. This meant that it would displace calcium and lodge in bones. It also had a half-life of twenty-five years, compared with the half-lives of days or hours for other radioactive elements found in fallout. As a result, S-90 would continue to lodge in bones, in decreasing amounts, for a quarter-century after each blast.[5] The *Lucky Dragon* incident and the new reports about S-90 left reporters in the United States clamoring for information. After a second, smaller test in late March, the AEC director Lewis Strauss tried to reassure reporters that there was no cause for alarm. The Marshallese were healthy and the fishers—who had only themselves to blame for what Strauss called minor illness—were recovering quickly. The amount of radiation that was released, he said, was too low to cause any harm to people, crops, or animals.[6] But during the question and answer period, he said that the hydrogen bomb could be made infinitely larger and could destroy a city as big as New York. Press coverage immediately focused on the enormity of the bomb, not on Strauss's reassurances.[7]

Fears about fallout and growing opposition to the arms race prompted small groups of pacifists to begin a public campaign against the tests in the summer of 1955. In June, members of the War Resisters League, the Fellowship of Reconciliation (FOR), and the Catholic Worker movement refused to participate in a New York City civil defense drill, sitting out the test in downtown Union Square. FOR and the Women's International League for Peace and Freedom (WILPF) circulated antiwar pamphlets, took out newspaper advertisements, and circulated the Einstein-Russell manifesto, which called for an end to the arms race.[8]

Scientists also began to express uneasiness about the tests. In September 1954, one of the fishers from the *Lucky Dragon* died, again focusing attention on the dangers of radiation. His death contradicted the AEC's assertion that the fishers had not been exposed to dangerous levels of radioactivity. Few scientists would assert anything definitive about the health effects of fallout, but some were beginning to raise questions about assurances of the safety of atomic testing. In November of that year, the geneticist A. H. Sturtevant, the Thomas Hunt Morgan chair of genetics at the California Institute of Technology and president of the Pacific Division of the American Association for the Advancement of Science (AAAS), published his presidential speech in *Science*. In it, he argued that radiation was dangerous for reasons other than its effects on people who were exposed to it: it would also, he said, cause mutations in future generations. No amount of radiation, he said, was safe.[9] The AEC responded a month later, asserting that there was little risk to the human population from this type of hazard. Other scientists joined in on both sides of the debate.[10]

The controversy grew in 1955. The AEC released a report in February designed to allay fears about the effects of fallout from the March 1954 test. Again, the agency argued that fallout posed little danger to human health. In March, two University of Colorado biologists challenged the AEC's assurances, arguing that fallout from the Nevada atomic tests that had taken place over the previous year posed a great danger to public health. The AEC's response, by now typical, was to dismiss the threat. Scientists launched another round of critiques of the AEC's stance. Curt Stern, of the University of California–Berkeley, worried that the AEC's reassurances of the safety of fallout, in the face of contradictory evidence from scientists, would "increase the existing distrust . . . by scientists . . . of the information coming from the AEC."[11] Throughout 1955, the debates continued, drawing in more biologists, including the Nobel Prize winners Hermann J. Muller and Linus Pauling. At a National Academy of Sciences (NAS) meeting in April, Muller argued that radioactivity from tests was likely to cause "tens of thousands of mutations."[12] Later that year, it was revealed that the AEC had denied Muller permission to travel to a conference on peaceful uses of atomic energy, igniting another storm of controversy between scientists and the AEC.[13]

In 1956, reports from science and government groups concluded that since any amount of radiation increased the likelihood of genetic mutation, steps should be taken to limit exposure. The first report, by the International Commission of Radiological Protection, urged setting lower limits on exposure by limiting atomic testing.[14] The second, a study of the biological effects of radiation carried out by the NAS and jointly sponsored by the Rockefeller Foundation and the AEC, was written for the nonspecialist and included reports from a number of subcommittees. The genetics subcommittee, led by the former AAAS president Warren Weaver, explained that any additional radiation would cause genetic mutation, but it was not possible to know exactly how much would cause mutations in present or future generations. Stopping short of recommending atomic testing limits, Weaver's committee report strongly urged that exposure to radiation before age thirty be limited.[15]

Neither journalists nor the AEC read the NAS report in this way, however. Headlines such as "Nation's Top Scientists Call for Radiation Controls; Fear Shorter Life Expectancy and Mentally Defective Babies" and "Guards Urged against Rise in A-Deformities" were typical.[16] The AEC responded to public alarm by trying to draw attention to the Genetics Subcommittee's conclusion that X-rays were more dangerous to human health than atomic fallout. Scientists from the subcommittee angrily responded in letters to the *Washington Post* and *Science*, asserting that this was not the main conclusion to be drawn and citing reports by the AEC itself that concluded that S-90 was genetically harmful. Other science groups were also unhappy with the NAS study, including the Federation of American Scientists (FAS), which thought that the report underestimated the dangers of fallout. FAS quickly commissioned its own study.[17]

This controversy was not simply a debate about which statements were true in the material sense. It was also a debate over the political status of scientific claims about fallout. The context for this debate was the 1956 presidential election campaign. The two candidates, Adlai Stevenson and Dwight Eisenhower, held different positions on atomic testing. Stevenson favored a test ban that would not require direct inspection because he believed it was technologically possible to know when any nuclear test was undertaken anywhere in the world. Eisenhower, on the other hand, favored continued weapons development and testing as means of making nuclear war so risky that it could never happen.[18] In October 1956, a month before the election, two contradictory reports on the health effects of fallout were released. The AEC released a report showing that the fallout from Nevada bomb tests was concentrated at higher levels in Illinois and Wisconsin than in any other area of the world, but it assured readers that the levels were still well below the maximums set by the U.S. government.[19] The other report was produced by the FAS. Far more sobering than the AEC's report, it concluded that radiation from fallout, especially from S-90, was a health hazard, perhaps especially for children, and called for a test ban and more study.[20] Prominent scientists lined up on either side of the debate. Some concluded that more knowledge about the effects of fallout was needed; others thought the AEC's study clearly demonstrated that fallout levels were within a safe range.[21]

In May 1957, the biologist E. B. Lewis published a study of the effects of radiation on survivors of the atomic bomb in Japan, children irradiated as infants for clinical treatment, radiologists, and patients suffering from joint disease. He showed that there was a direct, linear relationship between the amount of radiation exposure and the development of leukemia.[22] Two months after Lewis's article appeared, Ernest C. Anderson and his collaborators wrote in *Science* that although many studies had demonstrated that the absorption levels of radiation through food were deemed safe by many scientists, "if the rate of weapons testing continues to increase, . . . this margin may eventually disappear."[23] Nonetheless, the U.S. government and other individuals and groups continued to promulgate the idea that atomic radiation was harmless.[24]

At the national level, then, the three years following the March 1954 test were marked by controversy about the effects of fallout. Clearly, scientists did not know as much as they, or the American public, would have liked, yet decisions about testing were being made anyway. Like scientists at other universities, some biologists, physicists, and health researchers at Washington University in St. Louis became interested in these debates, formulating their own views and sharing them with their peers and with the public.

Precursors to CNI: Washington University Scientists Mobilize against Atomic Testing

CNI was cofounded by scientists from Washington University (WU) in St. Louis. The university's leadership in the decade after World War II was somewhat unusual. Unlike some other universities that were firing faculty who had been accused of disloyalty by state and federal investigators, WU leaders espoused a civil libertarianism that, if not exactly encouraging dissent, did not suppress free speech. This atmosphere helps explain why CNI was founded there.

The political culture of WU after the war was greatly influenced by Arthur Holly Compton, who was chancellor from 1945 until 1953. The winner of the 1927 Nobel Prize in physics, and the former director of the Manhattan Project's Chicago Metallurgical Laboratory, Compton began his career as a professor at Washington University in 1920. He came from a family deeply committed to public service and civil liberties. During his stint at the University of Chicago, for example, Compton welcomed black scientists to the Metallurgical Laboratory, and he was an outspoken antifascist.[25]

When Compton returned to WU to serve as chancellor, he brought his friend Joyce Stearns, a physicist and the former director of the Chicago Metallurgical Laboratory, with him to serve as dean of faculties. Stearns was a vocal participant in the 1945–1947 "scientists' movement." Compton and Stearns recruited both young, promising scientists and those who were already eminent and well established. In doing so, they continued to build the already world-class medical school faculty, and added prominent new faculty members in the physical sciences, zoology, and botany. Among them were two physicists who had worked with Glenn T. Seaborg on the discovery of plutonium, Joseph Kennedy and Arthur Wahl.[26]

One example of the political tolerance of WU during the 1950s involves the case of Martin D. Kamen, a codiscoverer of carbon-14 and a participant in the Manhattan Project. Kamen had been a target of FBI security investigations that led to the revocation of his passport, problems with grants, expensive lawyers, and hearings before various loyalty commissions.[27] He considered the university and St. Louis "a haven against the tide of cold war hysteria rising in the country . . . [that] had evoked latent anti-intellectualism."[28] Compton regularly defended Kamen from the periodic attacks on Kamen's political loyalty.[29] Compton also formally desegregated the university's undergraduate division in 1952 and appointed the university's first female full professor. Although he was criticized then, and by historians today, for the slow pace of his integration efforts, by the time Compton retired as chancellor in 1953, WU was a vibrant and active research center well known for its physics faculty and increasingly recognized for its biology faculty.[30]

During 1955 and 1956, Compton, who had resigned from administrative work and was now Distinguished Professor of Physics, organized a series of seminars on science and human responsibility. Focused around whether the pursuit of scientific knowledge was drawing society away from consideration of moral values, Compton pursued questions about the effects of science on social life.[31] No populist, Compton considered scientists, not citizens, to be the proper judges of the policy implications of scientific research. He also believed that since there was no world government in place, nuclear weapons testing and development were essential to peace.[32]

Compton's successor as chancellor, the attorney and civil libertarian Ethan A. H. Shepley, continued to promote academic freedom and civil liberties at the university. Shepley attracted national attention in 1956 when he hired the eminent physicist Edward U. Condon. Condon was a former president of the AAAS and a well-respected scientist. Although he had been the subject of multiple federal loyalty investigations, largely because he was a prominent liberal scientist who espoused internationalist views, Condon was never found guilty of disloyalty. Still, he had difficulty finding a permanent position.[33]

In this atmosphere of tolerance, WU scientists were able to engage in the critical debates about atomic testing with little fear of losing their jobs. In 1956, shortly after the conflicting FAS and government reports about fallout were released, the physicists Kamen and Condon and the biologist Barry Commoner held a panel discussion at WU on the effects of atomic fallout. Not advocating an atomic test ban per se, they discussed the problems of detecting fallout and the likely effects of radioactivity from fallout on genetic mutations.[34] Shortly afterward, twenty-four WU scientists, joined by scientists from the California Institute of Technology, Columbia University, City College of New York, the Federation of American Scientists, members of religious organizations, and Eleanor Roosevelt, among others, took out an advertisement in the *New York Times* warning readers of the dangers of testing and of atomic war. They advocated support of a ban on hydrogen bomb testing.[35] Condon, widely respected at WU for his courage in the face of attacks on his political loyalties in the previous decade and his principled political positions, was an especially powerful figure in motivating other scientists to take the risk of publicly declaring their opposition to atomic testing. Other WU scientists who were involved in the testing debate included Evarts A. Graham, a cancer specialist at the WU medical school. With the pediatrician Benjamin Spock and two other physicians, Graham had formed a committee to support Adlai Stevenson's presidential campaign.[36] In the last days before the election took place, Graham wrote to Stevenson about his belief that the principal potential sufferers from fallout, particularly S-90, were children, who were likely to ingest the radioactive element in milk. Stevenson used Graham's claim in his press releases and speeches up until the election.[37]

The WU scientists who spoke out against atomic testing were joined nationally by the eminent biologist H. J. Muller, the future AAAS president Bentley Glass, and scientists from Cal Tech, Columbia University,

City College of New York, Argonne National Laboratory, Yale University, and FAS, all of whom called for an end to atomic testing.[38] Lining up on the other side of the debate was an equally impressive group of scientists who advocated continued atomic testing, the most prominent of whom was the physicist Edward Teller, the major intellectual contributor to the development of the hydrogen bomb.

If before the election the antitesting scientists felt confident of their ability to find like-minded scientists, their capacity for attracting more scientists was diminished by Eisenhower's landslide defeat of Stevenson, and by renewed government suspicion of those who opposed atomic testing. One key reason for this concern was the State Department's release of a letter from the Soviet leader Marshall Bulganin that renewed an earlier offer to ban nuclear weapons testing. In it Bulganin noted that "certain prominent public figures" in the United States also supported the ban.[39] The Republicans did nothing to discourage the public from interpreting this to mean that support for a test ban was tantamount to supporting the Soviet Union. While not directly implicating scientists, it did make antitesting statements, whether backed up by research or not, politically suspect.

Raising the Stakes: Women's Activism and the Atomic Testing Controversy

While scientists at WU were focused on the effects of fallout nationally and internationally, a small group of politically experienced St. Louis women were raising questions about the effects of fallout in their own community. This group of women would eventually cofound CNI with other community members and scientists from WU. The women played an important role in raising awareness about fallout in the St. Louis area and nationally, and in convincing the scientists to formalize their work in providing the public with information about atomic testing.

Women were longtime visible participants in St. Louis's civic life. At the center of many twentieth-century projects was Edna Gellhorn. Born into an upper-middle-class family of social reformers, a graduate of Bryn Mawr, and married to an esteemed professor of gynecology at WU, Gellhorn was in many ways the grand dame of St. Louis's civic activities. A founder of the League of Women Voters and the St. Louis Urban League, she was devoted to women's rights, school reform, and, to some extent, civil rights. In the 1950s, Gellhorn was a member or officer of dozens of local, state, and national women's groups, and could count Eleanor Roosevelt among her friends.[40]

As the atomic testing controversy grew, Gellhorn and some of the twenty-odd other women with whom she would work became concerned about levels of S-90 in the milk that their children drank.[41] Many of these women were members of the St. Louis Consumer Federation, the International Ladies Garment Workers Union (ILGWU), and the WILPF. Some, like Gellhorn, came from old St. Louis families; others were married or otherwise related to prominent politicians and local leaders, including the commissioners of the Board of Education and the Police, and the presidents of the Missouri Bar Foundation and the State Board of Regents.[42]

The women activists first became concerned about atomic testing through the publicity about atomic fallout produced by the Stevenson campaign and the public controversy it generated.[43] Locally, they heard talks by the pathologist Walter Bauer and the physicist John M. Fowler, both of WU, on what was known about the health effects of atomic fallout. Neighbors and fellow Quakers, Bauer, Fowler, and the physicist Alec Pond had become concerned about the effects of fallout as a result of the scientific controversy surrounding it. In a series of self-instructed seminars in 1956, Bauer and Fowler began to learn more about what was known—and especially what was not. They began to speak to local groups about fallout. As Fowler recalled, Bauer and Fowler began to speak in "unsponsored ways to PTAs, women's clubs, Rotaries—to anyone who would listen. We [Bauer and Fowler] were all struck by the lack of knowledge, the confusion of myth and superstition which characterized our audiences and the sometimes overwhelming eagerness with which they sought clear answers."[44]

At the end of October 1956, after the release of the contradictory reports on the health effects of fallout by the AEC and the FAS, Gellhorn and twenty-three other women wrote to the St. Louis Health Commissioner and the St. Louis Dairy Council asking them to test St. Louis's milk supply for S-90. Both agencies responded that they did not have the equipment to carry out milk testing.[45] Stymied but undaunted, the women went into action. Using their extensive national network of labor, women's groups, and consumer groups, and relying on their well-honed mobilization skills, the women began to contact other people who might be interested in working on an atomic test ban with them. Marcelle Malamas, a member of the ILGWU, wrote a letter that the other women signed. It went out to national peace, labor, women's, and other activist groups and individuals, urging them to support a test ban.[46] Among the recipients of the letter were Norman Cousins, a peace activist and the editor of the *Saturday Review of Literature*, and Eleanor Roosevelt.[47] Mindful that the Dairy Council was unlikely to respond without pressure from politicians, the women sent similar letters to their congressional delegation, to the leaders of the AEC, and to the U.S. surgeon general, asking them to intervene by beginning a nationwide milk testing program.[48]

Meanwhile, WU scientists began to take more high-profile positions regarding atomic testing. In November, Condon delivered a well-attended campus lecture in which he criticized government officials for treating the public as if they were incapable of understanding the science and politics of atomic testing.[49] In addition to twenty-four WU scientists who had issued a statement in October opposing more testing, another 106 campus physicians and scientists joined them to petition the Joint Committee on Atomic Energy to investigate the health effects of fallout.[50]

The women's strategy of using politicians to provide the political leverage they needed to instigate milk testing paid off quickly. After the intervention of Missouri Congressional Representative Frank Karsten, whom the women had contacted, the AEC agreed to include St. Louis in a study of S-90 in the milk supplies of four cities.[51] And in December, the Senate Foreign Relations Committee's Subcommittee on Disarmament, chaired by Hubert Humphrey, came to St. Louis to hear public testimony about disarmament and weapons control. Humphrey's committee was unusual in its openness to hearing multiple perspectives on the issue. In the twelve previous public committee hearings, groups that were only recently accused of being communist, such as WILPF and the American Friends Service Committee (AFSC), were allowed to testify.[52] It was at the St. Louis hearings that many members of the women's group and the scientists who had been working on related antitesting activities met face to face.[53]

Speakers included Mayor Raymond R. Tucker and Arthur Holly Compton, both of whom supported continued atomic testing. Opponents of testing who testified included Theodore Lentz (a WU peace activist and author of the 1953 book *The Science of Peace*), Edna Gellhorn, and Gertrude Faust, a parishioner of the Reverend Ralph Abele, the leader of the progressive St. Louis Metropolitan Church Federation and a member of the St. Louis Consumer Federation.[54] Faust based her call for a halt to atomic testing on her experiences as a mother and a consumer. She argued that traditional "consumer methods" such as boycotts or widespread milk testing would not work, and reported that her group would continue to educate people about the dangers of fallout. "Many members share my personal feeling" she said, "that, in the interests of the life and health of our own people and those of the entire world, we should cease poisoning our soil, polluting our food, and exposing our children to disease, genetic dangers and possible death from Strontium-90" by undertaking a unilateral atomic test ban. Gellhorn made a much stronger case for the science behind her request for a cessation of atomic testing. She noted that she and the other members of her group were not in need of information: "I am here today," she said, "because many of us feel that the information already available so strongly indicates increasing danger and points to the fact that each H-bomb exploded increases the danger to such an extent that we cannot afford to wait." She concluded, "Do H-bomb tests add so greatly to our security that we are justified in risking the health of our children?"[55]

In the following few months, the women activists continued their letter-writing campaign, and scientists from WU took a much more visible role in the debate about atomic testing. In doing so, they began to test the viability of different strategies for attracting attention and persuading others to oppose atomic testing.

THE FORMATION OF CNI

Among the most politically engaged scientists at WU was the biologist Barry Commoner. Commoner was no stranger to politics. As an undergraduate at Columbia University and then as a graduate student at Harvard, he was involved in a variety of social and political causes, including supporting the Scottsboro Boys, addressing poverty in Harlem, supporting the Spanish Civil War, and serving on the executive committee of the Boston chapter of the American Association for Scientific Workers (AASW). After service in the Navy during World War II, Commoner served as Navy staff representative to the (Kilgore) Senate Committee on Military Affairs. His assignment was to gather the views of scientists on legislation about federal control over scientific research. This task put him in contact with many members of the FAS and other scientists who were lobbying Congress.[56]

Once he was out of Washington and at his job in St. Louis in 1947, Commoner turned to the AAAS, hoping, like the SSRS, to use the association as a vehicle for encouraging scientists to become more engaged in the public political debates of the day. In 1953, Commoner met with Warren Weaver, the president-elect of the AAAS and the director of the National Science Division at Rockefeller University, which funded Commoner's research, to see if the AAAS was interested in these kinds of activities. Weaver suggested that Commoner try to develop a concrete program of action, in line with the AAAS's new orientation toward social issues that was set out in the organization's Arden House statement of 1951.[57] Encouraged by this conversation and by the enthusiasm of Condon, the outgoing president of the AAAS, Commoner began to look for ways to introduce social issues into the AAAS.

In 1955, he helped to organize the AAAS Interim Committee on the Social Aspects of Science. The committee proposed to investigate the state of the relationship between science and the public. Its report was released in 1956. At a time when there was more dependence on science in government, industry, the economy, and social life, they wrote, science had failed to "attain its appropriate place in the management of public affairs." Calling for more discussion of the negative *and* positive effects of science on social life, the environment, and health, and urging that citizens and government gain greater knowledge of science, the committee recommended that the AAAS serve as a source of scientific facts that would allow citizens and government to make political decisions. The committee emphasized the dangers to democracy if scientists did not speak out on these issues: "If others express their opinions and scientists do not, a distorted picture[will emerge] in which the importance of science will be lacking and the democratic process will become to some extent unrepresentative."[58]

Although the report did not make any specific recommendations for action, it was met with mixed responses. The AAAS leadership, particularly its executive officer, Dael Wolfle, believed that it was the business of professional associations to participate in public political debate. But neither Wolfle nor other AAAS leaders offered much in the way of concrete ideas or direction for the Interim Committee.[59] The FAS was enthusiastic about the committee's recommendations, but cautioned that it might "raise a few eyebrows," since scientists had not been highly involved with the public in recent memory.[60]

Reactions were diverse. *The Nation* applauded the report, interpreting it to mean that scientists would not make policy recommendations, but would only provide scientific facts to the public and political decision makers. The editors thought that this might have been "the biggest story of 1956." Praising the "sturdy and challenging tone" of the report, they approvingly remarked that "the scientist becomes more of a politician."[61] *The New York Times* argued that the new role for scientists might be constructive "if done properly."[62] Detractors had stronger views. The *Wall Street Journal* dismissed scientists' claims to self-importance and rejected the implication that their involvement in politics was somehow nonpartisan. Their involvement in public issues, the paper editorialized, was "a clear call to the political barricades if we ever heard one." As for scientists' assertion that they had an "obligation" to serve the public, the editors dismissed it as mere self-promotion that placed scientists on a pedestal where they did not deserve to stand. "There is no more reason for scientists to feel 'obliged' to accept this role [to shape how their products were used] than there is for businessmen, soldiers, or college professors. Or," they mockingly wrote, "for that matter, comic book publishers, who have an important 'social effect.'"[63] Others thought that engagement with the public was hopeless, for the public was interested only in the magical powers of science and the contraptions that it could build, but would never have much appreciation for the scientific process. The assistant editor of *Science* Joseph Turner, for example, wrote in March 1957 that the public unfortunately had "a wish to manipulate the course of nature" that did "not necessarily imply a wish to understand the natural laws on which such control is based."[64] Other writers worried that the specialized language of science and scientists' lack of skills in communicating with the public, the lack of sophistication among the public, and security issues made it less than likely that citizens could be educated about science. Too much popularization, some believed, could trivialize science.[65]

As I noted in the previous chapter, despite the interest of some of the leaders of the AAAS, such as Weaver and Wolfle, in public political issues, there was no mandate for partisan involvement in political issues except those that directly affected scientists, especially issues of academic freedom. Thus, while the AAAS Council endorsed a statement of support for Hungarian scientists who suffered as a result of the brutal Soviet invasion in November 1956, it shied away from the proposals of the SSRS and the AASW for addressing issues related to militarism.[66]

The debates in St. Louis and Commoner's attempts to find a vehicle to express his political views were taking place just as greater public attention was being drawn to nuclear war and testing during the spring and summer of 1957. The film *On the Beach* was attracting serious public attention.[67] In April, radical pacifists and liberals, including Bayard Rustin, Robert Gilmore of the AFSC, and Norman Cousins, met in Philadelphia to discuss the formation of a committee to stop atomic weapons testing, which they called the National Committee for a Sane Nuclear Policy (SANE). SANE's first major action was in November 1957, when the group took out a full-page advertisement in the *New York Times* calling for the abolition of nuclear testing. Signed by a diverse group of liberals, including Pitirim Sorokin, Lewis Mumford, Eleanor Roosevelt, Norman Thomas, and Cleveland Amory, the advertisement stated that "because of the grave unanswered questions with respect to nuclear test explosions, especially as it concerns the contamination of air and water and food" atomic testing should cease.[68] Despite the growing national and international popular opposition, testing continued throughout the spring of 1957.

In May, Commoner and Condon invited the chemist, Nobel Laureate, and peace activist Linus Pauling to give a speech on disarmament at WU. Pauling was well respected by other scientists for his brilliant contributions to understanding the nature of chemical bonds, and increasingly well known nationally and internationally for his opposition to atomic testing and to war in general. Pauling began his speech "Science and the Modern World" by enthusiastically describing recent discoveries by physicists and biologists. He then turned to an extended analysis of the dangers of radioactivity to human genes. Pauling went on to advocate a full test ban and an end to the arms race. Making a special appeal to the scientists in the audience, he urged them to educate their fellow citizens on the dangers of radiation so that decisions about atomic testing could be democratically decided. Scientists, Pauling argued, had to "realize their obligations to enlighten mankind."[69] He ended his speech by reminding his audience that the problem to be solved was not merely technical, but also moral. Quoting a 1780 letter from Benjamin Franklin to the English scientist Joseph Priestly, Pauling said, "O that Moral science were in as fair a way of improvement, that men would cease to be wolves to one another."[70] Pauling invited his listeners to see the testing dilemma in a historical context, and to understand science itself as an activity with moral value.

The audience response was overwhelmingly positive. Pauling, Commoner, Condon, and Bauer rewrote the speech as a petition. Within two weeks, twenty-six hundred scientists had signed the petition; within two months, more than nine thousand had done so. The petition made the specifically scientific claim that all additional radiation "causes damage to the human germ plasm such as to lead to an increase in the number of seriously defective children that will be born to future generations."[71]

Unlike the SSRS, whose members based actions on the idea that they were no different than anyone else who had to make moral decisions regarding war on moral principle, signers of the Pauling petition understood scientists to have a special moral role to play. Expressing a "deep concern for the welfare of all human beings," the petition read, "We have knowledge of the dangers involved and therefore a special responsibility to make those dangers known."[72] They warned that if "testing continues, and the possession of these weapons spreads to additional governments, the danger of an outbreak of a cataclysmic nuclear war . . . will be greatly increased."[73] This call for scientists to serve as mediators to educate citizens about the dangers of atomic testing and to advise the government harkened back to the scientists' movement that took place after World War II. The National Committee on Atomic Information (NCAI) and FAS educational campaigns were decidedly more populist, and had as their goal something much closer to what John Dewey meant by an "educated" public than one that was merely informed about facts.[74] What was different about the Pauling petition is that it did not call for the sort of education that the earlier campaign had emphasized, but rather, for scientists to provide the public with information about the dangers of fallout.

Within a month, more than two thousand American scientists had signed the petition. It quickly generated controversy.[75] One of the reasons for this is that the petition was being circulated at the same time that a special AEC commission, convened earlier that year to investigate the biological effects of testing, was presenting its findings to Congress. One of the results of the controversy over testing was that scientists' disagreements further confused Americans, for it was unclear whom to believe. An exasperated editorial in the *New Republic* explained the growing feeling that scientists were hardly presenting facts. The debate, the editors wrote, "is not enlightening, because it is an argument, principally, between two groups of extreme partisans, both guilty of asserting scientific claims unconfirmed by the evidence."[76]

That summer, direct actions against testing increased. In August, members of the pacifist group Nonviolent Action against Nuclear Weapons attempted to occupy the Nevada Mercury Proving Grounds, in the hope of stopping tests. They were unsuccessful but gained national attention for their efforts. Local organizations from around the United States that had been picketing, holding vigils, and taking other types of action joined the Committee for a Sane Nuclear Policy, which was to be one of the most influential peace groups in the late 1950s and early 1960s. Around the world in Japan, Great Britain, Canada, New Zealand, and elsewhere, activists petitioned, wrote letters, and held demonstrations against atomic testing and the arms race.[77]

By the fall of 1957, American scientists were becoming worried about how the disagreement among scientists over the facts about atomic fallout shaped public opinion of science. In *Science*, the assistant editor Jonathan Turner wrote, "The possibility that advances in science may have important social consequences is not new. Nor is it novel to find scientists in dispute about the meaning of a particular finding. But the recent striking increase in the first possibility has made the general public forcibly aware of the second." The result, wrote Turner, was "a distrust of science arising from honest perplexity, for one may ask who shall decide when scientists disagree." What scientists needed "was something better in the way of public relations."[78] Another editor of *Science*, Graham DuShane, breathed a sigh of relief in reporting that perhaps a solution had been found to the problem of how members of Congress were to judge the testimony of different scientists who were called before congressional committees. He lauded those who called for panel, rather than individual, testimony before Congress. He believed that a panel system would allow scientists to question one another before they were questioned by Congress, and therefore the differences and similarities in the opinions of scientists would be known to scientists and worked out ahead of time. This, DuShane believed, would increase the legitimacy of the testimony.[79] In a letter to the editor of the *New York Times*, Ralph Lapp reassured readers that "disagreements in science are often essential to progress and understanding."[80] The earlier worries about whether the public could understand science were now multiplied by the concern that disagreements among scientists lowered their social status and the political value of their work.

A startlingly ferocious debate between Linus Pauling and Edward Teller in February 1958 on a San Francisco television program revealed that these debates were no longer gentlemanly disagreements about technical issues. Teller was the foremost scientific opponent of a test ban, a close confidant of President Eisenhower, and a staunch anticommunist who had testified against Robert Oppenheimer in his 1954 security investigation by the AEC. Teller warned that a test ban would give the Soviets a chance to carry out tests that would eventually harm the United States. Pauling warned that thousands of children would be defective if atomic testing continued. They made ad hominem attacks on each other, and each accused his opponent of misleading people about the facts in the debate.[81] Pauling's book *No More War!* and Teller's *Our Nuclear Future*, both published in 1958, codified each scientists' scientific and political interpretations of what to do about atomic testing. Neither book pretended to be a cool analysis of fact; each author put forth a strong political argument that was buttressed by scientific claims.

At the same time that the controversy about atomic testing was gaining much more attention, the United States was confronting the reality of Soviet superiority in missile development. The launching of Sputnik in 1957 worried many leaders and citizens about the capabilities of U.S. scientists, for it had long been assumed that the Soviet Union followed the U.S. lead in technological and military development, but did not originate many of its own new technologies. The worries about whether the United States was behind, and whether American scientists could keep the nation safe through weapons development, only added to the atmosphere of tension and uncertainty about atomic testing.

During this period of heightened interest in fallout, and with the momentum of the AAAS Interim Committee, Commoner continued to work on the development of methods to promote more interactions between scientists and citizens. One of his most important ideas was published in "The Fallout Problem," which appeared in *Science* in May 1958. Building on the arguments first put forth by the AAAS Interim Committee Report in 1956 and on Pauling's arguments, Commoner laid out a political strategy that he believed would resolve the problems of credibility that scientists were facing, and would make use of scientists' special skills without undermining democratic practice. Concerned, as the editors of *Science* had been the year before, that "the public has been confused by disagreements among scientists regarding the biological danger of present and anticipated radiation levels from fallout," Commoner advocated that scientists share more, not less, information about what was known about the effects of fallout with the public. It was scientists' "natural task," he wrote, to use their "professional organizations to bring the necessary facts and the means for understanding them to the public."[82] He was careful to assert that although scientists had a special role to play in this process by virtue of their knowledge, if the public were not involved in decision making once they had the facts, scientists would later bear responsibility for the future effects of nuclear testing.[83]

The sense in which he meant *information* is different from what the atomic scientists meant by *education* in the 1940s. Earlier scientists intended their project to be something more along the lines of what John Dewey had in mind when he wrote about the need for reason-based learning. *Information* was developing a particular meaning in science, and it was that meaning that came to be incorporated into CNI's project. In describing the meaning of *information* in the mathematical theory of communication in computer science, Warren Weaver wrote, "The use of 'information' in this theory [mathematical theory of communication] is used in a special sense that must not be confused with its ordinary usage. In particular, it must not be confused with meaning."[84] It was this meaning of *information* that Commoner appropriated. If scientists were able to provide citizens with distinctive bits of information that were divorced from evaluative statements, they would be providing citizens with the basic tools they would need to participate in political decision making, Commoner believed.

For Commoner, the principle behind his involvement in information provision was the similarity between democratic and scientific practices and goals. Opposing the top-down vision of a government elected by the people but not beholden to them, Commoner had great faith that ordinary people could use the tools of analytic thinking and consideration of evidence to arrive at decisions that were consistent with data. This vision assumed that political decision making was essentially a rational process dependent on an informed electorate, and that many political problems could be resolved with recourse to fact. Harkening back to the development of seventeenth-century liberal political theory and rules for judging scientific claims, Commoner's view treated the public sphere as a neutral space in which all positions are placed in full view, and universalistic rules are used to come to an agreement. Both systems depend on the capacities of individuals to interpret, classify, and manipulate nature using a set of universal principles that guide collective action and decision making. Problems are resolved through debate about the relative effectiveness of various solutions for different participants in decision making. In the American model, interest groups are key links between individuals and government. In a political system corrupted by the misrepresentation, misuse, and withholding of scientific information, scientists could restore the integrity of science and democracy by providing citizens and other scientists with accurate, unbiased, freely exchanged information.

Commoner's ideas strongly influenced the direction that CNI would take. In late March 1958, Edna Gellhorn suggested to John Fowler that they and other scientists and nonscientists form an organization that would coordinate the activities of the scientists, the women antitesting activists, and other community groups. In part, Gellhorn and other women activists were motivated by the failure of the women's national petition drive for milk testing,[85] and their inability to gain access to be heard by government groups. They were denied the right to testify at a hearing on atomic testing before the Joint Committee on Atomic Energy because the committee decided that only scientists were qualified to speak. This was the same committee to which the SSRS member Herbert Jehle was unable to speak because he did not represent a group. These new rules were a significant change from those of Humphrey's 1956 committee, which had enabled scientists and nonscientists to express their views.

The women decided to continue to focus on local issues.[86] In March 1958, Reverend Ralph Abele (the head of the St. Louis Metropolitan Church Federation), John Fowler, Edna Gellhorn, and Barry Commoner met at the home of Mrs. Ernest Stix, one of the women involved in the petition campaign and a cofounder of the St. Louis League of Women Voters. They met in order to form a new organization dedicated to educating the public about fallout. They named a steering committee composed of scientists, physicians from the medical school, women activists, and an attorney.[87]

In the next month, the group debated what sort of strategy they would use in their campaign. In an April 14 memo to the steering committee, Commoner wrote, "Mrs. Birk and Mrs. Faust . . . feel that it might be a mistake to commit the organization definitely not to take a stand at this time on continued testing, etc."[88] Faust later recalled that the group understood that to inform was not politically neutral. "If you're trying to educate people, you have some kind of agenda in your head why you want to educate, you have a philosophy that you're educating them for. So, I'm not so sure that you can be nonpolitical in something that involves your government in this way, and yet, we really did have to try and take that stance."[89] Faust said that the pressure to adopt a neutral stance came from their fears of red-baiting. Walter Bauer recalled in a 1973 interview that avoiding political positions was also considered a way to include a wider range of people: "A lot of people had come there for different reasons, and we could see that if we were going to do a good job we had to have support. And some of us were interested in the peace issue primarily, and some of them were interested in the quality of milk and so forth. And we didn't want anyone to join or not to join our organization on the basis of being a Republican or a Democrat or for the administration or not being for the administration or whether you were in big business or little business."[90]

Still, participants in CNI were not known for their pro-testing views. Virginia Brodine recalled, "Lots of people who came in to C.N.I. came in because they thought it [the bomb] should be banned. They were against it, but they weren't against it on—I mean they were against it for other very good and sufficient political reasons, maybe . . . of which what information they had played a role, you see. I think it would certainly be wrong to say that the people who founded the organization were not worried about it and didn't have political ideas about it, they certainly did."[91] Thus, rather than a timeless, natural position to take, information provision seems to have been a politically expedient tactic.

The week before the official founding of CNI, the group of scientists and women activists wrote that the proposed organization was necessary because "the public needs to decide whether it would approve the use of possibly suicidal process [of using nuclear weapons] for the purpose of 'settling' conflicts between nations. . . . A nation that conducts tests of nuclear weapons must somehow weigh the military or political advantages which it expects from these tests against the harm that they will cause to the health of people in all parts of the world, now living and expected for some generations to live on earth."[92] This was hardly a politically "neutral" statement, given the contrast it draws between suicide and "military and political advantages." Later that week, the group drafted another set of principles that proposed, instead, engagement in public speaking and use of "self-education groups" led by nonscientists who had mastered the basics of the science of nuclear power and weapons. This strategy was almost identical to what the education groups after World War II had done. "The issues of modern life," the statement read, "require a confrontation between ascertainable scientific facts, and ethical, religious, and political principles."[93]

With a strong push by Commoner and some others, the founders of CNI agreed that the group "will accept the responsibility" to form a scientific and medical advisory group to "collect and analyze available information, stimulate discussion about information and policies, and to express informed public opinion on these matters." Of the three goals, only the first was starkly expressed in terms of what the group would actually do; the other two were framed in terms of what the group might do, with majority consent. The statement was signed by 182 scientists, physicians, attorneys, activists, and others. Within weeks of its formation, at the urging of Commoner, CNI decided to wait a year to make a decision about whether to advocate a particular position. In the meantime, the group would simply present facts to the public.[94]

At about the same time, other scientists across the United States were forming local groups that were undertaking related activities. A group of California scientists organized to oppose atomic testing, but they decided against the popular education method that CNI was developing. Their leader, E. Richard Weinerman, thought that "while the reasoning that leads one to reject the nuclear weapons security thesis are almost entirely biological and humanitarian, the resulting lines of action must be almost entirely political. Hence, our concentration on the Democratic Party level, instead of the usual community education activities, important as these are." The group's main objective was to provide the California Democratic Party with information on the biological effects of fallout and the effects of nuclear weapons use on worldwide populations. At Rockefeller University, another group, whose founders were friends of Commoner's, formed to create a public education program on the biological effects of radiation. Both the California and New York groups sought advice and scientific information from CNI.[95] Clearly, there were other models of political action available for scientists aside from the neutral information-provision model that CNI had developed.

During its first year, CNI scientists were researching what was known about fallout and health, and gathering their own information about levels of fallout in St. Louis. Their findings, published in CNI's new newsletter, *Nuclear Information*, indicated that atomic testing was continuing to increase the levels of radiation in milk, but they stopped short of recommending that children stop drinking it. What gained them more publicity, however, were their speaking engagements. CNI scientists spoke to seventy-five groups that year. They also widely circulated a document prepared specifically for the public, on the effects of radiation on human biology.[96]

Even though speakers from the group were careful to distinguish their own views from the facts, it is clear that this was, at first, unconvincing to other scientists. The most important opponent of CNI at the local level was Compton. Beyond his opposition on scientific grounds, Compton took issue with CNI's populist stance.[97] Compton's view was supported by other scientists at WU. Leslie Drews, for example, had once considered the group a great boon to communists, and concluded that although he vacillated between "Jeffersonian Democracy and Hamiltonian Republicanism . . . at this particular time I think that Hamiltonian Republicanism should be our guiding principle."[98] Another critic concluded that CNI was misguided and a waste of time from a scientific standpoint, and that university faculty ought not to be involved in "expressing their political feelings."[99] At a national level, prominent scientists raised questions about the wisdom of the provision of information to the public, fearing that it would undermine scientists' detachment. Rather than seeing the provision of information to the public as a natural and obvious activity for scientists, H. Bentley Glass argued that "our first responsibility is really to be biologists." Although Glass himself was one of the most active participants in the debates about atomic testing, served on numerous government and NAS panels, and was an ardent defender of academic freedom, he was worried about what would happen if scientists spent time engaging in political debate. Those who urged scientists to take a more public stand, he said, were shortsighted because "the possibility of conflict between detachment and evangelism is ignored."[100]

Another pressure toward the provision of information divorced from policy recommendations was that audiences in St. Louis were more likely to react favorably to talks that presented what was known (and unknown) about the effects of radiation, rather than presentations that sought to convince the audience to support a test ban or milk testing. The success of the group in its first year helped to solidify Commoner's and others' belief that the group ought to eschew any policy recommendations. At CNI's first annual meeting, the group's constitution was changed to reflect this new policy.[101] Fowler, who was not initially in favor of the nonpolitical stance, said that the change came not because there "was any sense of being devious; we recognized that it was strongest place from which to proceed."[102]

There was another major reason for CNI to move away from endorsing political positions: the group needed money. A CNI subcommittee reported at an emergency board meeting in early 1959 that "CNI must now approach various foundations for money. . . . [T]he feeling was prevalent that foundations and some individuals would be more prone to give if CNI were strictly educational."[103] They decided to incorporate as a nonprofit educational group. Meanwhile, the women who had raised questions about fallout in milk decided to disband their group and join with CNI. The decision was not, however, enthusiastically made. Noting that CNI did not publicly oppose atomic testing, the women voted 5–4 in favor of dissolution, with several abstentions.[104]

The new, nonpartisan and formally apolitical approach that CNI espoused left most of the women activists in CNI in traditional, gendered roles. They served as secretaries and administrators, the "backstage" positions, while the scientists engaged in "front stage" public speaking, in addition to researching and interpreting scientific data. The work that women participants did was essential to the group, however. They managed most of the research projects, in addition to the conventional office work that was necessary for the group to run. Two women in particular were the backbone of CNI's administration: Gertrude Faust, formerly of the St. Louis Consumer Federation, and Virginia Brodine, a reporter for the ILGWU. Brodine worked with CNI for more than twenty years.[105]

CNI Receives National Recognition

CNI was already an active participant in public political debates about atomic fallout by 1961, but it became a household name through the publication of a report written by one of the few women scientists in CNI. Florence Moog, a biology professor at WU, began editing *Nuclear Information* in 1959. Under her guidance, readership increased monthly. Her most important contribution to the group, however, was her analysis of what would happen if St. Louis were under a full nuclear attack. A gifted writer, in 1959 Moog produced a fictional, almost clinical account of the deaths from radiation and fire, contaminated food and water, blindness, disease, and of the destruction of the city's infrastructure. Such reports were commonplace in the second half of the 1940s; NCAI and other groups produced many of them. This report brought home the experiences that the Japanese had endured, which had been described in detail in John Hersey's *Hiroshima* (1946) and Shute's *On the Beach*.[106] Reprinted in the *Saturday Review*, and then as an individual document with forty-five thousand copies in print in the next decade, the CNI report made the group a household name.[107] This was the document that young SSRS activists in the Boston area distributed.

In 1959, CNI drew more national attention for its participation in the Baby Tooth Survey, begun in December 1958. Carried out in conjunction with the WU Dental School, the project collected children's baby teeth, along with information about the child's age, geographic location, and diet, to analyze the extent to which S-90 had lodged in the teeth. Citizens contributed more than fifty thousand baby teeth to the survey. The results of the ongoing survey, widely publicized throughout the United States, provided information to the public about the levels of radioactivity in the bodies of the most vulnerable group in the population.[108]

One of the most important political aspects of the Baby Tooth Survey was that citizens were voluntary contributors to the data collection. Rather than being told little about the health effects of atomic testing and radiation, or hearing information from seemingly distant scientists, citizens were participating in the development of scientific ideas that affected them. The survey was so popular that other organizations participated in it. In 1961, for example, Women Strike for Peace activists from around the country urged women to send their children's teeth to the Baby Tooth Program, and to send the final report to their senators.[109]

CNI had an audience in part because of the new attention to issues of peace and atomic testing that had developed over the previous four years through the actions of SANE, Women Strike for Peace, the Student Peace Union, the AFSC, and the Committee for Nonviolent Action. Students for a Democratic Society, a small, new organization formed in 1961, was also critical of the arms race. But SANE was by far the most visible group. By the end of 1958, it could boast more than 150 chapters in the East, West, and Midwest, tens of thousands of American members who subscribed to its newsletter, and the capacity to use high-profile advertising and other forms of public communication so effectively that many Americans were aware of the group's message. These forms of public action were complemented by SANE's Washington, DC–based lobbying campaign.[110] Individually, none of these groups had a major effect on government policy, but their activities, which included civil disobedience, public speaking, advertisements, newsletter and information distribution, and delegations to Congress, raised the visibility of the nuclear testing issue. These groups, along with the scientist groups Pugwash and the SSRS, offered a number of different approaches to questions of peace and nuclear testing.

Yet the information provision model, with its claim to nonpartisanship and public service, would have been hard for CNI to abandon. In addition to the difficulties of changing strategy once one has been established—a common problem for most organizations—red-baiting was still common. The House Committee on Un-American Activities (HCUA) was still investigating individuals and organizations thought to be pro-Soviet, and organizations such as the American Legion continued to monitor and harass groups, including the SSRS, that they suspected of being communist. In July 1960, in response to a Senate Internal Security Subcommittee investigation of New York SANE leader Henry Abrams, SANE adopted a policy that effectively excluded communists.[111] The Student Peace Union and Women Strike for Peace were also targeted by security investigators.[112]

Like the SSRS, CNI was not entirely successful in its attempts to convince others of its political "neutrality." In October 1960, the *St. Louis Globe-Democrat* published an editorial reviving the question of the CNI member Edward U. Condon's political loyalty, reminding readers that HCUA had once called him "the weakest link in America's security chain." Condon asserted his innocence of the charges, as he had many times before to different audiences. In a letter to the publisher of the *Globe-Democrat*, Condon acknowledged that he was the subject of five loyalty hearings and investigations between 1947 and 1954, but reminded readers that he had never been found to be disloyal. He called the editorial an effort "to revive McCarthyism."[113]

This event shook CNI. The organization considered removing the word *Citizens'* from its name after the Condon controversy, and renaming itself the Committee for Nuclear Information. In November 1960, the Community Relations Committee (CRC) of CNI met to discuss the public image of the group. In order to create an image of nonpartisanship, the group agreed that members would speak to groups with varied political viewpoints about nuclear war and weapons. They even considered inviting a former member of the AEC, known to be a proponent of nuclear tests, to speak at a meeting sponsored by CNI because "it would make it clear that we are hospitable to a point of view different from that of CNI in the minds of some St. Louisans."[114] In January 1961, at a joint meeting of the executive board and the CRC, the CRC expressed concern that "some people in the community regard us as a 'left' organization." This, they argued, was derived from "the nature of our information. . . . It makes people uncomfortable. It raises questions in their minds about government policy," and from the simple fact that CNI "was founded by people who were and are deeply concerned with the dangers of the military energy." Fearing more red-baiting, CNI decided to "include more positive information . . . on the things to be gained from the use of nuclear energy, such as electricity and the medical use of isotopes."[115]

Changing the group's name and including more positive views of atomic energy did not stop the attacks. Four months later, a group calling itself the Committee to Ban the Communist Party picketed WU. They were protesting because Linus Pauling remained a close associate of CNI members Commoner and Bauer. Pauling was still the most vocal scientist critic of atomic testing. In 1960, he had given a speech at a liberal Jewish organization in St. Louis on the topic "Our Choice—Atomic Death or World Law." Very shortly after the speech, Pauling came under attack from the Senate Internal Security Subcommittee.[116]

Charges that CNI was spreading propaganda rather than scientific fact worsened when it turned out that some of CNI's scientific assertions and calls for caution were not supported by scientific evidence. New data gathered by universities and government agencies from studies of fallout in soil, in milk, in the ocean, and on plants indicated that it was possible to remove some of the radioactivity from fallout. As a result, fears about fallout were waning.[117] In St. Louis, new studies of safety from fallout in milk showed that using a more accurate testing method caused recorded S-90 levels in St. Louis milk to drop by 50 percent.[118]

The *St. Louis Globe-Democrat* was quick to use these new findings to attack CNI. The paper characterized CNI's data on fallout in milk as "phony scare propaganda" and assured readers that they need not be worried about "vigilante citizen committees who attempted to discredit reputable experts." The *Globe-Democrat* also made hay out of the appearance of CNI in a footnote in a Senate Internal Security Subcommittee report on communist groups. The communist magazine *Soviet Russia Today* had written an article on peace groups, and had mentioned CNI by name. The *Globe-Democrat* featured this fact on the front page of the paper.[119] At least among those who knew that many CNI members opposed nuclear testing, the supposed political neutrality of the group was unconvincing.

That fall, CNI faced more challenges to its supposed neutrality, this time from Compton. Compton believed that CNI was in favor of a test ban, despite the group's claim that it took no position on the matter. Walter Bauer recalled that "Dr. Compton, the Nobel Prize winner, was arrayed against C.N.I. and gave a number of prominent speeches through our town saying that we were all wet, we didn't know what we were talking about."[120]

These attacks caused CNI to redouble its efforts to provide the public with unbiased scientific data in order to call into question uses of atomic energy. One effort was the provision of information to opponents of one part of Project Plowshare, a proposed AEC study of the utility of nuclear power for "constructive" purposes. Among the proposed Plowshare activities were gathering useful isotopes from surrounding rock after a nuclear explosion, and generating energy by pumping water to turbines through underground rock melted by underground explosions. CNI assisted opponents of another Plowshare project, called Project Chariot, which proposed creation of a new harbor in Alaska by blasting away seventy million tons of soil and rock along its coast. The blasting would have two purposes: to provide scientists with information about the potential of harbor blasting as a "peaceful" use for atomic energy, and to build a harbor that would facilitate mineral exports from the state.[121]

The proposed Project Chariot test site was in an Inuit-populated area. At that time, the Alaskan Inuit had the highest S-90 concentrations of any group in the world. CNI provided detailed information to the Inuit, politicians, and scientists about the extraordinary amounts of S-90 that could be expected, if Project Chariot were to be carried out, to become concentrated in lichen eaten by caribou, which in turn would be consumed by the Inuit. In preparing its own reports, CNI explicitly contested the scientific claims of the U.S. government, whose report predicted that harmless levels of fallout could be expected from the blasts. Project Chariot never took place, partly because of strong opposition from the Inuit and their Alaska representatives, and partly because it had been scheduled to take place during a time when the United States subsequently committed to a test ban.[122]

CNI continued these kinds of information projects on subjects related to fallout. Commoner, CNI's charismatic leader, was becoming more involved in environmental science, specifically with how toxic wastes affected the ecology of the planet. Drawing attention to the ways in which the same products of modern science could harm the planet while also benefiting humans was to become an even more important theme in the work of Barry Commoner and other American scientists in the 1960s.

INSTITUTIONALIZING THE INFORMATION— SOCIAL RESPONSIBILITY MODEL

CNI occupied an important place in the politics of science in the 1960s not only because of its ability to attract public attention, but also because the information model became the most important form of public scientist engagement in the 1960s. Its success was in part due to Barry Commoner's efforts. He helped the AAAS to push forward with a similar model through the formation in 1958 of the Committee on Science in the Promotion of Human Welfare (CSPHW). This permanent committee replaced the earlier Interim Committee on the Social Aspects of Science. Commoner and his AAAS colleague Margaret Mead quickly pushed for the new permanent committee to create a report assessing the status of science in public life and the role that scientists ought to play.[123]

In the first report issued by the new committee, Commoner expressed his worry that close partnerships with economic and political institutions had eroded the integrity of science, and he implored scientists to face their social responsibilities. Noting that methods developed by other scientists, including members of the FAS, the Pugwash movement, and the SSRS, had not won broad acceptance by the scientific community, Commoner advocated that the AAAS assemble facts and prepare reports for public distribution, and develop partnerships between local communities and scientists.[124] As Margaret Mead recalled in 1963, the group decided not to take political positions, and to keep itself separate from other groups such as peace organizations or political parties, because one of the most "important things in the formation of this kind of activity is that it should be relatively—I can only say relatively—free from infiltration or the accusation of infiltration."[125] The committee sponsored numerous reports, symposia, and conferences during the 1960s, on topics ranging from civil defense to environmental issues. Commoner played an extraordinarily important role as a member of the CSPHW, establishing a way of engaging scientists in debates about public issues that did not require the AAAS to take a stand on any given issue, and of allowing scientists with different views on a subject to participate in the debate.[126] This would have been impossible in the AAAS a decade earlier, but the new activist position that the committee advocated was becoming established in the association.[127] Despite opposition from scientists who wished for a less populist and politicized stance, throughout the 1960s the AAAS pushed forward a program of spreading greater public understanding of science through science education and communication.[128]

This call for more engagement by scientists was ultimately a hopeful and optimistic expression of how scientists could use their ideas to promote human welfare. These same sentiments were expressed in the micro-biologist Rene Dubos's 1962 book *Reason and Utopia*, which was nominated for a National Book Award. Dubos valorized scientists as longtime contributors to the public good. Like C. P. Snow, Dubos believed that scientific ways of thinking had much to offer in the sphere of public decision making. In his review of the book, H. Bentley Glass praised Dubos for offering a hopeful and ultimately realistic portrait of scientists—and a realistic portrait was what Glass thought was needed. Science was under attack, he wrote, in part because the public did not understand the fundamentally human qualities of the scientist, including "his childlike curiosity and enthusiasm, his errors, and his blindness!"[129]

Dubos's and Glass's enthusiasm, however, would soon be tested. Thomas Kuhn's *The Structure of Scientific Revolutions* (1962) and Rachel Carson's *Silent Spring* (first published serially in the *New Yorker* in 1962) raised the stakes for scientists' engagement in public political life. Kuhn's book depicted scientists not as bold, objective thinkers, but mainly as followers of taken-for-granted "paradigms" that shifted only when massive amounts of evidence dislodged them. For Kuhn, scientific ideas were shaped by the contexts in which they were created, not independently given by nature. Carson's claims had nothing to do with the epistemology of science, but her harsh criticism of the chemical industry raised serious questions about whether science as a whole, and scientists individually, really understood—and cared about—the problems that technology created. Coupled with congressional calls for funding research based on nonmilitary issues, and dissensus among scientists about the proper way to engage public issues, Kuhn's and Carson's work further undermined the long-term possibilities for scientists to remain politically neutral and politically objective. By the end of the 1960s, the information-provision model—still powerful until the late 1960s—would come under fire from scientists and other activists for its failure to acknowledge the kinds of claims that Carson and Kuhn had made.

THE DECLINE OF ATOMIC TESTING AND THE FORMATION OF THE SCIENTISTS' INSTITUTE FOR PUBLIC INFORMATION

In April 1960, the New York Scientists' Committee for Radiation Information (the group founded by Halsted Holman in 1958) and members of CNI and of other fallout information groups located in Rochester, Cleveland, Hartford, Philadelphia, New Jersey, Palo Alto, and Missoula, Montana, met under the auspices of the AAAS. Each group was engaged in a different kind of activity. Some, such as the Missoula group, which was led by the zoologist E. W. Pfeiffer—who was also a member of the SSRS—were actively opposed to atomic testing, and published materials saying as much, while others provided information to the public, government, the media, or other scientists.[130] By late 1962, the number of fallout information groups from around the United States had grown to twenty-two, and many members thought that the time was right to centralize their activities. At the urging of Margaret Mead, these groups formed the Scientists' Institute for Public Information (SIPI) in February 1963.

SIPI's founding statement reflected goals that were nearly identical to those of CNI: the organization sought to provide "information unencumbered by political or moral judgments." The information, the statement said, "is prepared with scientific objectivity" and the "information is freely available to all."[131] Commoner's speech at SIPI's founding highlighted the importance of the information model for buttressing democracy. Scientists, he said, should place their hope in the method of "bringing reason and justice to the solution of our nation's and the world's tragic problems . . . because the public's need is identical with the commanding imperative of our own discipline, which is to seek and teach the truth."[132] Unlike CNI, SIPI's goals were to provide scientific information on a variety of subjects to the public.

In July 1963, the Atmospheric Test Ban Treaty was signed by the United States, Great Britain, and the Soviet Union. The effects of the treaty on CNI were mixed. On the one hand, such a ban was what the group had been working for since 1958, and since they had played an important role in shaping public opinion and the AEC's arguments about the dangers of fallout, they could, and did, feel satisfied with what they had accomplished.[133] On the other hand, as with many other social movement groups, the achievement of a specific goal left CNI without a clear direction. The St. Louis group considered dissolving, but most members felt strongly that it should not. Noting that the public was not especially interested in underground nuclear testing and that a broader scope would likely bring more members and subscribers to their newsletter, now called *Nuclear Information*, they considered a variety of issues to which they could turn their attention.[134] Given that CNI had largely focused on the biological effects of testing (not, as physicists had done earlier, on the blast effects), the group decided to turn its attention to environmental issues. Commoner was especially interested in this subject; he was becoming a specialist in the new field of ecology and was actively working to focus legal and scientific attention on problems of pollution from industrial products. He continued to use the AAAS to launch discussions and projects on the environment, and he became more involved in developing what he called "a science of survival," the outline of which would appear in his 1966 book *Science and Survival*.[135]

In 1967, SIPI and CNI agreed that *Nuclear Information* would become the official newsletter of SIPI, in exchange for financial support for its publication. In 1964, the newsletter was renamed *Scientist and Citizen* and began to publish more articles about environmental issues and fewer about atomic issues.[136] The pattern that CNI had developed continued, however: scientific boards within CNI evaluated information about sociotechnical issues, adjudicated the science, and published their evaluations. CNI members continued to speak to groups directly and on the radio and television. They also continued to participate in exchanges with government agencies and other organizations about the soundness of the science behind policy proposals. The circulation of *Scientist and Citizen* was not as large as it had been in the past—1,200 in 1964, compared to a circulation of 1,700 for *Nuclear Information* in 1962—and it was increasingly a publication to which organizations and agencies, rather than individuals, subscribed.[137] Reflecting the increasing focus on environmental issues, in 1967 the CNI board voted to change the name of the group to the Committee for Environmental Information, and the newsletter title to *Environment*. This became a well-respected and widely circulated magazine.[138]

Scientists as Conduits, Knowledge as Information: CNI's Legacy

Like the SSRS, CNI did not explicitly raise questions about how political relationships might shape scientific knowledge. In both cases, the groups treated the content of scientific knowledge as politically unproblematic. It is not clear, however, whether individual members of CNI believed that the content of knowledge was utterly untouched by the political perspectives of the groups that produced it. Yet the decision to treat knowledge as information was not an unexamined decision or the default option given by the rules of "science." Commoner had spent years on AAAS committees, unsuccessfully attempting to persuade peers to address social issues, as the SSRS had also tried to do. There were other choices available to CNI. It could have followed the SSRS model and chosen to boldly espouse political viewpoints, as individual scientists such as Pauling and Leo Szilard did, or it could have continued the elite negotiation model that groups such as Pugwash were using. Ultimately, the choice that CNI made, to treat knowledge as information, was shaped by the political realities of the era rather than by a normative belief in the inherently apolitical quality of scientific knowledge. By treating scientific knowledge as information, free from its context of production and its producers, CNI was more insulated from charges of partisanship and particularism. As nonideological information, scientific knowledge fit into the broader intellectual ideas about the end of ideology and the confidence in technocratic decision making that was, at least until the early 1960s, part of the political landscape of the era.

Just as knowledge was packaged as a bit of information divorced from its makers, scientists were treated as conduits of information divorced from their social locations. Individual scientists need not have partisan motivations to participate in CNI's activities, since action was couched in terms of collective duty rather than individual choice. The idea of duty was not unfamiliar to scientists; it was used as a basis for explaining their participation in the Manhattan Project. Just as in that project scientists had a special task to complete by virtue of their technical skill, so, too, according to CNI, scientists had a special duty to promote democracy, based on a rare but important set of technical skills.

Underlying this role is a particular vision of political decision making, in which the public makes decisions on the basis of rational deliberation and facts, rather than ideological or moral preferences. It shares with the liberal view of how political decisions are made the notion that there is such a thing as "the public," the generic citizenry that is the basis of democratic political practice. Citizens' decisions, in this view, are filtered upward to representatives, who take these views into account when deciding on various policies. By providing citizens with information and envisioning them as rational actors, CNI challenged the idea that "the masses" were inherently irrational. When public participation is circumvented, according to CNI, democracy is undermined. Information provision could restore democracy by allowing citizens, not just the government and its experts, to participate in decision making based on reason.

"Information," "duty," and the notion of a generic "public," in the historical context in which CNI was founded, simultaneously shifted responsibility both away from scientists and to them. This framework— in its ideal form—reinsulated scientists and scientific knowledge from ideology. If collective duty, rather than individual choice, draws scientists into public political debates, then attention is deflected from the moral motivations of individual scientists, which could discredit the idea of scientific knowledge as information. Action comes from relationship to the collective, not from self-interest or individual motivation. Thus, because motivation is externalized, actors can distance themselves from defending their choices against charges of self-interest. Moreover, in the late 1950s and early 1960s, when red-baiting was still common, situating motivation in terms of public duty could signify detachment from particularistic interests that might draw the attention of anticommunists. The decontextualization of knowledge similarly drew attention away from the conditions of production, and thus limited the potential to discredit claims as mere particularism.

The most powerful challenge to scientific authority put forth by CNI was that "the public" ought to be able to participate in sociotechnical decisions. The liberal political theory espoused in the 1950s was suspicious of "the public" and centered authority in the executive branch, increasingly advised by a cadre of scientific and other experts. CNI challenged the idea that stable democratic systems were best run by keeping the masses from having too much input. CNI was able to revive the public engagement in debates over nuclear power that had been stunted by McCarthyism, but in this new version, CNI asked the public, not scientists, to come to their own conclusions. Of course, given that in 1960 only 41 percent of American adults had graduated from high school (compared to 85 percent in 2005), it is not surprising that *Nuclear Information* had limited readership.[139] But CNI did have influence on the media and on the decisions of the AEC. CNI members served as counterexperts who insisted on keeping scientific debates in public view and invited the public to critically engage in debates that had been kept within the state and the laboratory in the 1940s and 1950s. As such, scientists served as "conduits" for information. In this conception of their role, scientists collectively mediate the relationship between nature and the state.

There is little evidence to suggest that CNI convinced its readers and other observers that the group had no stake in the outcome of the debate over atomic testing, especially given that *Nuclear Information* strayed from the strict "neutrality" stance to draw conclusions about policy. Nor was CNI able to completely separate the political perspectives of scientists within the organization from the organization itself. Yet CNI was scrupulous in its research, and rarely were other groups or agencies able to contest its scientific claims, even if they disagreed with the policy implications of those claims.

What CNI accomplished was to routinize the distribution of information and, through the Baby Tooth Survey, a particular form of voluntary public participation in knowledge production. This new system was novel in the United States. Previously, the government, particular scientists, or reformers would have distributed such information, or it would have been distributed on a temporary basis, as in the atomic scientists' public education movement at the end of World War II.

As the test ban movement declined, U.S. involvement in Vietnam spawned what would become the largest mass-based antiwar movement in American history, incorporating pacifists, African American and Latino groups, liberals, intellectuals, professionals, and, above all, students. Unlike in the test ban movement, in which scientists played the role of information providers, in the anti–Vietnam War movement, they were not only sources of information about weapons and opponents of the war but also targets of activists because of their role in the production of weapons and delivery of advice that perpetuated the war. By the late 1960s, the antiwar movement was becoming more radical, using direct action and calling for wholesale reform of American institutions. Scientists who were part of this radicalization denounced the liberal information-provision and advice model of interaction among scientists, citizens, and the government, as well as the moral individualist model. They called for more attention to the way in which science was subjugated not only to a military state, but also to capitalism and the interests of economic elites. In the next two chapters, I turn to an examination of the role of the anti–Vietnam War movement on college campuses and the development of "radical" science activism.

Confronting Liberalism: The Anti–Vietnam War Movement and the ABM Debate, 1965–1969

> Over the past 25 years the scientific community has grown very large in numbers and in resources, but we have become complacent with our prestige. As scientists have become more and more dependent on the government for research funds and for their very livelihood, speaking out on public issues has been done more and more cautiously. We must therefore strike to regain our full intellectual and political freedom. We shall work for change within our present affiliations (professional society, university, laboratory), but foremost we shall strive to present our opinions as an independent body of socially aware scientists free from the inhibitions which abound in the established institutions we now serve.
> —Michael Goldhaber, Martin Perl, Marc Ross, and Charles Schwartz, "Announcing the Formation of a New Organization Dedicated to Vigorous Social and Political Action," February 1969

In 1969, four physicists who were frustrated by the failure of the American Physical Society (APS) to publicly oppose the war in Vietnam called for the organization of a new group of scientists. More than three hundred physicists attending the annual APS meeting in February 1969 came together to discuss the formation of the new group. First called Scientists for Social and Political Action, and then Scientists and Engineers for Social and Political Action, and, more colloquially, Science for the People (SftP), the organization was the most important radical science group in the United States in the mid-twentieth century. The term *Science for the People* referred to both the organization and the cluster of related ideas that undergirded it. *Science for the people* meant, above all, that science should be created and used in the service of "the people"—the working class, oppressed people, or nonelites, depending on the speaker—instead of in the interests of elites, the military, and capitalism.

From the founding of the group in February 1969 through the winter of that year, many of its activities were similar to the liberal model of information provision and lobbying, embracing the idea of "social responsibility" among scientists. Yet by early 1970, a new, vocal group of SftP members was engaged in public political action and claims making that rejected the liberal model of information provision. They condemned the model for its elitism and its failure, in their view, to adequately critique the economic and political relations that shaped knowledge production and use. The conscientious objection model was also rejected by many SftP members for its failure to direct attention to the economic and political contexts of knowledge production and use.

Radical scientists argued that economic, race, and gender systems, rather than a temporarily corrupted government or a misinformed public or insufficiently moral individuals, were the source of the problems caused by science. The problem, as radicals saw it, was that systems of knowledge production and distribution in the United States mainly benefited ruling groups, and were as much a source of harm as of benefits to humans and the environment. The systemic changes that radical scientists called for required more fundamental reorganization of power relations, not simply more public education or personal renunciation of associations with the military.

Because the radical science movement centered around SftP was decentralized, local chapters and individuals were free to choose whatever actions they wished. Some called for university-based changes, such as demands that universities renounce military funding and add new "relevant" science curricula. Others called for more far-reaching changes that included joining other workers in a struggle against imperialism, and an end to discrimination against women scientists. In the first two years, SftP activists used "direct action" to force their peers, professional associations, and employers to take action against the institutional arrangements that implicated science in these problems. Direct action included political theater at professional meetings, public and face-to-face denunciations of their peers, and refusals of professional honors. After 1970, the group developed a broad range of activities at the local, national, and international level.

As in the case of the information and conscientious objection models, the development of the radical model was shaped by the contemporaneous political activities of groups and movements on the left, and by the relationship between scientists and the military. Two major changes were especially important. The first was the congressional call for scientists to attend to a broader range of social and political issues and to spend less time on basic and military research. The era of free-flowing federal research dollars—and scientists' capacity to shape where they went—was disappearing. Although many scientists readily signed on to the new regime, it also meant that some scientists would lose power. Among them were the physicists, who were used to being listened to on the numerous military advising panels on which they served. They and other scientists were especially angered by the failure of the Congress to reject on technical grounds the antiballistic missile (ABM) system. This issue created a breach between scientists and the government.

A second key factor was the radicalization of the student and antiwar movements. The hearts-and-minds methods of persuasion that had been so successful in the civil rights movement no longer seemed viable to many activists, and there was a growing belief that personal relationships and the major institutions of society would have to be fundamentally reorganized, not simply reformed. It was a moment when faith in liberalism was declining among activists and intellectuals. Yet it was unclear what would replace it. The debates that scientists had about the Vietnam War and about their relationship to the government and other parts of society were thus not only about substantive issues. They were also about the status of science as an ideological system to serve as a basis for organizing political life. Could information and technical advice be considered politically neutral, as liberal and conscientious objector scientists had claimed? Were scientists simply technicians with little moral culpability for the uses made of their work? These were questions with which earlier scientists had also struggled. In the second half of the 1960s, these questions were addressed via debates about an unpopular war and, in particular, about the relationships among the war, professions, and universities.

At MIT and at other campuses in the Boston area and nationally, science students and faculty involved in the student and anti–Vietnam War movements were especially critical of military research on college campuses. This posed a direct challenge to university scientists with military contracts, and to those who considered military contracts and advising to be a personal choice that scientists had a right to make. In the fall of 1968, students at MIT organized what became known as the March 4 movement, as a way to critique such ties and to explore what alternatives there might be for a socially responsible science that addressed the myriad social and environmental problems of the day.

This chapter shows how the antiwar movement, the growing radicalism on the left, and the debate about the ABM system shaped the emergence of something that organizers of SftP called "radical," without quite knowing what that would mean in practice. When SftP was formed, it was clear that the older methods of advising and personal responsibility would not satisfy a new generation of activists—but what would, they did not know. In this chapter, I explore the roots of the rejection of the liberal and conscientious objection models, and the embryonic attempts by scientists and science students to articulate an alternative.

BEGINNINGS: THE ANTI–VIETNAM WAR MOVEMENT AND THE ABM DEBATE IN BOSTON, 1965–1968

In the words of former SftP member Jon Beckwith, "the origins of this [radical science] movement can be found in the political convulsions which were taking place around the Vietnam War and the civil rights movement. Many people working in science initially developed a social conscience and became politically active, not around scientific questions, but around these nonscientific issues."[1] Similarly, Donna Haraway argued in her comparison of SftP with the Marxist scientist movement in 1930s Great Britain that the development of radical science in the United States in the 1960s was driven by concerns about scientists' relationship to the war in Vietnam.[2]

Although the United States had been militarily involved in Vietnam for more than a decade, in 1962 and 1963 intellectuals, members of Congress, and a few peace groups began to articulate their opposition to U.S. involvement. By 1965, U.S. involvement was growing, and so too was a movement against the war. It was increasingly coordinated by groups such as Students for a Democratic Society (SDS), the American Friends Service Committee, and the Student Peace Union. Although dispersed across many different constituencies and settings, the antiwar movement was most centrally situated on college campuses. Subject to the draft and morally appalled by the war, students targeted their government and their universities for their roles in the war. The beginning of significant university-based critique was the 1965 "teach-in" movement, in which more than one hundred and twenty campuses held panel discussions on the war and other political issues of the day. In October, antiwar activists organized the International Days of Protest, which involved more than one hundred thousand people in more than eighty cities and several nations.[3] In 1966, widespread student opposition to the draft developed. Protests against the presence on campus of ROTC and recruiters for war-related companies such as Dow Chemical (which made napalm) escalated.[4] Students also began researching the extent to which military-sponsored research was taking place on their campuses. As I discussed in chapter 2, at campuses ranging from Penn State to Stanford, students found that contracts for weapons and surveillance systems were plentiful.

By 1967, students and other activists were beginning to call for dramatic changes in U.S. involvement in Vietnam and in the organization of public and private life. In 1968, the assassinations of Robert F. Kennedy

and Martin Luther King Jr. and the relentless escalation of the war outraged many activists, who concluded that fundamental social change required more than polite acts of civil disobedience and teach-ins. Student
and ethnic- or racial-based movements in the United States and elsewhere
began to call for revolutionary acts, and to question the organization of
institutions ranging from universities to the art world to the medical system. In May 1968, the Mobilization to End the War in Vietnam staged
the largest antiwar rally in American history. The mood of the protestors
at the August 1968 Democratic National Convention in Chicago was
angry and defiant. The black power movement, as well as the emergent
Puerto Rican, Asian American, and Chicano movements, called for radical action and revolution, not compromise or piecemeal change. In Mexico, Brazil, France, Italy, Japan, and other countries, students demonstrated in huge numbers against war, political repression, and economic
injustice. A vocal segment of the American left, with growing confidence
in its own power, was rejecting liberal strategies of piecemeal change. The
American university was one of its major targets.[5]

In the Boston area, faculty and students were becoming involved in
activism against the war. Even though MIT was one of the largest recipients of Department of Defense funding of any university in the United
States, and MIT and Harvard faculty members were among the most important science advisors in the nation (often referred to as the "Charles
River" scientists), significant numbers of scientists in the area opposed
the war in Vietnam. In 1964, scientists formed Scientists and Engineers
for Johnson-Humphrey. The Massachusetts chapter, centered at MIT and
Harvard, was the most active of the local chapters, with more than three
thousand members. Much of the impetus behind scientists' support for
Johnson was their belief that Goldwater, the Republican candidate, would
escalate the war.[6] In January 1966, twenty-nine prominent Boston-area
scientists took out a out a full-page advertisement in the *New York Times*
opposing the U.S. government's use of chemical weapons in Vietnam.
John Edsall, the director of the Biological Laboratories at Harvard University, drafted the letter. The authors "emphatically" objected to the use
of "chemical agents for the destruction of crops, by United States forces
in Vietnam." Whether or not the agents were poisonous to humans, they
wrote, the use of such weapons was "barbarous and indiscriminate" and
showed "a shocking deterioration of our moral standards. These attacks
are also abhorrent to the general standards of civilized mankind."[7] Eschewing the language of fact and information that was the lingua franca
of science advisors and liberal groups such as the Committee for Nuclear
Information (CNI) and Scientists' Institute for Public Information (SIPI),
these scientists used emotion and morality as the basis of their arguments.[8]

Beginning in 1967, students and faculty from MIT and other Boston area universities were involved in another political activity: opposition to the ABM system. In the previous five years, scientists, political leaders, and American citizens had been hotly debating the larger question of how to protect Americans from potential bomb attacks by the Soviet Union and China. Some favored a shelter-based civil defense campaign, others verified arms control agreements, and others a strong defense system.[9] The ABM system was designed to destroy incoming Chinese intercontinental ballistic missiles by placing missile silos near major cities, where the missiles could be deployed to attack incoming Chinese weapons. At an estimated cost of five billion dollars, it would be the most expensive weapons program ever developed by the U.S. government, and it was proposed at a time when funding for basic science research was being cut.[10]

In 1965, Mack Newell, a physics graduate student at the University of Washington, became interested in the technical feasibility of the ABM missile. Newell asked Senator George McGovern to debate the proposed ABM in Congress. McGovern declined, but Newell pressed on. He prepared a technical paper raising doubts about the feasibility of the ABM, and distributed it at a University of Washington talk by visiting physicist Hans Bethe. Bethe, who had been active in the post–World War II scientists' movement in favor of civilian and scientist control of scientific research and nuclear weapons, was also active in international arms control debates. He agreed with Newell's analysis, and around them the two men gathered a group of University of Washington scientists who were concerned about the siting of the proposed missiles. Newell discovered, through his analysis of documents and correspondence with other scientists in the areas where the missiles were to be sited, that many missiles would be placed dangerously near major cities. Through the Seattle group's correspondence and the chance discovery by Argonne National Laboratory scientists that a missile was to be sited near Chicago, more scientists came to know about the proposed system.[11]

In April 1968, Bethe and the Columbia University physicist Richard Garwin wrote in *Scientific American* that the proposed ABM system, soon to be debated in Congress, would be useless as a defense because the Chinese would soon develop more sophisticated missile systems that the ABM system would not be able to stop. They were soon joined by other scientists who also argued that it would be unwise to expend billions of dollars on a system that had a low probability of shooting down incoming missiles.[12]

Despite the pleas of scientists and strong opposition from Democrats and some Republicans, Congress approved $1.2 billion for a scaled-down version of the ABM system. Citizen opposition began to grow. Local scientists were among the sources of the public warnings about the dangers of the system: they asserted that siting the missile silos close to urban areas might help protect cities, yet it also made those cities inviting targets of an attack. When the Army sent public relations teams to communities where ABM sites were scheduled to be built, scientists sounded the alarm. Soon citizen-scientist anti-ABM organizations were formed. In Chicago, physicists from the Argonne National Laboratory west of Chicago publicized the intended siting of the missile system near Chicago and began to describe the dangers of the missile systems to area residents, forming an informal citizen-scientist alliance.[13] Other groups, composed of combinations of nuclear scientists, homeowners, conservationists, pacifists, and real estate developers, formed in Los Angeles and Detroit.[14] In the Boston area, the New England Citizens Committee on ABM recruited local scientists and journalists to participate in a meeting with the Army at Reading High School Auditorium on January 29, 1969. It turned out to be less a staid public information event than an angry confrontation between fifteen hundred citizens and scientists and a handful of Army public relations specialists. The meeting was a public relations disaster for the Army.[15]

Despite opposition from the public and scientists, the Nixon administration did not abandon the ABM system. By March 1969, it was clear that scientists who opposed the system were being shut out of the decision-making process. As the liberal magazine *The Nation* reported, "in the crucial matter of the Sentinel anti-ballistic missile, the scientists see that the decision is not being made by scientists, and the country . . . should count itself fortunate if the most authoritative of them manage to get a hearing."[16] This was a new situation for scientists, given that for the previous twenty years their scientific advice had been courted by the government, not ignored. The knowledge that scientists were, at least in this case, not the valued advisors that they had once been became fodder for the debates that took place during the March 4 movement, especially at MIT.

The anti-ABM campaign marked a significant public breach between science advisors and the president, at a time when federal funding for basic research was being cut. The symbiotic relationship between scientists and the federal government had been based on a quid pro quo exchange in which scientists claimed status by virtue of their association with national security and prosperity, and the government benefited by having an on-tap source of ideas and technologies and the prestige that approval by "objective" outsiders could bring. The Johnson and Nixon administrations, through their failure to attend to the widespread opposition among scientists—especially those such as Bethe who were longtime advisors to the government—gave scientists less reason to continue their political relationship than in the past. Moreover, the ABM controversy brought scientists into contact with citizen groups, in much the way that CNI scientists had been in contact with citizens groups concerned about atomic testing between 1958 and 1963. The ABM controversy helped to organize scientists in areas such as Boston, Los Angeles, Berkeley, and Chicago, where anti–Vietnam War activism was growing. The networks of scientists who participated in the anti-ABM debates were among those who became participants in the challenges to the APS that led to the formation of SftP.

THE MARCH 4 MOVEMENT

The antiwar and anti-ABM movements converged at MIT in the spring of 1969, during what was called the "March 4" movement. Although MIT did not have a history of widespread student or faculty involvement in political issues, it was far from politically quiescent in the mid-1960s. As I showed in chapter 3, between 1963 and 1965 the students in the Boston-area chapter of the Society for Social Responsibility in Science (SSRS) organized a series of well-attended talks and debates on science and social responsibility. In the spring of 1968, the MIT chapter of the SDS, together with the MIT chapter of SSRS and the MIT Committee to End the War in Vietnam, produced the pamphlet *M.I.T. and the Warfare State*. In it, the authors documented a range of relationships among the Department of Defense, MIT, and military contractors. Their special emphasis was not on the sciences, but rather on Defense Department funding for the MIT Center for International Studies, which was sometimes distributed to members of the political science department. The authors demanded that MIT become a "truly free and responsible university, free of vested interest, free of classified research, free of government interference," and urged students to join them "in demanding that MIT accept responsibility for its present collaboration with an unjust and imperialist war, and that future collaboration cease."[17]

In calling for an end to such collaborations, *M.I.T. and the Warfare State* diverged from the liberal critiques of the misuse of science by the state that had characterized earlier science activism. Reflecting the New Left's interest in the subtle forms of political influence, the authors argued that even if one did not accept monies from the Department of Defense, the existence of defense contracts on campus had the effect of normalizing what the authors of the pamphlet thought was abhorrent: "The atmosphere of the [political science] department is one which constantly encourages graduate students and younger professors to get involved in these [military] studies, and set their sights on becoming prominent government advisors. After all, study is only part of what goes on in a department: equally important in orienting the students are informal topics of conversation. In MIT's political science department they talk about keeping security clearances and how to get government contracts."[18] For these authors, the problem to be solved was not merely how to end the war in Vietnam, but how to disengage the university from the military. Going beyond the liberal critique of the misuse of science, these antiwar activists argued that students became involved in military research through a subtle process, rather than through an explicit choice. Writing about the problem was one way that student activists hoped to begin resolving the problem. Ultimately, they hoped to convince members of the MIT community to end military-based research. This perspective influenced younger science activists who would begin to organize around antimilitarism in the fall of 1968.

Antiwar sentiment on the campus spread in October, when the MIT chapter of the SDS and Resist, a Boston war resistance group, housed an AWOL soldier at the student center for several weeks. His presence had an extraordinary effect on the campus community. Students and faculty heatedly discussed the war in dorms and in laboratories, at lunch tables and in classrooms. As the linguist and leftist intellectual Noam Chomsky recalled, the effect on the campus was electrifying. "MIT had practically shut down," he said. "Practically the whole student body was over there [at the student center], thousands of people, twenty-four hours a day. There was an endless stream of everything from political seminars and meetings to rock music. . . . It just turned the whole Institute around."[19] Suddenly, it seemed, the entire campus was engaged in the debate about the war in Vietnam. Among those who participated in the discussions and activities were the MIT physics graduate student Ira Rubenzahl, the visiting Cornell University physics graduate students Joel Feigenbaum and Alan Chodos, and Jonathan Kabat, a graduate student in the MIT Microbiology Department.[20]

In early November, after the soldier left the campus, Rubenzahl and Chodos had dinner at Feigenbaum's. The three continued to discuss the war in Vietnam and MIT's relationship to it, as they had been doing for several months. Inspired by the discussions and ideas raised while the soldier was on campus, they decided to embark on a campaign to draw faculty and students into these discussions. Their first idea was to ask students and faculty to sign a petition against the war. After their initial success with the petition, they began to draft a statement of their views, and shared it with physics professor Kurt Gottfried, who, like Feigenbaum, was visiting from Cornell for the year. Professors David Frisch and Bernard Feld were initial supporters as well, providing the students with comments that the students integrated into their statement.[21] The new group called itself the Science Action Coordinating Committee (SACC).

The group's first public statement was relatively moderate in its tone. It advocated that scientists explore their responsibilities, and in many ways it was consistent with the liberal and moral individualist models of public political action among scientists. It called for critical examination of governmental policy, and urged scientists to look for ways to turn research applications away from military technology and toward the solution of environmental and social problems. Calling on students to "devote themselves to bringing the benefits of science to mankind" and to "scrutinize the issues raised here before participating in the construction of destructive weapons systems," the group wished specifically to express its opposition "to ill-advised and hazardous projects such as the ABM system, the enlargement of our nuclear arsenal, and the development of chemical and biological weapons."[22] In mid-January, Chodos, Rubenzahl, Feigenbaum, and Kabat wrote a letter to the President's science advisor, Lee A. DuBridge, stating that they were "deeply disturbed" about the increasing influence of the military on society. Such claims had been made for the previous twenty years, of course. Echoing the earlier *M.I.T. and the Warfare State* pamphlet, the students wrote that the influence of the military on the university was so profound that it endangered the university as "a center for scholarly research and productive social criticism," and fostered the development of "a group of people and institutions whose interests cannot lie in arms limitation." The letter was mentioned in an article in the *New York Times*.[23]

The same month, the students came up with another method of publicizing and exploring the close relationship between the military and science: a one-day research stoppage, during which they would discuss the relationship between science and the military. The idea was agreeable to faculty supporters, so they wrote a press release announcing the event, to take place on March 4. As a former member of SACC recalled, "Joel [Feigenbaum] set up this [research stoppage] on March 4th. We sent [the announcement of the research stoppage] to the [Boston] *Globe* with a press release. And by the way, this is where Alan [Chodos] came in, 'cause Alan knew the press, he worked on a paper [at McGill University] so he knew how to do that stuff." The group of students "began meeting with Chomsky, we'd go over and talk to Chomsky about what we should do, we'd run back and talk to the physics professors and organize the graduate students."[24] Chodos and Rubenzahl modified the statement they had sent to the press, and brought it to Kurt Gottfried. Gottfried was enthusiastic, as was Chomsky. The document that SACC produced became an organizing tool.[25] The *Boston Globe* carried a front page story about the research stoppage, as did the *New York Times*. The *Times* called the proposed event a "strike."[26]

After the MIT group received inquiries from other campuses and spread the news of its action, groups on other campuses began to plan their own March 4 events. SACC distributed five thousand copies of its literature kit, consisting of a seven-page information sheet on MIT and the military-industrial complex, and a three-page statement of SACC's positions. The information sheet provided information about military funding for research projects related to war, including chemical and biological warfare research at Ft. Detrick, Maryland; the military research sponsored by the Institute for Defense Analyses (IDA) that was carried out on college campuses; and military research programs at Washington University, MIT, the University of California–Berkeley, and Harvard.[27]

In early January, SACC began to argue strenuously for a new form of "counterexpertise," "organized against that of the government and corporations that dominate society . . . specifically antimilitarist, and specifically directed at those aspects of militarization that dominate and direct scientific research, particularly at the university." They took aim at MIT itself. Calling for an end to credit for classified theses, a board to help students find nonmilitary work, and the discontinuance of cooperatives in laboratories heavily dependent on the military, SACC looked to transform the university, not simply change the government or individuals' consciences.[28] They were not alone: at colleges and universities all over the United States, student demonstrations, sit-ins, and other actions—and threats that they might engage in other actions—forced schools to sever ties with ROTC, to establish black studies programs and hire black faculty, and to give student protestors more lenient sentences.[29] In the spring of 1969, more than three hundred American universities experienced demonstrations and disruptions of classes and administration. One-fifth of the actions were accompanied by bombs, property destruction, or arson.[30] Student organizers of March 4 had reason to feel empowered: all over the country, challenges to universities were resulting in changes in policies and the institution of new programs.

The students' new focus began to make faculty involved in planning for March 4 uneasy about what they perceived as SACC's potential to embarrass them or undermine their intellectual and political reputations. As one physics faculty member recalled, "The students had another agenda, which was not known to us, which was to move on to MIT itself, attacking the existence of the I-Lab [the Instrumentation Laboratory, a laboratory on campus that was largely funded by the Department of Defense] and classified research . . . which took me certainly by surprise."[31] Another dimension of this rift was the research stoppage. The physicist Jerrold R. Zacharias declared that he and other scientists at MIT objected to the proposed events because they were "an act of protest with an implied prejudgment of the questions at issue." His view was echoed by Nevin S. Scrimshaw, the chair of the Department of Food and Nutrition Sciences, who argued that "to imply somehow that the research is at fault and should be stopped is naive and hardly useful."[32] As Chodos recalled, faculty were concerned about their reputations, while students thought that they were too timid. "I think that the faculty were suspicious of the students in some sense. There was hairsplitting, as there is in any academic situation. Students were a little bit, contemptuous is too strong a word, but they felt they were dragging these faculty people along. You know, they [faculty] should be wanting to do the right thing, yet they have to be persuaded. And they were very timid; they wanted to put the brakes on everything. . . . The faculty didn't want to have their reputations totally ruined . . . and the students didn't want to be held back."[33]

The divisions between the students and the faculty—and the lack of clarity about exactly what March 4 was about—came into bold relief as a result of the March 4 announcements in *Science* and the *New York Times*. The original plan for March 4 was that it would be a day of protest in which students and faculty would stop work to discuss issues of public political concern. The story in *Science*, however, suggested that organizers had more radical goals by calling the event a one-day "strike," much to the consternation of some of the faculty.[34] For some of the faculty, a "strike" implied an association with labor and communism; for those who had lived through the red-baiting of the 1950s, such associations were to be avoided. Taken in a larger context however, the interpretation of the event as a strike by the *Times* and *Science* can been seen as one of many attempts by journalists to understand this novel form of action among scientists and the uncertainty the organizers themselves felt about the meaning of the event. Other journalists, for example, called the event "a research stoppage," a "form of protest," a "research strike," a "practical and symbolic expression of apprehension by scientists," and a "convocation."[35]

The uncertainty about what exactly the events of March 4 would be and how they fit into larger questions about the university and American society are central to understanding the political significance of the March 4 movement. In the process of organizing, science students were raising questions about two related political issues: whether the university could or ought to be autonomous from political issues of the day, and whether scientists' technical advice could be considered amoral and apolitical. The concern with how institutions were related to systems of power followed the lines of thinking of New Left intellectuals such as Marcuse and Paul Goodman, as well as the SDS. The May 1968 takeover of Columbia University, spurred by demands that the university cut ties with the IDA and cease plans to build a gym, marked the beginning of what Nancy Zaroulis and Gerald Sullivan argue was a new pattern of student protest on college campuses, whose object was "not just to protest certain university policies, but to force the university to become an institution that reflected their views—to become a 'revolutionary political weapon' with which they could attack the system."[36] Although in this earlier period SACC did not call for a revolution, it raised a question similar to that raised by student activists in the United States, France, Mexico, and elsewhere between 1965 and 1975: what was the role of the university in larger economic, political, and social systems? By calling for changes in the university and the actions of individual scientists, SACC expanded the scope and breadth of the critiques of science that were made by liberal and conscientious objector scientists a decade earlier.

Since few faculty were eager to make the kinds of demands for immediate institutional changes in the relationship between MIT and the military that SACC was calling for, and there was now considerable ambiguity about the nature of the event, faculty who were involved in planning for March 4 decided to form a separate group, called the Union of Concerned Scientists (UCS). UCS was composed of forty-eight faculty from many disciplines, but the majority were physicists and engineers.[37] Many of the physicists who were actively involved in planning for March 4 and who later joined UCS had been involved in the post–World War II political debates about the control of atomic energy and the place of science in public life, and so were not unfamiliar with the process of criticizing the government.[38] Francis Low, a member of UCS, distinguished UCS from SACC by saying that UCS was "willing to go slower. We have the same concerns but it's a difference of activism and style." UCS was interested in "how to apply our fantastic technical aptitude to solving our fantastic technical problems—poverty, undernourishment, urban decay, the breakdown of transportation, and environmental pollution."[39] While SACC and UCS might agree on the problems that needed to be solved, UCS members were more interested in solving existing problems through technical advising than in making fundamental changes in social institutions or the university itself.

SACC continued to push for a more critical format for the March 4 events. It announced, "The research stoppage is a protest and a political act. . . . [T]he research stoppage is a strike in the European sense—a one day strike to dramatize discontent with the conduct of affairs in our country."[40] UCS and SACC eventually compromised on the title of the event, calling it a "Day of Reflection" in order to make it more inclusive of diverse political viewpoints.[41] Alan Chodos wrote shortly after March 4 that "in the considered judgment of SACC leaders at the time, this [the use of moderate language] was the best way to maximize support while still retaining a useful political focus for March 4."[42]

The March 4 movement encouraged scientists to consider not only what kinds of social and political responsibilities scientists had as a group, but also what specific ethical responsibilities members of their own disciplines had. Small groups of MIT biology faculty, students, and postdoctoral students, for example, issued a statement advocating that their fellow biologists examine how they might contribute to solving health, social, and environmental problems. They cautioned that to do otherwise could place biologists at risk of becoming morally culpable for any ill effects of biological research. "The nuclear physicists," they cautioned, "awoke too late to the horrible potential of their discoveries. Let us not repeat their tragic error."[43]

In a letter to the editor of the *New York Times* on February 27, Rubenzahl and Feigenbaum explained that participants in the March 4 events were likely to have different motivations. Some, they said, saw March 4 as a "strike" in which participants withheld their services to express a "vote of no confidence in the government to make wise and humane use of science and technology." Others were engaged in a "research stoppage" to make a "symbolic personal commitment toward reforming a set of government policies that have resulted in the growing power and influence of the military-industrial complex." Other scientists saw the event as less a protest of any kind, but as a day to discuss government support of scientific research at universities.[44] Clearly, the scientists who were participating in the March 4 events were both inspired by the experiences of other activists, past and present, and also struggling to differentiate and articulate what place they ought to have in the foment over the war and American society, and in the longer term, what relationship they should have to public politics.

By the time that March 4 actually arrived, SACC's tone was sharply critical of American society, particularly of the role of technology in consumerism and war. The group was disillusioned with promises to apply technology to solve problems, instead seeing its uses as confined.

> As young scientists we have waited for America to apply her technological resources toward solutions of international social and economic problems. . . . We were persuaded that more sophisticated weapons systems were needed for our national well-being, and our elders helped to provide the facilities to build them. In the Sixties . . . America awoke to the congestion of her cities, and to the sterility of the ever-increasing advertisement and consumption of 'technological miracles.' We awaited action to these problems. . . . But instead America undertook to bring democracy to Viet Nam. We watched in increasing disbelief as America brought technological expertise to an underdeveloped nation [Vietnam] in the form of napalm, B-52's, anti-personnel weapons, and strategic hamlets.[45]

UCS, on the other hand, issued a statement at the March 4 meetings that it was a group of "counterexperts" who would use science to correct problems of public concern. Like CNI, UCS based its participation in public life on a set of special skills. It called for "a concerted and continuing effort to influence public policy in areas where your own scientific knowledge and skill can play a significant role." Its main areas of concern were what it called " 'survival problems'—where misapplication of technology literally threatens our continued existence. These fall into three categories: those that are upon us, as is the nuclear arms race; those that are imminent, as are pollution-induced climatic and ecological changes; and those that lie beyond the horizon, as for example, genetic manipulation."[46]

UCS was not simply looking for a way to strategically maintain that scientific knowledge was "apolitical," nor were they cynically attempting to protect their own reputations. The open-ended nature of their efforts to find a method of engaging in politics can be seen in one of the last ideas that they circulated in the informational document at March 4: they called for "an exploration of innovative forms of action."[47] It would be a mistake, then, to see UCS (or SACC, or any other group I examine) as merely efforts to protect their members' status.[48] That view cannot explain the exploratory nature of both UCS and SACC's arguments, or their development over time. To be sure, each group's members were in some sense "acting in their own interest," but such a simplistic idea does not convey that what scientists' interests were, precisely, was uncertain at that moment.

The events on March 4 were a mix of calm debate and heated exchanges about topics including the responsibilities of intellectuals, the conversion of military to nonmilitary research, student concerns, arms control, and technological solutions to urban problems.[49] For the students, the association between the military and science was anathema to the idea that science was essentially a force for good. In his talk at the March 4 meetings, Joel Feigenbaum said that news coverage of the war in Vietnam always showed "a commingling of technology and death" in which cold, logical body counts were mixed with horrific descriptions of dead bodies. The relevance of the war to MIT, he said, was that students "must perceive the relationship between our sparkling, expensive laboratories and the instruments of death produced by the fellow next door. We [students] cannot live comfortably in a place such as MIT, which declares it is 'apolitical' while producing MIRVs [Multiple Independently Targetable Reentry Vehicles], ABMs, and weapons for Vietnam."[50]

Some faculty were less concerned about the substantive issues than they were about what they considered the bullying and uncivil behavior of the students. One physicist recalled that the accusatory tone that faculty felt students used in exchanges with invited speakers was reminiscent of the tactics used by anticommunist government investigatory committees in the 1950s. "[They] had Jack Ruina [the dean of the faculty and a former director of the Department of Defense Advanced Research Projects Agency] up in front of a big lecture hall and they were saying, 'Is it true that on such and such a date you met with so and so who was Deputy Secretary of Defense in Charge of Evil Things?"[51] The political scientist Thomas C. Schelling, responding to a call by the SDS New England regional director Eric Mann for community members and students to contest what was being taught in college classrooms, argued that one could "no longer speak of an academic community" because students did not use standards of civility and politeness in or outside the class-room.[52] The growing rift was between those who embraced the liberal model of reform, using accepted norms of civil debate, and those who saw a crisis in science that was so dramatic that uncivil actions were legitimate and necessary.

The March 4 movement was the most widespread discussion of the proper relationship among scientists, "science," and public life since the end of World War II. Not only at MIT, but at more than thirty universities around the country, students and faculty held meetings similar to those at MIT to discuss their relationships to the public, the military, and the government. The significance of the movement is that it exposed and helped to develop different ideas about what that relationship ought to be. At some universities such as Stanford, student organizers faced the same faculty concerns that SACC did at MIT, and therefore organized an event that included a broad range of scientists from the Bay Area, including those in industry and the university, and scientists who ranged from luminaries to students. At the University of Chicago, faculty and students organized a series of conferences on the topic "science and society." Less concerned with what scientists could do to solve public problems or with the legitimacy threats posed by those who were critical of scientists for failing to address political issues of the day, conference participants expressed concern that external political controls over science would limit scientists' ability to freely choose topics that interested them. The presidential science advisor Lee A. DuBridge explained this idea shortly after March 4. "We [scientists] do not have, and do not want, any organized social controls over the active pursuit of basic science—the pursuit of new knowledge—other than the inherent controls which exist in the minds of and hearts of scientific investigators themselves," he wrote. "Scientists must be free to pursue the truth wherever they can hope to find it."[53] A group of computer programmers at Rockefeller University handed out literature to passersby on Fifth Avenue asserting that "we will not program death."[54]

The mix of views among scientists represented a number of concerns—at the most obvious level, the war in Vietnam and the ABM system. But scientists were also concerned about whether individual scientists or "science" in some general sense was responsible for the ill-use of science. Concerns about freedom of speech represented, at a superficial level, debate about rules of discourse. But they were also a microcosm of the much larger problem: could scientists serve as neutral providers of information if they did not use rational and disciplined discourse? If one of the hallmarks of science is supposed to be the use of reason in a "public" forum to arrive at a collective answer, the new incivility threatened to weaken that foundation. The public face of scientists as cool, collected, and unemotional was belied by the heated debates during the March 4 movement.

The March 4 movement was one of the first direct attacks on earlier forms of joining science and public political issues. It did draw thousands of scientists into organized debates about the proper relationship between science and relationships of power in the United States and elsewhere. The organization for the events helped to create a network of physics graduate students and faculty who began to question the liberal information provision model and its assumptions about public life.

CHALLENGING PROFESSIONAL ASSOCIATIONS: THE SCHWARTZ AMENDMENT

A month before March 4, three physics professors, Martin Perl of Stanford University, Charles Schwartz of the University of California–Berkeley, and Marc Ross of the University of Michigan, and a physics postdoctoral student, Michael Goldhaber of Rockefeller University, announced the formation of Scientists for Social and Political Action (SSPA) at the 1969 meeting of the APS. Although SSPA was not the brainchild of one person, Schwartz helped to galvanize physicists to take this collective political action in opposition to the war because of the failure of his efforts in 1968 to convince the leaders and members of the APS to oppose the Vietnam War.

The idea of using a professional association to undertake explicitly political or moral projects was, of course, not new among American scientists, as I have shown in previous chapters. The 1960s, though, marked a dramatic change in how many leftist professionals saw their professional associations. Professional associations are traditionally associated with the promotion of the professional interests of a group, undertaking tasks that include certification of members, sponsorship of meetings and reports, and advocacy for funding and public recognition. Although there have always been political struggles among members with different visions of the association, most of these were carried out through genteel debate and bureaucratic mechanisms for electing new representatives or introducing new programs and panels.

Not content to have the professional associations promote only the interests of their members in the narrow sense, leftist academicians in the 1960s wanted the associations to serve as political vehicles. The kinds of challenges that Schwartz and others made, which would rock professional science associations, must thus be seen in the context of the activities that took place in other professional associations and the concomitant debates about whether and how professionals could also be radicals.

Some associations, including the American Sociological Association and the American Political Science Association, were taking formal political positions opposing the war, and in the process sharply dividing their members. The American Historical Association was bitterly divided over how the association should approach political issues. In 1968, radical historians supported Staughton Lynd's bid for the presidency of the association against Robert H. Palmer of Yale University. Lynd was defeated by a vote of 1,040 to 396, following an acrimonious and divisive meeting in Washington, DC.[55] Members of associations of microbiologists, public health professionals, philosophers, psychiatrists, linguists, and modern languages scholars raised formal political propositions for members to consider and forced political debates about war and feminism in business meetings. They organized sessions on "radical" subjects. This usually meant analyses of underrepresented groups in research, and topics that would help solve problems of concern to the less privileged. American Philosophical Association radicals, for example, argued that the discipline ought to bring its talents to bear on the issues of the day. More conservative members argued that philosophers' main contribution was to advance thought dispassionately aloof from politics.[56] Modern Language Association (MLA) members had voted to move their 1969 annual meeting from Chicago to Denver to protest Mayor Richard Daley's handling of the protestors at the 1968 Democratic Convention. Radical MLA members also charged that the Department of Defense's Defense Language Institute used expertise in foreign languages as a means of oppression, and accused the Department of stockpiling language experts, especially in Chinese, Vietnamese, and Russian, just as it did weapons.[57]

Moderate and radical women were also actively organizing within professional associations. Between 1968 and 1971, twenty women's caucuses or committees were formed in professional associations ranging from the MLA to the American Society for Microbiologists. Women also formed organizations outside their professional associations, either because there were too few women within a particular association (as was often the case in the sciences) or because women felt that they would not be coopted if they were outside the control of a professional association (e.g., Sociologists for Women in Society, Association for Women in Science). The earliest to organize women's groups were the MLA (1968) and the National Vocational Guidance Association (1968). Women scientists were relative latecomers; of the women's groups formed within scientific associations or organized outside of them, most were formed after 1971.[58]

At an even broader level, professionals were considering a range of different actions they could take to change their relationships to political arrangements. At a 1967 conference sponsored by the Ann Arbor Radical Education Project of the SDS called "Radicals in the Professions," the organizer Ted Steege suggested three ideal types of action models for the individual radical professional: the "community organizer, the spy who channels information to radical groups, and the radical subversive who tries to break down the barriers between his radical activity and his professional field. He attacks both his profession and the larger society."[59] At the same conference, the longtime SDS activists Barbara Haber and Al Haber described another professional role in radical politics, this time emphasizing its collective nature: "it is autonomous, yet responsive to a group of comrades; it is self-defined . . . it can change . . . it provides financial and status gain that are most rewarding to movement people, and at the same time most morally acceptable."[60]

In 1965, the American Association for the Advancement of Science (AAAS) began to feature more discussions of and debates about formally political issues than it had previously. That year the American Association of Scientific Workers (AASW) submitted a resolution to the AAAS urging an end to "hostilities in Vietnam."

This statement was followed in the next year by a series of other activities that explored the relationship between science and social problems in a more aggressive fashion than the AAAS had done in the 1950s. In 1966, the association resolved to study the long-range consequences of the use of herbicides, and proposed to have a Canadian-U.S. conference on social responsibility in science in 1970. At the 1967 annual meeting, panels on "Is Defense against Ballistic Missiles Possible?" and "Crime, Science, and Technology" were held. In 1968, the AAAS sponsored symposia on "Unanticipated Environmental Hazards Resulting from Technological Intrusions," cosponsored with SIPI. Other symposia focused on "public considerations" of genetic technology, and on the global effects of environmental pollution. Even the physicists, long considered the most resistant to integrating social and political concerns, sponsored a session on the social relevance of physics and symposia on "The Physiology of Fighting and Defeat" and "Arms Control and Disarmament."[61]

Throughout 1969, the AAAS Board of Directors, under the leadership of Dael Wolfle, devoted a great deal of attention to the responsibilities of scientists for the values and uses of science. They were especially concerned about disaffection among younger scientists. In June, the Harvard University historian of science Gerald Holton, who was a member of the AAAS Board of Directors, proposed that the AAAS begin to decide which social problems the association should address. The board also approved a motion that each of its members bring a young colleague to the next meeting in order to exchange views about the responsibilities of scientists for the values and uses of science. At the board meeting, the student group met and proposed to study the problem and report back to the committee in March.[62]

Professional associations in science were clearly moving toward engaging public issues. Yet one in particular, the APS, was especially resistant to engaging public issues. This is not a surprise: the APS member Charles Schwartz proposed not a discussion or an investigation, but an explicit political statement against the war in Vietnam. To Schwartz, the APS was especially complicit in the war in Vietnam because its members were so heavily funded by the Department of Defense.

THE ATTEMPT TO REFORM THE AMERICAN PHYSICAL SOCIETY

As was the case with many of the people who became involved in SftP, Schwartz's views about the Vietnam War developed slowly. He (like the UCS chair Henry Kendall) received his doctorate from MIT in 1955. Like most other physicists of that era, he was removed from engagement in political action. In 1962, Schwartz spent one summer in Washington, DC, working on projects for the elite Jason Program of the IDA. As he recalled, "There was a sense of glamour attached to it. I had to get a security clearance, go to Washington, get paid, have a bunch of briefings . . . the headiness—wow, you were really getting into it!" When "the problem was presented of things like better infrared detectors, you know, we understood that was how to find bad people in a jungle at night and kill them. But you know, just another technically challenging problem. I mean, I recognize that, and I say, my God, how amazing! I see that now as part of the familiar cloistered environment that physicists fall into—are encouraged into, and probably want—in getting into the profession."[63]

After his stint with the Jason Program, Schwartz returned to Berkeley in the fall of 1962. By 1966, the Bay Area antiwar movement was in full swing, and Schwartz was becoming more involved in it. First, he signed a petition against the war, and then attended demonstrations. Students at Stanford and Berkeley were beginning to expose the relationship between physics research and the war in Vietnam. Antiwar activists in the Bay area were uncovering defense contracts at Berkeley and Stanford, and at federally funded research centers, such as the E. O. Lawrence Laboratories in Berkeley and Livermore and the Stanford Research Institute (SRI), that were affiliated with them. Students were especially focused on the financial ties between SRI, affiliated with Stanford University, and the development of weapons systems, including infrared detectors used to locate insurgents in Southeast Asia. In 1967 and 1968, Bay Area students occupied the Applied Electronics Laboratory at Stanford, which was the recipient of more than two million dollars in funding from the Department of Defense.[64] These challenges to SRI were made as part of the highly mobilized and increasingly aggressive antiwar movement in the Bay Area.

Frustrated with his inability to persuade his peers to take action against the war, Schwartz came up with idea of having the APS oppose the Vietnam War. He proposed an amendment to the APS constitution that would have allowed members to "express their opinion, will, or intent on any matter of concern to the society by voting on one or several resolutions formally presented for their consideration."[65] His request was denied. The APS bylaws allowed only officers to generate propositions on which the membership could vote. After Schwartz protested, the APS board indicated that if 1 percent of the APS membership would agree that a vote should be taken, the board would then allow the membership to vote on a second proposition that APS members should be able to "initiate a vote of APS members on any issue of concern to APS and on the public stand the society should take on these issues."[66]

In the January 1968 issue of *Physics Today* (PT), Schwartz argued that rules preventing members from proposing actions amounted to "censorship completely alien to the principles of free discourse upon which a scientific community is built." To those who argued that the APS would be destroyed if it became a debating club on issues of the day, he replied that physics was already "intimately tied up with political decision making." The APS was thus obligated to express concern over a public issue, said Schwartz, when "there exists an external crisis of such magnitude that we fear a general catastrophe of a political, military or cultural nature."[67]

When the ballots were sent out in May 1968, the board included a statement of its opposition to the amendment. The amendment failed by a margin of three to one. Throughout the summer, PT published letter after letter from APS members condemning the board for electioneering at the ballot. In August, the board published a defense of its actions in a statement titled "What We Are Not Against." In it, the board argued that its opposition to the proposition was based on its belief that PT spoke "to" physicists, not "from" them. Further, it argued, scientists ought not to make use of the "halo effect" that they had by virtue of their social status to legitimate political arguments.[68] The board's modesty about the halo effect had a certain irony, for most of the board members held or had held prestigious advisory positions in government. If there was to be a routine means for physicists to speak as a group about their political positions or issues of the day, it would have to be a separate vehicle because the APS and its magazine were, in the board's view, a means of promoting the narrow interests of physics, and nothing more. Although the amendment failed and APS remained resistant to the kinds of activities in which other organizations engaged, Schwartz had ignited a storm of protest. Throughout the summer and fall, physicists continued writing to PT to urge physicists to engage in political activity against the war.[69] The insularity that was the dominant model for most professional associations before the middle of the 1960s was clearly losing its legitimacy.[70]

THE FORMATION OF SCIENTISTS FOR SOCIAL AND POLITICAL ACTION

Undaunted by his experiences with using existing institutions to lead scientists to oppose the war, in late 1968 Schwartz and the Stanford University physics professor Martin Perl discussed other possibilities for mobilizing physicists against the war. Like Schwartz, Perl had been involved in Bay Area antiwar activism. He suggested that they form a new organization, tentatively called Scientists for Social and Political Action, that would gather together physicists and other scientists who wished to express their political views.[71] The organization would be independent of any existing group, and would therefore allow scientists to express themselves freely. Joining Perl and Schwartz in their effort to organize the new group were Michael Goldhaber, who had been Perl's graduate student at Stanford before receiving his Ph.D. and who had been involved in antiwar activism in the Bay Area, and the University of Michigan physics professor Marc Ross. Ross was involved in antiwar activism on the Michigan campus and was a participant on a panel on "Alternatives to U.S. Policy in Vietnam and Asia" at the 1965 University of Michigan teach-in about the war in Vietnam.[72]

In two feverish weeks, Perl, Ross, Goldhaber, and Schwartz worked with thirty or so other scientists to publicize the formation of the new group, which they planned to announce at the February meeting of the APS. Rather than relying on the well-established practice among physicists of providing technical advice and criticism of government actions, the call to organize stated that the new group would be a forum for exploration and would be a break with past assumptions about science and progress. "We reject the old credo that 'research means progress and progress is good.'"[73] Like SACC, they intended the group to be a forum for debate "where all concerned—especially students and younger members of the profession—may explore the questions, Why are we scientists? For whose benefit do we work? What is the full nature of our moral and social responsibility?" Organizers hoped that participants would "seek new and radical solutions for long-range problems and immediate issues" and would relate their activities to those of "similar groups (radical caucuses) now forming in other professions." They would not, as "the dominant professional associations"—such as the APS—had done, remain "aloof from desperate problems" of the day.[74] Schwartz explained that he considered the group "New Left" and "radical" but that radicalism did "not mean destructiveness," but rather "a distinctly new direction and we feel that is what is needed."[75]

More than three hundred physicists packed the room for SSPA's first meeting. Debates about what the organization should do were lively and contentious, and many participants wore "Stop ABM" buttons. Participants debated the meaning and significance of the Schwartz amendment, how to organize for March 4, the ABM, and the formation of radical caucuses in other professional associations. When the meeting ended, more than thirty individuals from locations including Chicago, Madison, Woods Hole, Massachusetts, and West Lafayette, Indiana, agreed to organize local chapters. Members were supposed to define what the group's major issues and methods would be.[76] Despite the enthusiasm for reform among a small number of physicists, not all physicists were ready to jump into political debates. At the APS meeting, members voted 8,500–6,400 against moving the 1970 annual meeting from Chicago in protest against police behavior at the 1968 Democratic National Convention.[77]

Like SACC, the founders of SSPA were not sure initially what the new ideological and material relationship between science and politics would look like. They were clearer about the organization's structure. It was to be "an independent and loosely organized assembly" in which a "national framework" would exist to "coordinate projects formed upon the initiative of their members" rather than to set policy.[78] Indeed, in the first few weeks of the group's life, many possibilities for how to act as a physicist in political debates were offered, yet only some were taken up by the group. Through interactions with one another, participants would have to discover what their interests were, how to act on them, and with whom to organize.

Two weeks after the initial meeting, operating out of Perl's Stanford office, the group added *Engineers* to its name in response to numerous expressions of interest in the organization by engineers. By late April 1969, the group was coming to be known as Scientists and Engineers for Social and Political Action. Most participants and observers referred to the group as "Science for the People," a term that also represented the group's philosophy.

Perl's office produced a newsletter that began to elaborate the group's identity. After the "fantastic success" of the initial meeting at the APS, the group announced that it was a new organization devoted to "radical redirection in the control of modern science and technology." Very quickly, more than one hundred people from academia and industry joined the group. Enjoying a level of self-determination uncommon in more hierarchical and homogeneous organizations, locals developed varied forms of research and action that responded to local needs.[79]

The formation of SftP was taking place at the same time that the March 4 movement and the concomitant debate about the ABM system were picking up steam. The Committee for Environmental Information and SIPI, CNI's successors, were flourishing, and even though the SSRS was not as widely known, it, too, continued to draw in participants interested in debating the proper role of science in public life. Other established groups, such as the Federation of American Scientists (FAS), were also becoming more active as well. It is clear that scientists' engagement in public political debates was becoming more institutionalized, although the models for action were highly diverse. Walter Sullivan, writing in the *New York Times*, explained the extent to which scientists were becoming politically mobilized. "A deep groundswell of discontent is rolling through communities of scientists from Moscow to New York and, perhaps even to isolated Peking." To those who wondered what these scientists had in common with student radicals at Berkeley, Columbia, and the Sorbonne, or with the Soviet dissident scientist Andre Sakharov, Sullivan wrote that they all shared the conviction that "the various societies of our planet are trapped in ideological dogma, nationalism, and power politics. They are unable to adjust to rapidly changing circumstances." To those "self-styled realists" who would consider calls for more political engagement by scientists naive, he wrote that "the reply that many of the restless scientists would have given last week [at the meeting of the APS] could be summed up by the words of Dr. John R. Platt, professor of physics and biophysics at the University of Chicago: 'The world has become too dangerous for anything less than utopias.' "[80]

The involvement of leading physicists in military advising and research, coupled with the apparent lack of interest in the political issues of the day among the majority of physicists, drove many early members' involvement in SftP. Like Charles Schwartz and early members of SACC, SftP cofounder Michael Goldhaber was critical of scientists' failure to understand the moral gravity of advising the military. In his view, a leading physicist at Stanford University who had served as part of the Jason Program "was very, very excited by his contact with power. It just seemed to me to be really nauseating, to put it mildly. They [leading physicists who served as government advisors] weren't thinking about what they were doing so much as they were glad for the glamour of it, and that bothered me."[81] If some physicists were too closely connected to the military, others, Goldhaber felt, were wrongly disconnected from the political problems of the day. Shortly after Martin Luther King Jr. was shot in April 1968, Goldhaber recalled that he was upset because a physicist who was scheduled to give a seminar didn't cancel it. "I remember marching in this memorial march that was hastily set up in Palo Alto and just feeling like, you know, why aren't these people [scientists] realizing that this incredibly horrible thing just happened and just carrying on as if it was business as usual? I mean it was sort of a defensive, remote, complete removedness from politics that went on. I got very involved at that point."[82]

For others, the cauldron of challenges to the university sparked their interest in how politics and science intersected. One biologist, an early active member of the Chicago SftP chapter, began to question the relationship between politics and knowledge during the 1968 sit-in at the University of Chicago:

> In 1968 there was a major sit-in at the University of Chicago . . . and around this issue a number of student power issues started up. I was one of the people that attended meetings and listened, but I learned a tremendous amount during that time; I had endless discussions with people. I was particularly interested in how all this related to science, and this related to what I was doing. In other words, was all the science that I was learning as socially conditioned as the political ideas that we were combating? And was there some knowledge that was pure knowledge and other knowledge that was sociologically conditioned? This was a major thing for me.[83]

Similarly, Britta Fischer, a member of the Boston chapter of SftP, first became involved in antiwar activism, and later connected her activism with the politics of science: "[My] radicalizing experience was the Vietnam War and the protest against the Vietnam War. . . . So I was definitely radicalized by that experience, as well as by the exposure to some fellow students who had clearly studied some Marxism. By the end of the sixties, I was turned around from the wishy-washy liberals. So in '69, I got, through some friends, a friend of a friend type of a thing, involved in Science for the People."[84]

Early members of SftP thus included scientists with little previous experience in activism and those with years of experience, physicists and non-physicists, men and women, and people of all academic ranks. Few, if any, came to participate in SftP because they had a clear template for how to act on their political beliefs and values as a scientist. But what they shared was an interest in discovering new ways to join their political and scientific concerns.

This uncertainty is evident in the first SftP newsletter. Few of the projects that the editors in the Bay Area suggested broke with the liberal models of reformist politics and information provision. None urged scientists, as the SSRS did, to forgo certain kinds of research. The newsletter suggested that SftP members work with anti-ABM groups to lobby against the missile system, create a new APS division on the problems of physics and society, make presentations at congressional hearings about the spread of weapons systems, develop March 4 events, organize seminars or colloquia on the concept of science for the people, and circulate anti-ABM and antiweapons petitions.[85] These projects suggest that at least among the Stanford and Bay Area editors, ABM issues were at the forefront of their thinking.

Not all SftP participants wished to abandon the familiar, well-established liberal role of scientist as government critic. In fact, the earliest SftP participants saw existing organizations and science leadership as contributors to the problems of the day rather than as resources for their resolution.[86] SftP members were encouraged by the Stanford SftP newsletter editors to reach out to the liberal FAS, then reviving after being moribund the previous twenty years, and to SIPI. The FAS leader Jeremy Stone was enthusiastic about the new group, and invited SftP to become a branch of the reviving FAS. Some members of SftP were not enthusiastic about collaborating with FAS. Schwartz, for example, did not see FAS's political interests and those of SftP as similar.[87] Michael Goldhaber said that SftP did not "speak the language of FAS." Goldhaber found FAS's lobbying efforts for changes in weapons policy "an insane way of speaking," because it never addressed the problems of why weapons were being developed at all and in whose interest they were developed. Moreover, association with FAS would be a bad idea for attracting young scientists to SftP, because younger scientists, Goldhaber said, saw FAS as "big fat partners of the Pentagon."[88]

Goldhaber's criticism of FAS exemplifies a liberal/radical divide that would grow over the next three years. On the liberal side of the divide, professionally staffed "public interest" advocacy and educational groups were growing rapidly. These organizations, which included UCS, SIPI, and the Committee for Environmental Information, distributed information to "paper members" and urged them to take a conventional form of action such as writing a letter or signing a petition.[89]

Although the liberal public interest model was gaining strength among moderate scientists and Americans, it was not triumphant. On the radical side of the divide, all the talk about "radicalism" would take material form at the AAAS meetings in December 1969 and December 1970. The radicalism espoused by the scientists who took action at these meetings was shaped by the new radicalism on the left.

RADICALISM DEVELOPS ON THE LEFT: THE NEW COMMUNISM,
THE BLACK PANTHER PARTY, AND FEMINISM

In the eight months before the 1969 AAAS meeting where SftP would make its first national appearance, the group, as a compilation of local chapters, as a cluster of ideological perspectives, and as a form of action, became more clearly defined and shaped by developments among activists on the left. SftP's development coincided with the development of the "New Communism," and the embracement of confrontational, sometimes absurdist direct action.[90] After a decade of being overshadowed by the New Left, American communists influenced by Mao Zedong, V. I. Lenin, and Che Guevara reinvented themselves as supporters of third-world revolutionary movements and socialist revolution in the United States. The intersections of class, race, and gender were more important for New Communists than they were for their communist predecessors. Some New Communist groups saw race and class as intersecting in ways that made it impossible not to organize around race and class simultaneously.[91] Other New Communist groups, such as the October League, tried to organize primarily along class lines but encouraged women and minorities to participate on an equal basis with white men.[92] Central to the "New Communism" was the idea that imperialism, or the extension of political and economic rule over colonies and colonylike nation-states, had to be overturned in order for a communist revolution to take place.

The turn toward radicalism came also from the deaths of Martin Luther King Jr. and Robert F. Kennedy in 1968, and the continuation of the war in Vietnam. Many student activists were frustrated and disappointed with the peaceful, reformist methods that had marked the civil rights and anti–Vietnam War movements earlier in the decade.

The largest leftist student group in the nation, the SDS, was collapsing, and in the process its (largely white) members were trying to invent a new role for themselves as revolutionaries. The key questions for them were: Were students workers? Or intellectuals? What role should they play in a movement to overturn imperialism? And how should they relate to race- and ethnic-based movements, many of which were hostile to whites?[93] After protests at Columbia University in April and May 1968, more student activism against universities took place the following year. Students at Berkeley walked out of graduation, and hundreds of campuses experienced sit-ins, takeovers, and other disruptions of routine activity. Black students used direct action to force Cornell University to attend to their demands for more black students and African American curricula on campus. Harvard University students went on strike in the fall of 1969, and MIT students engaged in a series of demonstrations and direct action campaigns at some of the university's most lucrative research laboratories that were funded by the Department of Defense. The December 1969 killing of the Black Panther Party leader Fred Hampton and the Peoria Black Panther organizer Mark Clark by state authorities in Chicago, and the highly publicized harassment of the group, were especially galling to leftist activists, who took on the Black Panther Party's repression as a cause célèbre.[94]

The women's movement was also gaining strength in 1969. Feminists challenged the ways in which writers, artists, professionals, and the law had constructed ideas and images of women's inferiority.[95] Although there were few women in the sciences at this time, small groups and networks of women scientists began to raise questions about the validity of the idea of objectivity and, later, even its desirability.

By late 1969, the New Left, which had helped to generate so much campus-based and antiwar activism during the decade of the 1960s, was collapsing. In its place were a series of fragmented groups, some of whom were intent on revolution but had serious disagreements about the role of students, intellectuals, workers, blacks, and women in this revolution. Other groups turned toward local community building. The first eight months of SftP's existence were, thus, during a historical moment in which mass-based nonviolent organizing was being replaced by mass disruptions on college campuses and new assertions of power by combinations of blacks, women, workers, and students. It was in this context of excitement, range, and possibility that SftP would fire its first salvo into the larger scientific community: the disruption of the meeting of the largest professional science association in the United States.

CHAPTER 6

Doing "Science for the People": Enactments of a New Left Politics of Science

> Traditional attempts to reform scientific activity, to disentangle it from its more malevolent and vicious applications, have failed. Actions designed to preserve the moral integrity of individuals without addressing themselves to the political and economic system which is at the root of the problem have been ineffective. What is needed now is not liberal reform or withdrawal [from science] but a radical attack, a strategy of opposition. Scientific workers must develop ways to put their skills at the service of the people and against the oppressors.
> —Bill Zimmerman, Len Radinsky, Mel Rothenberg, and Bart Meyers, "Science for the People," February 1972

The "Science for the People" manifesto, published in *Liberation* in 1972, captures two key features of the "science for the people" project: its explicit rejection of the reform of political institutions and the moral individualist strategy, and an open-ended approach to the development of ways of using scientific ideas and skills in the service of "the people." Scientific knowledge, according to this perspective, was not politically neutral, but the product of political and economic arrangements that led it to be most beneficial to elites, rather than to ordinary people. Liberals and moral individualists had already asserted that scientific knowledge was not always beneficial. Radicals went further, arguing that interlocking systems of capitalism, racism, sexism, and imperialism would have to be eliminated in order for the benefits of science to extend beyond the ruling classes.

Science for the People (SftP) was a diverse organization, with chapters and individual members carrying out a variety of activities all over the United States. Yet SftP might have remained relatively unknown had they and other scientists not used unconventional and disruptive tactics at the 1969, 1970, and 1971 American Association for the Advancement of Science (AAAS) meetings to draw attention to their ideas. At these meetings, SftP activists disrupted panels, used political theater, harassed scientists who had associations with the military and industry, held "alternative" sessions, and distributed literature. Their methods and claims disrupted a key organizational representation of the social integrity of "science": the professional meeting. Such meetings are traditionally ordered around staid debates over the validity of scientific claims, in which scientists take turns presenting evidence and argue, usually in a genteel fashion, over interpretations and assumptions. SftP activists refused to observe these rules of etiquette. They made it impossible for meeting participants not to hear their perspective, and disrupted a key moment in the ritual reproduction of science as a social practice. These disruptions, moreover, helped to draw more attention to SftP's assertions about the relationships between scientific research and the reproduction of inequality.

Although SftP was perhaps best known for these activities, the majority of the group's activities looked nothing like this. Unlike the Society for Social Responsibility in Science (SSRS) and the St. Louis Committee for Nuclear Information (CNI), which had templates for liberal and pacifist action, SftP did not. By 1969, the New Left, which served as a reference point for many SftP members, was becoming an amalgam of loosely connected forms of action and ideas. The diverse directions in which the New Left went after 1969—"opting out" of professional life and moving toward revolutionary action, community organizing, and other activities— were mirrored in the diverse activities of SftP nationally, and among its local chapters. Within the broad contours of an ideological framework that called for disconnecting science from the reproduction of existing power relations, group members produced a continual flow of ideas and actions in which they rethought what was unique (or not) about science and scientists, how knowledge and power were related, and how best to redirect science for "the people." Above all, SftP was a contentious organization, in the sense that its members continually challenged other scientists, both within and outside the group, to examine how their work contributed to inequality. Few of SftP's targets likely welcomed these challenges, but they had the effect of laying bare the very different ways that scientists thought about the proper relationship between science and public political issues.

SftP, like many other professional-based, anticapitalist organizations from that era, never collectively resolved the problem of how best to assist the working class without resorting to the use of expertise. SftP members worked, instead, on projects such as providing assistance for agriculture, creating teaching modules for secondary-school teachers, carrying out health studies for unions, supporting Vietnamese scientists, and countering the claims of sociobiologists. SftP chapters often carried out these projects with other New Left organizations and networks, including the

Medical Committee for Human Rights, Computer People for Peace, the War Resisters League, and Health/Pac.

In an analysis of the ways in which scientific knowledge can serve ordinary people instead of elites, the sociologist Brian Martin articulates four theoretical possibilities:

1. "Science for the people, (rational version)": an enlightened government directs scientific resources into areas of greatest benefit to society;
2. "Science for the people (pluralist version)": research agendas, "rather than being dominated by corporate and government imperatives and thinking, are shaped by wider social priorities as articulated by individuals and groups who are in touch with genuine social needs";
3. "Science by the people": by participating in the scientific research process itself, citizens shape the subjects and content of scientific knowledge; and
4. "Science shaped by a citizen-created world": society is reorganized so that citizens directly determine what is produced and how it is produced.[1]

Since SftP was a decentralized group, it is no surprise that members had different views on which of these versions they aspired to, and which they could achieve. They perhaps all preferred the fourth choice, but most chapters actually engaged in the second. This chapter explains how the actions and claims of SftP members, and of other scientists and engineers who felt similarly, created and laid bare fundamental disagreements about the political and moral status of scientific knowledge, through their disruptive and confrontational tactics at professional meetings in 1969 and 1970. The chapter then turns to the less well-known activities of the group to illuminate varied ways that SftP members interpreted and acted on the idea of "science for the people" in local settings by an examination of the activities of three influential chapters, in Chicago, New York, and Boston. Here I show that the political networks in which scientists were involved at a local level shaped the development of innovative ways to engage in "science for the people."

The Turn toward Radicalism: February–December 1969

After months of uncertainty about the direction the group would take, between March and December 1969, SftP chapters began to turn away from liberal reform strategies.[2] As the organizers had planned, local chapters began to explore different ways of joining scientific identities and ideas with radical political action. Chapters sometimes split into groups focused on problems of class and race, and those who wanted to focus on liberal reform issues such as the antiballistic missile (ABM) system.

During the fall of 1969, local groups did not coordinate national actions, so the organization existed as many separate units exploring options for taking bold action. SftP was especially active in Boston. A dense network of war resister, labor, peace, feminist, and anti-imperialist groups was spread throughout the Boston area, and many of the scientists who were part of Boston SftP were also participants in some of these groups. In the late spring of 1969, the Boston chapter began to publish the *Science for the People* newsletter. The beginning of the 1969–1970 school year in Boston saw the continuation of activism that had been generated by March 4. The main focus of the Science Action Coordinating Committee (SACC) continued to be MIT, but overlapping networks of SACC and other science activists from around the area continued to look for ways to broaden the challenges to the institutional arrangements of science. A group of students who had been asked by the AAAS to organize a youth-oriented symposium, members of SftP, and SACC activists came up with the idea to use the December AAAS meeting in Boston to promote their call for more radical action. During these discussions, they formulated the slogan "science for the people."[3] Organizing under the name Scientists' Action Group (SAG), a "coalition of Boston area radical science groups who work with community action groups to promote 'science for the people,'" they announced that they had requested that the AAAS allow them to hold a two-day symposium called "The Sorry State of Science" at the upcoming annual meeting in Boston. They also requested that the AAAS allow the public to attend the meeting. The AAAS board originally refused their requests, but when SAG threatened a demonstration, the board relented.[4] The inclusion of nonscientists at the meeting was not entirely new (spouses of scientists, overwhelmingly women, had long attended the sessions); the "public" now included any person who wished to attend. They would have a chance to hear scientists speak directly, instead of having scientists' views mediated by journalists.

The call for the inclusion of debates about urgent political issues of the day was not limited to the students who planned "The Sorry State of Science." In the previous year, the AAAS board had responded to members' and other scientists' requests to take action on or debate political issues ranging from the use of herbicides in Vietnam, to the lack of student involvement in the association, to the more general question of the responsibility of scientists for the values and uses of science.[5] In November and December, various groups and individuals submitted resolutions on political issues that they wished the Council to consider. The Scientist's Committee on Chemical and Biological Warfare requested that the association sponsor a study of the use of defoliants in Vietnam. Roman Skorski, council member from the University of Alabama, called on the AAAS to encourage more East/West scientist exchanges, in part because "complete world destruction is feared by many persons," and the United States and the Soviet Union were "engaged in fruitless negotiations concerning disarmament of chemical, biological, nuclear, and conventional warfare." Another resolution insisted that the AAAS denounce the "ruthless repression" of the Black Panther Party, two of whose members were killed in a police invasion of their headquarters in Chicago in early December.[6] In the previous two years, the AAAS had sponsored an increasing number of panels on political issues such as population growth and environmental destruction.[7] Clearly, the concern about how science was implicated in public political issues was not limited to radicals. Yet it was the radicals who would go beyond the use of routine bureaucratic procedure and professional practice and dramatically force these issues to be considered by more than just a handful of organizational committees. Radicals' actions drew nearly all participants into the debates, whether they wished to be included or not.

The two-day "Sorry State of Science" symposium resembled a typical symposium in many ways. Panelists prepared papers and debated with the audience the actual versus the proper relationships between science and politics. Participants raised the possibility of disrupting other meeting sessions, but did not move beyond the usual bounds of discourse and presentation. The session did, however, break with professional norms by beginning with a slide show of the war in Vietnam, pollution, and dead soldiers, accompanied with music by Bob Dylan and the Beatles. Allen S. Weinrub, a graduate student organizer of the symposium, encouraged the audience to "interrupt the speakers" and "ask challenging questions." He sounded the "misuse of technology" theme voiced by earlier groups of liberal science activists, but his emphasis on who benefited from such misuse and who lost went beyond simply stating that the government ought to redirect funds toward pressing social problems. Misuse, he argued, took the form of "militarism, economic exploitation, and psychological domination."[8]

Outside their own symposium, however, SAG activists parted company with most of the other attendees by using confrontational and disruptive tactics to make their points. At a televised and heavily publicized session on the future of the space program, activists held signs that read "Rocket on the Moon, Slums on Earth," "Redirect Society to Redirect Science," "Science for the People," and "The U.S.: First in Space, Sixth in Infant Deaths." Two women presented Charles Stark Draper, the head of the beleaguered Instrumentation Laboratory at MIT, with a papier-mâché moon rock with a $1.49 price tag on it. The scientific value of an actual moon rock, they argued, was equivalent to that of their faux moon rock.[9] Melvin Marguiles, a member of the Columbia University Ecology Group, the least organized, least sophisticated, and most disruptive group of radical scientists in attendance, refused to leave the speaker's platform and refused to let panelists speak.[10]

Feminists involved with the SAG activities drew up a list of eight demands for improving the status of women in science, including equal pay, access to admission to graduate programs, changes in high school counseling, and an end to sexist curricula. They disrupted the women's honorary Sigma Delta Epsilon meeting to present their list of demands, but were prevented from delivering them.[11] At a session on developmental psychology, feminists charged that a panelist's argument that children were healthier if mothers stayed home was "discriminatory."[12] Going well outside the routines of civil discourse, a group of feminists from the Women's International Terrorist Conspiracy from Hell (WITCH), a short-lived group founded in 1968 by members of New York Radical Women, used political theater to draw attention to women's subordinate status generally, and to condemn the use of science for war making. WITCH "hexed" the science honorary society Phi Beta Kappa–Sigma Xi at the society's luncheon "because it represents and embodies the elitist and exclusionary nature of science today. . . . [I]nstitutionalized science has set up a hierarchy that has excluded and oppressed white women, as well as black people, poor people, and all but a select group of white males. When asked about this . . . they [scientists] invoke the utterly nonsensical fallacy that white males are the only ones qualified. This is not science; this is verbal magic to maintain an oppressive system. Since in this matter scientists are resorting to magic, we are countering with our own effective magic."[13]

Audience members were divided on the appropriate response to activists' disruptions. Many were especially upset by SAG's disruption of the televised session on the space program. Some AAAS members argued that "making a circus out of the televised sessions of Americans' best known forum for issues important to society" had to be stopped. Others, by contrast, argued that viewers would distinguish between foolish and serious comments, and would give the AAAS high marks for allowing free discussion. And journalists were quick to distinguish between those activists who followed the usual rules of dress and hairstyle and those who did not. Those who were discredited in journalists' stories were those who were "shoeless," "bearded," "wearing a tam-o-shanter," or wearing a "wool shirt and work pants." By contrast, journalists remarked favorably on those men who were "clean cut" with "short hair," who waited patiently to speak, and came prepared with "thoughtful questions."[14]

Antiwar opponents—who included those using disruptive tactics and those who did not—scored a political victory when in a vote of 114 to 51, the AAAS Council officially opposed the continued use of defoliants in Vietnam. The vote was based in part on the evidence gathered by Harvard professor Matthew Meselson in an AAAS-sponsored study that showed the harmful effects of defoliants on the environment in Vietnam.[15] Yet it was not only data that was changing how AAAS leaders and the rank and file thought about their public political role. Howard O. McMahon, chair of the 1969 Boston meeting, said, "In previous years, the organization would not have been willing to take formal action on a moral basis. In the past, the AAAS has said 'We're scientists, we shouldn't get involved in moral or political questions.'" According to the *Boston Globe* writer Richard Knox, the new trend among "young radicals and not-so-young members of the AAAS establishment, was one of increasing concern for dealing with ethical as well as scientific concerns, and a new dedication on scientists' part to emerge from their ivory towers."[16] The degree to which the AAAS board was willing to engage moral and political issues had its limits, however. Although the association elected its first female president, Mina Rees, a mathematician at the City University of New York, the AAAS Council refused to debate the feminists' eight demands or to act on them. They also refused to consider the two resolutions that SAG submitted for consideration at the business meeting: the call for withdrawal from Vietnam, and for the withdrawal of technical support for the "repression of the Black Panther Party."[17] After waiting through the business meeting, SAG introduced its proposals, but the AAAS president H. Bentley Glass immediately adjourned the meeting because none of the resolutions had been submitted according to AAAS procedure.[18]

Despite the failure of the AAAS leadership to take action on specific SAG demands, the radicals' activities and claims were an opening salvo that challenged the routine, ritual reproduction of science as usual. As many scholars of social movements have demonstrated, the use of tactics that undermine the moral continuity of institutional arrangements is a powerful means of creating crises—and leverage—that can be used to force change. But disruptive tactics such as sit-ins, the appropriation of public and private spaces for political purposes, refusals to leave, vigils, and street parties, to name just a few, are not always effective. Many factors are involved, including the capacity to neutralize tactics without escalating their dramatic power, the inclination of those in power to grant challenger's wishes, and the power of opposing groups of activists.[19] SftP's main victory at the AAAS meeting was to raise the stakes of ignoring an increasingly vocal subgroup of scientists who wished to redirect science.

THE 1970 AAAS MEETING: THE ESCALATION OF DIRECT ACTION

Although "the sixties" are often remembered as a time of extremism and revolutionary fervor, it was in 1970 that *revolution* became a watchword for activists who were fed up with the hearts-and-minds approaches to social change that had characterized some activism in the 1960s. No event was more emblematic of the new revolutionary spirit on the left than the spring 1970 student strike. At thousands of high schools and colleges around the country, students demanded more power over curric- ulum and an end to the war and military research on their campuses. Students at New York University took over Kimball Hall, and faculty and students voted to suspend classes until the end of the semester. At Washington University, students rioted and burned the ROTC Air Force Building. At the State University of New York–Stony Brook, students lit fires across the campus and firebombed an art exhibit. Other schools held peaceful vigils and marches, workshops, and town meetings. By far the largest concentration of striking schools was in the northeast, particularly the Boston area, although schools in the mid-Atlantic re- gion, California, Illinois, Indiana, Michigan, Minnesota, and Wisconsin were also rife with activity. In every region, public and private schools, small and large, religious and nonreligious, were involved. In some areas, high school students were active participants, and at most schools there was considerable faculty support for the strike.[20] The feeling of power and possibility was palpable. Revolutionary social transformation did not seem utopian, but realizable.

The AAAS board prepared for the possibility that the organization's annual meeting in Chicago might be disrupted by those calling for revolu- tionary action. After canvassing the leadership of other professional asso- ciations about how best to prepare for this eventuality, they concluded that disruptions were unlikely. Only one annual meeting, that of the American Psychiatric Association in San Francisco, had been severely dis- rupted by activists. The board recommended the following rules: that every session be open to any person, registered for the meeting or not; that care should be taken to include the full expression of all views; and that when a disruption did occur, the session chair should request a show of hands concerning continuation.[21] Despite these preparations, the 1970 meeting was more disruptive than the 1969 meeting had been.

In this spirit, at the AAAS's 1970 meeting in Chicago, SftP's actions were confrontational, uncompromising, and insistent. As one SftP mem- ber warned at the beginning of the 1970 meeting, "[AAAS attendees] are not here to educate us. We're here to educate them."[22] SftP went well beyond the neutral distribution of scientific information, cool logical argument, and gentle moral discourse that previous scientists had used in their engagement with public political issues at AAAS meetings. To draw attention to their concerns and force discussion of the power relations that shaped scientific knowledge, SftP made it impossible for the meeting to proceed in a genteel fashion. As the *New York Times* science journalist John Noble Wilford argued, the 1970 AAAS meeting starkly contrasted with earlier meetings. The AAAS meeting, he noted, used to be "a quiet, dignified and optimistic affair" where scientists gathered to "renew fraternal ties, to report in an arcane language on their research and to speak of 'endless frontiers' and the wonder and elegance of science."[23] This sort of clubbiness and old assumptions about technology and science were exactly what SftP wanted to undermine.

Like the activities at the Boston meeting, those at the Chicago meeting were organized by SftP members and others who were active in local-area activist groups, including the New University Conference and feminist groups.[24] As in 1969, SftP activists held placards with "Science for the People," "We Are a Death Oriented Culture: Reverse the Priorities—All Power and Life to the People," and other slogans emblazoned across them.[25] Sometimes shouting and interrupting, SftP members demanded that speakers discuss their funding sources and the uses that were made of their work.

Audience members who were not radical scientists contributed to the disorder at the meeting. When radicals disrupted a session on crime, violence, and social control, an audience member, the wife of a scientist on the panel, jabbed the activist Frank Rosenthal with a knitting needle after stealthily moving forward to sit behind him. She said she was frustrated because Rosenthal refused to "keep quiet and let the symposium continue."[26]

Although SftP's analysis of science and power was fundamentally structural, members also used public shaming of individual scientists whose professional lives they believed represented the sorts of alliances they deplored. For example, SftP presented Edward Teller, the anticommunist, hawkish physicist who had played a major role in the development of the hydrogen bomb, with the "Dr. Strangelove Award," a wood and chrome soldier with the words "I am just following orders" written on it.[27] Teller refused the statue and told its presenter that he should be ashamed of himself.[28] Recounting his experiences as a Jewish refugee, Teller said that he refused to be subject to tactics he had experienced in Nazi Germany.[29] He was, he said, "now under attack from thoughtless individuals who do not know what they talk about and whose acts are about to induce violence and lack of reason."

Three days later, SftP activists used the same tactic to embarrass Glenn T. Seaborg, the chairman of the Atomic Energy Commission.

They read an indictment of him for "the crime of science against the people," but Seaborg walked out before activists could confront him. The indictment read:

> We, scientific workers and students, here in Chicago at the annual meeting
> Of the A.A.A.$., do hereby on this, the 30th day of December, 1970, indict
> You, Glenn T. Seaborg, for the crime of SCIENCE AGAINST THE PEOPLE.
>
> . . . You are guilty, Glenn T. Seaborg, of a conscious, major, self-serving and ruthless role in establishing, organizing, maintaining and developing institutions of science and government for effective use by the ruling class. . . . Glenn T. Seaborg has . . . performed this function, of coordinating and strengthening the dependence of science and universities on war and profit in a unique criminal history of responsibility in many of these institutions, for example: Livermore Radiation Laboratories . . . University of California . . . Atomic Energy Commission . . . Department of Defense . . . National Aeronautical and Space Agency . . . The President's Science Advisory Committee.[30]

They continued that Seaborg's administrative positions in these agencies and institutions had allowed him to direct both the building of weapons and "the most outrageous forms of waste" at NASA, where "dishonest appeals to the noblest traditions of science [are] seen to divert attention from the obvious neglect of people's needs." Herb Fox, a member of the Boston chapter of SftP, condemned Seaborg for his role in "establishing, organizing, maintaining, and developing institutions of science and government for the effective use of the ruling class."[31] John Froines, a chemist and a defendant in the 1969 Chicago 7 trial in which he and six others were accused of inciting riots at the 1968 Democratic National Convention, had harsh words for scientists who sold their knowledge for profit and professional status. Scientists, he argued, had "sold their souls to the Defense Department and the Federal Government for grants and status in the intellectual community. Others have sold their souls to industry." In "the black community," he said, "we have a word for those who sell their soul for a 'pittance.' We call them Uncle Toms."[32] The biologist Richard Lewontin urged scientists to give up their belief that science was pure and separated from other aspects of social life. "It is time to stop saying that science stands outside of society. Science is a social activity just like being a policeman, a factory worker or a politician."[33]

The ideas that radicals presented seemed to elicit less comment from journalists covering the meeting than the radicals' *activities*. *Science* editor Philip H. Abelson wrote that if the radicals aimed to "bring notoriety to themselves and to tarnish the image of AAAS [they] were to some extent successful." He noted that the "militants did not advance their cause; nor did they advance the cause of science."[34] Other participants were angry at the radicals' disruptions, and that the AAAS did nothing to stop them. Peter Suedfeld, a psychologist from Rutgers University, asked if "innocent bystanders" had rights. "Must a large number of people be frustrated in their desire to hear a symposium, an address by the president-elect of their association, because a small number of people prefer their own histrionics?" he asked.[35] *Science* reporter Philip M. Boffey reported the various activities in which the activists had been involved and tried to understand what activists meant by "science for the people." Ultimately, he concluded, the radicals "frequently 'turned off' as many people as they 'turned on.'"[36]

SftP conveyed its message not only through disruption, but also through the distribution of literature, a practice it continued at other scientific conferences for almost a decade. It is difficult to tell in retrospect how rank-and-file participants felt about the groups' views as opposed to their actions. Given that SftP members were articulating different perspectives—some calling for anti-imperialist revolution, some for the inclusion of women in science, others for more attention to the problems of the poor—it is likely that there were different responses.

What is clear is that AAAS leadership responded to radicals' demands—or at least pursued like-minded activities in which the AAAS was already engaged. The 1970 program reflected the association's engagement with scientists' role in public life to a greater degree than ever before. Panels on women in science and the contributions of minorities to science were held, and there were a number of special student-led sessions as well.[37] A Youth Council, appointed in June 1970, reported that they had planned, among other things, a series of activities for young people at the annual meetings. A founding member of the committee was Alan McGowan, who would go on to lead the Scientists' Institute for Public Information (SIPI) in the 1980s.[38] At the 1970 annual meeting, the AAAS Council endorsed all eight of the demands that feminists had introduced at the 1969 meeting.[39]

The disruption of the AAAS meetings in 1969 and 1970 disorganized the collective ritual representation of the "community" of scientists. Radicals' incivility challenged the image of the public meeting as an encounter of rational equals. The debate about the form the meeting should take mirrors the seventeenth-century debate between Thomas Hobbes and Robert Boyle, in which Hobbes denied Boyle's assertions that the laboratory was a "public" place and that experiment was a valid way to constitute philosophical knowledge. For Boyle, the laboratory experiment, with witnesses vouched for by other witnesses, was the best way to constitute a social community of natural philosophers and of valid facts about nature. SftP's actions and claims, from their attacks on the moral credibility of individual scientists, to their rejection of the idea that scientists alone were the proper witnesses and judges, echoed the challenges that Hobbes had posed.[40] To the extent that the rules for participating in such public witnessing are derationalized, the notion of collective agreement through witnessing is weakened.

Refusing to participate in acts of ritual solidarity denies acceptance of the values of the community. For groups that depend on conformity, rituals are even more important, thus making refusal a more significant act of deviance. SftP's activities at the AAAS meetings were akin to hurling the bread and wine to the floor during Communion in a Catholic Church in order to protest the treatment of homosexuals. Not only are these acts condemnations of policy, they are affronts to the practices that reproduce the organization. At an organizational level, activists were playing out the heresy that comes from orthodoxy, real or perceived.

The disruptions of the AAAS meetings did not precipitate a crisis in science in and of themselves, nor were they the main cause of the weakening of scientists' capacity to speak as neutral experts. But they did rattle the easy, civil, and gentlemanly structures that made it possible to see scientists as one-dimensional, cognitive, rational beings rather than complex moral individuals. SftP's use of unconventional tactics did not end with the AAAS meetings. They also drew attention to prominent scientists' ties to the Jason Program of the Institute for Defense Analyses (IDA). Unlike the pacifists, whose plan of action would require all scientists to renounce military funding, SftP activists saw their targets as representatives of a particular kind of relationship with the military. By insisting on a public accounting of Jason scientists' moral choices, SftP hoped both to draw attention to the military-science relationship and to raise the costs for those who wished to accept funding from the military. In the exchanges between the radicals and their mainly liberal targets, the very different visions of how each group understood the role of the individual scientist, how ideas were related to political systems, and what features of scientific investigation each sought to protect are revealed.

THE RADICAL CHALLENGE TO JASON SCIENTISTS

The Jason Program was created in 1958 to streamline weapons development by offering scientists the opportunity to explore basic and applied physics and engineering problems of a military nature. To attract the very best scientists in the nation, the program offered participants generous salaries and nearly unlimited funds, the prestige of working on problems of national importance in conjunction with the nation's military leaders, and summer workshops in vacation areas such as Woods Hole, Massachusetts. Salaries and grants went directly to individual scientists rather than to universities, turning them into private contractors who carried out their Jason work separately from their university-based work. Thus, Jason scientists were essentially private contractors to the IDA, which was itself a contractor to the Department of Defense.

Before 1967, few people outside the scientific community knew much about the Jason Program. That changed when student and antiwar activists, largely at the instigation of Students for a Democratic Society (SDS), turned their attention to examining how their own universities were involved in the war in Vietnam. In an effort that overlapped with the campaign to prevent Dow Chemical from recruiting on campus, activists called on their institutions to sever all ties to the IDA. As I noted in earlier chapters, because Jason recruited elite scientists, it is not surprising that anti-IDA demonstrations and actions took place at elite science institutions, including Cal Tech, the University of Chicago, Columbia, Berkeley, Stanford, and Princeton.[41]

To publicize their demands, most early anti-IDA activists used methods that included petitions, demonstrations, letters to administrators, and the distribution of documentation about the IDA's presence on campus. Anti-IDA activists were especially critical of the program's secrecy requirements, arguing that the interests of neither the students nor of other scientists were served, since IDA research was not publicly disseminated knowledge. Their other complaint was by now a familiar one: that participants and their defenders did not take into account the kind of knowledge the IDA produced and the purposes for which it was used. This, activists charged, was contra the spirit of science.[42]

SftP's campaign against the Jason scientists began in 1971, following the *New York Times'* publication of what became known as the "Pentagon Papers." These documents, originally titled "The History of U.S. Decision-Making Process in Vietnam," were commissioned by Secretary of Defense Robert McNamara.[43] They showed what the historian John Morton Blum calls "a depressing record of mistaken assumptions, prevarications, and flawed judgments" that characterized governmental decisions about U.S. involvement in Vietnam.[44]

Included in the *Pentagon Papers* was evidence of forty-seven Jason scientists' participation in the development of what was called "the electronic fence." In 1966, having condemned the Johnson administration's plans for more saturation bombing in Vietnam, Jason scientists proposed that the area stretching from the demilitarized zone between North and South Vietnam into the Laotian panhandle be covered with new kinds of weapons that could be set off by movement, heat, or light. Among them were new and more deadly land mines, acoustic sensors, and nail bombs.[45] The weapons were designed for what the Jason scientists envisioned as a long-term, open-ended war of "cat and mouse" rather than conventional, battle-based warfare.[46] The Jason scientists hoped that this system would help cut the supply lines between North and South Vietnam, thereby ending the war more quickly.

In early 1972, SftP published a long booklet titled *Science against the People*. It documented the history of the Jason Program, and contained descriptions of Jason scientists' views about their work, gleaned from "encounters" with them. These "encounters" took the form of scheduled interviews and letters. SftP's descriptions of the Jason scientists' views were interspersed with comments about what was "generally believed" about these scientists. The descriptions do not present a flattering portrait, in most cases making them appear as yes-men concerned with their own power rather than self-aware and thoughtful people. One key theme that SftP emphasized was that the Jason scientists were not politically disinterested, but allowed their political views to shape their decisions about weapons. SftP asserted that Jason scientists were often part of research groups that included political scientists, government officials, and other "interested" parties. For example, of one scientist they wrote: "He admits that politics was not a small and incidental part of their considerations."[47] Another, they argued, had stated that "the human element—the personal relations between the adviser and advisee [the Jason scientist and the government]—is very important to the success of the advising process; yet he continually stressed that the advising was strictly objective, non-political, and related only to technical evaluations."[48] In a description of a professor of physics at Berkeley, SftP wrote, "At a faculty meeting during the time of the Cambodian invasion, in 1970, [the professor] was heard to comment, 'Why is everyone getting so upset about a little war?' "[49] In response to a later letter to this scientist asking him to comment on the notes that the interviewer had taken before the material was published, the scientist responded, "This report contains several misrepresentations and/or quotations out of context. More significantly, it violates the conditions under which I agreed to meet with SftP, which were that I would listen and you people would talk."[50]

In a separate chapter titled "Why They Do It," SftP articulated, and presented its refutations of, what it thought were the main motivations of Jason scientists: (1) Jason's work must be harmless because it is so often ignored; (2) liberal scientists' advice counterbalances that of the government; (3) Jason provides accurate information that is not available elsewhere; (4) Jason scientists do not fully realize the consequences of their work; and (5) they are seduced by the "thrill of making history."[51]

SftP's activities were paralleled in Europe in the summers of 1971 and 1972 by student activists in Italy and France. These campaigns were led by younger scientists who, like their American counterparts, wanted the

United States to withdraw from Vietnam. Jason scientists at the Varenna, Italy, summer school on the history of physics and the Trieste (Italy) International Physics Symposium were prevented from speaking by disruptive activists who heckled and demanded that they speak about their involvement in Jason rather than about technical issues. In June 1972, the physicist Murray Gell-Mann was chased from the Collège de France,[52] and the University of California–Berkeley physicist Charles Townes was prevented from speaking at two engagements in Rome. French activists created a poster that simply said "War Scientists" and listed the names of thirty-nine members of Jason. The poster was circulated by French, Italian, and American radical scientists, including SftP. Among those listed was the chair of the Union of Concerned Scientists, Henry Kendall.[53]

JASON SCIENTISTS RESPOND TO CRITICS

Jason scientists explained their participation in the program in three ways. First, they asserted that their motives were moral because they were acting as public servants who hoped to give the government more sound advice than it otherwise would have received from scientists on staff in the government. Second, some argued that SftP's attempt to pressure them to end their research was antithetical to free speech and academic freedom. Finally, some argued that they delivered facts, not policies, to the government and therefore they were not responsible for the uses made of their ideas. Their responses reveal that, like CNI members, they believed that individual choice to carry out public duty should drive scientists' engagement in public political issues and that scientists could and should act as conduits for information dissemination. This stood in stark contrast to the model of scientist as servant of the people that SftP put forth and also contradicted SftP's claim that it was not possible to provide "apolitical advice."

Altruistic Motivations

For many of the Jason scientists, the criticisms that SftP and European activists made were based on a failure to understand that participation in the Jason Program was a form of public service. Providing advice to the government would help avoid bad political decision making, analogous, they argued, to the kind of technical advice that Manhattan Project scientists had given in the spirit of ending the war. By pruning out bad projects based on bad science, one helped the government. As Sidney Drell argued, "The [Hans] Bethes, the [Wolfgang] Panofskys, [Eugene] Wigners, [Edward] Tellers, who got drawn into the war . . . came in and they had a task to build radar and to build an atomic bomb—and they did it. We in a sense were following in their footsteps . . . we were intellectually recruited by them . . . we saw them as our models."[54] The physicist Charles Townes echoed Drell: "We had a lot of men over there [Vietnam] who were fighting, whose lives were in danger, and we ought to push on trying to see that there are right policies." He framed his role as a military advisor in the same terms: "I felt a good deal of pressure to help the people who were there [Vietnam]. The country [United States] was involved, and it would be best to try and stay in and give good advice."[55] Their critics did not understand these motivations, some Jason scientists argued, because they were ignorant and irrational. Townes felt that activists were unable to differentiate, politically or morally, among different kinds of involvement with government or business. For activists, he said, "anything to do with General Motors was anathema." He saw them as unreasonably characterizing universities and corporations in the broadest terms, unable to see complexities.[56]

Jason members viewed their service to the government as a bulwark against irrationality and muddled thinking, in much the same way that CNI and SIPI thought that the provision of accurate information to the public would improve political decision making. Yet, at a time when trust in the government was eroding, the Jason scientists' allegiance to government service could easily be construed as complicity in a flawed system.

Free Speech versus Totalitarianism

Some Jason scientists considered the pressure to end their association with Jason to be totalitarian. Just as their motivations were shaped by an earlier generation, they saw their actions in terms of political conflicts that that generation had experienced. The attempts to force them to avoid studying military issues were the same as the security investigations of the 1950s that attempted to discredit scientists for their real or imagined political associations. Malvin Ruderman, for example, felt that there were "false accusations" made against him, including assertions that he had used the funds from Jason to engage in military research, which he had no chance to refute. "Very personal attacks have used underhanded innuendo," he said, "and entirely false accusations that dwarf the excesses of McCarthyism of two decades ago."[57] Similarly, Richard Garwin wrote that the Jason attacks signified that "freedom itself is under fire."

> What is under attack is the right of an individual, in his own time, away from his regular job, to engage in legal activity to which some individuals are opposed. Make no mistake—this is precisely the same right that allows some of us to be Democrats, some to be Republicans,

some to be Christians, some to be Jews, some to be agnostics, some to favor the liberalization of abortion, some to do research on brain disease, and some to attempt to forbid it as invading the seat of the soul. The techniques used by these protestors are those of blackmail and coercion. Having failed with words and arguments, even lies, they attempt to deny to completely uninvolved individuals [students] access to work or education until these individuals force the Jason members to resign. . . . [T]his is a powerful tactic, extortion.[58]

In the fall of 1972, the liberal magazine *Christianity and Crisis* published a symposium of the arguments of Frank Baldwin, an assistant professor of history at Columbia, William J. McGill, the president of Columbia University, and the Columbia physicists Garwin and Ruderman concerning the morality of working for Jason. Baldwin argued that weapons and warfare techniques of the electronic battlefield were "widely regarded as criminal under Nuremberg precedents." In response to McGill, who argued that the university "had no business attempting to dictate the political or private activities of its professors," Baldwin argued that McGill had inaccurately characterized Jason scientists' activities. "By classifying the Jason activity as a 'private activity,' a very doubtful characterization, and exempting the actions of Jason members from criticism, it becomes improper to challenge the behavior of faculty members who helped devise a system of indiscriminate killing—even when that role is documented by indisputable Government records and the technology is viewable on the nightly news."[59] The Jason scientists Garwin and Ruderman did not consider their work to be in the service of totalitarianism or genocide, nor did they agree with Baldwin's and SftP's argument that all scientists' work had been captured by powerful economic and political interest groups. Instead, they saw scientists in the classical liberal sense: as autonomous individuals pursuing what interested them—in this case, service to the nation.

Delivering Facts, Not Policies

Jason scientists shared the view that scientific ideas are independent of the character of their producers or sponsors. Although they did not always agree with the government's use of research, they did not see themselves as responsible for how their ideas were used, given that their advice was based on scientific fact, not political preference. For example, Drell argued that "many people . . . who got involved in the electronic barrier [between North and South Vietnam] went in with the best of motives, and saw some of the technical contributions they made used in ways that they felt quite unhappy about. But it's inevitable you know. The laws of physics are fixed. The laws of politics change."[60] The distinction between the technical and the moral is evident, too, in discussions about weapons that were abstracted from their effects on human beings. Critics had charged that scientists were coldly rational, avoiding the moral implications of technical decisions. For the Jason scientist, this distinction was essential, for it prevented facts from being distorted by ideologies and political preferences, and clearly distinguished the areas in which the scientist had expertise (science) and in which he did not (politics).

Not all Jason scientists held the same views, of course, and some left the program because they were unconvinced of the effectiveness and morality of some of the Jason group's work. Richard Blankenbecler, for example, joined the Jason Program for some of the same reasons that other Jason scientists did. He hoped to "save lives and do something for my country." He found the problems on which he worked as a Jason member, particularly the feasibility of the ABM system, interesting and challenging. But while working on the electronic fence he began to question government leaders' motivations for U.S. involvement in Vietnam and the effectiveness of the electronic fence. The work he was doing with the Jason Program seemed less and less congruent with his desire to help the country.[61] He came to the conclusion that the study of "fixed barriers" (the electronic fence) and the use of "climate control to try to affect things [the war] . . . just led to a senseless loss of life."[62] He left the Jason Program in 1967. In the 1970s, he found more satisfying ways to use his skills to assist the government: he helped develop a system for detecting clandestine nuclear tests, and served on an arms control panel under the directorship of Wolfgang Panofsky. For Blankenbecler, these projects were satisfying because they served to reduce the number of weapons, not increase them.[63]

Jason scientists' responses to critics were based on liberal political principles. They typically distinguished between scientific knowledge and its contexts of production and use; they understood scientists to be autonomous individuals whose political and scientific tastes, not states or economic systems, determined the research they did and its relationship to systems of power; and they treated science as a social system that was autonomous from the state. Their choices about what kind of research to engage in also came from a sense of political duty, much as members of CNI argued that their actions were based on duty. In asserting individualism, scientists preserved the credibility of the knowledge they produced, whether in the service of government or not. Since individualism is also a highly valued quality in liberal politics—individual people, not larger systems, are the key actors—claims of independence served as the means to indicate and defend the scientific *and* political morality of participants.

SftP's Translocal Actions, 1971–1975

Although SftP's activities at the AAAS meetings and their critiques of Jason scientists were highly visible, much of what the group did was less dramatic and confrontational. Their activities represented attempts to articulate a "science by the people," and different ways of challenging the liberal and personal responsibility models of activism.

In 1973, SftP organized a trip to China to examine science under communism. The trip included ten members from five different local SftP chapters. They visited research institutes, universities, agricultural communes, and factories, observing and discussing how scientific projects were devised and implemented. Inspired by the Maoist model of investing ordinary individuals rather than formally educated people with the power to create scientific solutions and ideas, members of the China trip group wrote *China: Science Walks on Two Legs*. The book detailed the Chinese model and laid out the basis of how it might be applied in the United States.[64] In Brian Martin's terms, they were advocating "science by the people."[65]

Participation in professional meetings continued to be an important activity for the group throughout the early and mid-1970s. They staffed informational tables, held sessions, and petitioned the leadership of the AAAS, the National Science Teachers Conference, the American Chemical Society, the American Physical Society, and other conferences through 1974. For example, for the 1973 AAAS meeting in Mexico City, the group wrote *AAAS in Mexico: Por Qué? Science and Technology in Latin America* and *Los Nuevos Conquistadores*. Both publications addressed the relationships among capitalism, science, and inequality in Mexico and Central America. The failure of the Green Revolution, the activities and intentions of the Ford and Rockefeller Foundations in Central America, and science and political repression were among the subjects covered in other booklets distributed at the meetings and through mailings.[66]

Another important translocal activity was a campaign against racist sociobiology. The campaign was sparked by the publication of two articles that asserted that intelligence was heritable and largely unchangeable. In 1969, the educational psychologist Arthur Jensen published an article in the *Harvard Educational Review* arguing that boosting I.Q., and thus scholastic achievement, was impossible. He concluded that it was a waste of time and money to spend money on schools and special programs for those with low I.Q.s.[67] Two years later, the comparative psychologist Richard Herrnstein published "I.Q." in the *Atlantic*. Echoing Jensen's arguments, he asserted that Americans should simply accept the reality of fixed, unequal levels of intelligence. To be sure, those who were not naturally blessed with intelligence should be treated with kindness and compassion, but it was hopeless, Herrnstein argued, to change nature.[68] Jensen's article was highly controversial; dozens of newspapers and magazines published comments and essays in support of or in opposition to Jensen's claims. In March 1970, the SftP member Richard Lewontin published a refutation of Jensen's claims in the *Bulletin of the Atomic Scientists*, in which he asserted that Jensen's solution, to encourage black children to capitalize on the skills they had, was "so clearly at variance with the present egalitarian consensus and so clearly smacks of racist elitism, whatever its merit or motivation, that very careful analysis of the argument is in order." He concluded that Jensen had no evidence to support the claim that black-white differences on intelligence tests were caused by heritable, unchangeable genetic differences.[69] Lewontin and other members of SftP wrote and spoke out against other sociobiological arguments in the early 1970s, and in 1975, after the publication of the biologist E. O. Wilson's *Sociobiology: The New Synthesis*, Lewontin and other biologists, some of whom were members of SftP, formed the Sociobiology Study Group (SSG). The SSG was one of the major players in the debates over sociobiology in the 1970s, publishing reports and articles, presenting papers, writing letters to editors, and speaking at conferences.[70] In 1983, the group became a nonprofit organization called the Council for Responsible Genetics, a staffed nonprofit organization that fosters public debate about the social, ethical, and environmental implications of genetic technologies. Among those who have been active in the group are former members of SftP.

Inventing "Science for the People" at the Local Level

By early 1970, local SftP chapters began to take on issues that were generated by the interests of local members and by the places in which they lived and worked. The group had a national and especially regional focus as well, but these local groups were incubators and experiments in how to do "science for the people."[71] SftP had approximately ten chapters in 1970 and approximately five in 1975, but here I use the New York, Chicago, and Boston groups to illuminate how "science for the people" was carried out in practice.[72]

New York City: Scientists as Workers and the Politics of Domination

The New York City SftP was one of the least cohesive local groups, with at least three overlapping clusters of SftP activists who did not always know one another personally. Three major issues for the New York chapter between 1970 and 1974, the heyday of SftP, were the war in Vietnam, the use of technology for political oppression, and the labor process.

In 1971, SftP shared office space at the Dolphin Center, on East Fourteenth Street in New York City's East Village, with six other activist groups. These groups included the Committee for Social Responsibility in Engineering (CSRE) and Computer People for Peace (CPP). CPP, formed in 1968, was a New York–based but nationally networked group of activists who worked in the computer industry.[73] CSRE, a radical engineering group, was formed in 1971 by electrical engineers at Brookhaven Laboratories on Long Island to oppose the war in Vietnam and redirect engineering away from military applications.[74] Many New York SftP activities overlapped with those of CPP and CSRE.[75]

As in other cities, such as Boston, Chicago, and San Francisco, where the antiwar movement was especially strong, in New York there was a palpable feeling that revolutionary changes were possible. As Barbara Ehrenreich, one of the founders of the health activist group Health/Pac, a group with which SftP membership overlapped, recalled, "I wish I could convey to you some of the excitement of those days, the swirl of people and activity coming through Health/Pac." She goes on to describe the lively interactions and projects that went on among worker's groups, the Black Panther Party, medical professionals, and feminist organizers.[76]

For example, SftP, CPP, and CSRE organized a three-session "counter-conference" timed to coincide with the 1971 meeting of the Institute of Electrical and Electronics Engineers (IEEE) in New York City. Speakers included Seymour Melman, a Columbia University industrial engineer whose research interests included converting industry to peacetime activities, William Davidon, a Haverford College physicist and Quaker peace activist who was a former president of the SSRS, and Victor Paschkis, a Columbia University engineer and the main organizer of SSRS. Paschkis spoke about the old problem that had confounded SSRS activists: how to predict secondary effects of technological advances.[77] A second CPP/SftP/CSRE campaign involved challenging corporations that produced technologies used in war or for political repression. IBM, Polaroid, Honeywell, and Litton Industries were among the major targets.[78] Anti-IBM/Polaroid actions were intended to pressure those companies to cease supplying technologies used to track nonwhites in South Africa. IBM's computers tracked nonwhites' whereabouts through a national database, and Polaroid's ID-2 system created the passbooks that every nonwhite South African had to carry or else face imprisonment.[79] The campaign against the company included demonstrations, a letter-writing project, and efforts to pressure shareholders in both companies.

At a New York City conference, the three New York groups verbally attacked the Polaroid Corporation president and engineer Edwin Land.

They saw him as the "human embodiment of the Polaroid Corporation," and thus responsible for the political repression of black South Africans. Land, they argued, was responsible for the "misuse of our technology for political oppression," but not merely because he was the head of Polaroid. They refused to "separate his role as a scientist from his role as the father of a repressive corporation."[80] His actions were not simply the actions of an immoral person, they argued, but those of a scientist who headed a profit-making firm that helped perpetuate political oppression. This method of personal attack had something in common with the old SSRS strategy of encouraging each person to be true to an inner, morally good self, for it treated personal actions as important sources of political change. Unlike liberals, who encouraged one another to take responsibility for the outcomes of scientific work but were rarely critical of individual scientists, some SftP activists used shame as a political weapon.

CPP and SftP jointly printed and nationally distributed seven thousand copies of their thirty-page booklet *Data Banks: Privacy and Repression*, in which they explained how computers were used to track individuals' whereabouts and their spending patterns, and to provide national and international police with information about suspected political agitators.[81] Their concerns with technology and privacy also found more local expression. SftP and CPP worked closely with a group of Mt. Sinai Hospital health care workers to publish analyses of the costs and benefits for patients of computerized health care. Among the publications were *Health: Big Business for Computers*. The group predicted that in the future, patient information would be bought, sold, and exchanged without patients' knowledge.[82]

Somewhat disconnected from the CPP/CSRE/SftP network was another group of New York City SftP activists whose main interest was in labor politics. They began a campaign to organize faculty and staff at City University of New York (CUNY). Many of the members of this network were also members of the New York New University Conference (NUC), formed in May 1968 in Chicago as a way for young faculty and professionals to continue the kind of work they had done as students in the SDS. SftP members Bart Myers and Bill Zimmerman of the Brooklyn College Department of Physics, authors of the "Science for the People" manifesto that appeared in *Liberation* in 1972, were leaders of the organizing campaign.[83] In 1973, the CUNY SftP network wrote a pamphlet, widely circulated among SftP locals and graduate students, in which they argued that although scientists had once considered themselves invulnerable to unemployment because of the supposedly rare and valuable skills that they possessed, those days were over. The fiscal crisis of the state, they argued, had "forced reduction in support for scientific research, resulting in the 'proletariatization' of science." Although scientists still had "significantly higher salaries and better working conditions than the vast majority of workers," the "integration of more technology into production has placed their self-image into contradiction with the realities of their working conditions." Their pamphlet urged scientists to strengthen their position as workers. Science done in "an anti-elitist manner and by careful and imaginative choice of research" could in some way "serve the people" by contributing to a movement in which human values rose above monetary values.[84]

The New York SftP chapter's concerns were shaped as much by the local political networks and activities in the city as they were by ideological perspectives that were shared by SftP members in other chapters. In Chicago, too, "science for the people" was shaped by local concerns as well as a shared ideology.

Chicago: Science for Vietnam and the Politics of Epistemology

Like the New York City group, Chicago SftP had a membership that overlapped with the NUC. Medical Aid for Indochina, which provided medical supplies and expertise to North Vietnam and National Liberation Front zones in South Vietnam, was active in Chicago, and it was with this group that much of the Chicago SftP membership overlapped between 1970 and 1972. SftP became more focused on the needs of Vietnamese scientists specifically after a trip to Vietnam by the SftP member Richard Levins, a professor in the Department of Biophysics at the University of Chicago.

Levins had long been involved in the struggles of third-world people. In 1970, Levins visited Hanoi as part of a delegation of scientists sponsored by the World Federation of Scientific Workers. In Vietnam, Levins visited laboratories and schools and met with Vietnamese scientists. He found that as a result of the war, the North Vietnamese were in dire need of material assistance for scientific research. These scientists were also unaware of the latest research because they lacked textbooks and had only limited means of communication with scientists outside Vietnam. In the summer of 1971, Chicago-area and other SftP members met and decided to create Science for Vietnam (SFV).[85] Afterward, NUC, SFV, and Chicago SftP became an interlocking group and set of activities.

One of the Chicago group's SFV projects was a study of the effects of herbicides on plants and animals in Vietnam. The SFV project provided a detailed history of Operation Ranch Hand, the official U.S. defoliation program in Vietnam. Begun in 1962, the program increased dramatically in the following years and reached a peak in 1967. Other biologists were also concerned with this issue. When SFV began its investigation, the National Academy of Sciences (NAS) had just published a report that concluded that herbicides were not, in the long run, harmful to people, plants, or animals.[86] SFV's study was of the extent to which plants and animals could recover from herbicides. Contra the NAS study, they concluded that there had been significant and long-term damage caused by the herbicides. The U.S. government ignored SFV's findings. As a result, many SFV scientists revived the moral individualist policy of noncooperation with the government that Norbert Weiner and the SSRS had pioneered after World War II. They refused to share research with the government or to serve on government committees.[87]

SFV had a much longer-term project as well. The group worked with Vietnamese scientists to develop curricula and research programs so that biology would not "follow the traditional sequence that characterized spontaneous development of biology in the more industrial capitalist societies." Rather, they sought to develop it in a more "integrated fashion with systematics, evolution, distribution, physiology and ecology, etc. studied together" and with an "understanding of environments as dynamic and constituted by the organisms within them, not fixed and external."[88] Rather than considering individual units, such as genes or people, as passive subjects shaped by a fixed economic, political, or biological "environment," this view was based on the idea that reductionism was a poor way to understand human or organismic processes.[89] To further their goal of assisting Vietnamese scientists, they gathered science textbooks, course outlines, lab manuals, and scientific and medical supplies and sent them to Vietnamese scientists and doctors. They also raised money to buy and ship other items needed by North Vietnamese researchers.

The idea of providing supplies to Vietnamese scientists quickly spread to other cities, and within four months of the group's founding in Chicago, there were at least a dozen SFV chapters throughout Canada, Europe, and the United States.[90] In 1971, SFV conferences were held in Chicago, Boston, Madison, and Berkeley. Participants came from Northwestern University, Windham College, the University of Kansas, the State University of New York–Stony Brook, Indiana University, Berkeley, Washington University, and the University of Montana. The titles of sessions, including "Local War and Counterinsurgency Research" and "The Strategy of People's Science," were as likely to be found at any New Left conference as they were unlikely to appear on the program of a professional science conference.[91]

Separately from SFV and NUC, Chicago SftP and other organizations provided assistance to the Black Panther Party. Chicago SftP raised money for bail and for lawyers for Party members who were accused of crimes. More specific to their role as scientists, they and other locals created a Technical Assistance Program (TAP), designed to teach radical groups means of subverting surveillance systems, to help them acquire free electricity for clinics and "free schools," and to teach adults science and math that they could then pass on to other poor or oppressed groups. The program was never successful, however. TAP members often ended up doing the work themselves, rather than being able to teach others how to do it.[92]

Boston: The New Communism and the Politics of Biology

The Boston group bears extended discussion because it published the magazine *Science for the People*. The magazine symbolized SftP chapters'—or sometimes, the Boston chapter's—ideological perspective. Until 1975, the magazine served as the main organ of interchapter organization, and as a widely circulated subscription magazine, which, at its peak in 1974, had more than four thousand subscribers.

The Boston chapter was riven by conflict between 1970 and 1977, but still managed to engage in innovative actions. The main line of dispute was whether the organization ought to have a specific, formalized set of principles to which members would adhere. No other chapters had formalized principles, nor did the national organization.[93] Britta Fischer and Herb Fox gathered around them a group of people committed to Leninist ideology. From their perspective, the most important task at hand was to form a vanguard party that would support an anti-imperialist revolution in the United States and abroad.

At regional conferences in 1974, 1975, and 1976, SftP chapters in the Northeast gathered together to try to create a set of "Principles of Unity." The discussions were tortured and relentless, but there was little resolution. On the question of whether scientists were part of the working class, SftP members had very different views. One faction, influenced by the October League and calling itself the "Unity Caucus," pushed for the adoption of principles that would encourage the development of "revolutionary science as it is being demonstrated in China, as opposed to pseudo-science."[94] Debating whether to mention capitalism specifically or to discuss the current economic situation or other issues kept the Northeastern chapters busy. Many academic members were uncomfortable with the Caucus's desire for working-class leadership of the group. Others found the Unity Caucus's methods heavy-handed and alienating, while still others were disappointed that their own views were not considered. Ultimately, the group never produced any principles of unity.[95]

Despite the draining effect of the struggles over a formal political position, the Boston chapter worked tirelessly in the early 1970s. Among its activities were organizing factory workers and holding sessions at professional science meetings. An active science teaching subgoup of SftP developed science curricula in partnership with primary and secondary school teachers, held conferences for science teachers to share skills and ideas, and provided models for including analyses of gender, science and war, intelligence testing, and ecology in course materials. The teaching group also held sessions at professional association meetings.[96]

All SftP chapters were composed mainly of men. In part this reflects women's underrepresentation in science, but it also reflects the inability of women to make feminism a central part of the Boston chapter's (or other chapters') agenda. One of the members of the Boston chapter, Rita Arditti, a researcher in the laboratory of Jonathan Beckwith of the Harvard University School of Public Health, was one of the most vocal feminists in the Boston chapter. In December 1969, Arditti was an organizer of the feminist actions at the AAAS meeting, and an author of the eight demands that women presented to the board that year.

Women's concerns became more central to the Boston chapter as they sought to develop "principles of unity." Six women proposed a set of Principles of Unity that explicitly identified gender as an important concern: "Science for the People is an organization of women and men in science . . . who are part of an international struggle for freedom from exploitation based on class, sex, or race." Like other members of the group, they argued against "the myth that science is neutral." They insisted that science and technology were "ideological and practical weapons used by the existing power structure both here and abroad, to justify racist and sexist oppression."[97]

Like advocates of a class-based analysis of science, the feminists in SftP were part of a broader network of activists. In conjunction with other Boston-based women scientists, feminist activists from the NUC, and women's health groups, some of the women from SftP anonymously published and distributed a hundred-page document titled *How Harvard Rules Women*. Although it mainly focused on the subordinate role of women in academic and social life at Harvard, it also included a section on science. Like the articles in *Science for the People*, it focused on the politics of women's biology. In the view of medical professionals and researchers, they argued, "women's capacity for childbirth is their biology," making them no more than "machines for reproduction."[98]

In spite of the interest in feminism, some activists felt that gender was only discussed as a token nod to SftP's professed commitment to fight sexism, but not taken seriously by the group. Rita Arditti, who had left Beckwith's lab and was an assistant professor of biology at Boston University in 1971, had become more involved in the women's movement. The overlap between her professional activity in science and her involvement in women's activism made it "impossible not to reflect about women in science, my own personal experience . . . how hard it was to try to be a woman scientist." Yet participating in SftP did not provide the opportunity to explore those issues in the way Arditti would have liked. As a member of SftP, she recalled, she "started to feel the difference . . . between having to struggle to be heard about women's issues in mixed groups, the difference being [that] in an only women's group, they were [so] much more open to these ideas, it was possible to think and move forward much quicker."[99] Although she never experienced any overt opposition, Arditti never felt that the majority of the group shared her concerns. Focusing on feminism in SftP "was a constant struggle. . . . To bring a feminist perspective was always a risky thing to do . . . and one had to be on the defensive . . . to keep pounding [in] that it was meaningful. It was like they wanted something on women because they knew they had to have something on women, but it had to be limited to what they wanted and it had to be something that they [the editorial collective] liked."[100]

The problems with integrating feminism into the group were not exclusive to the Boston chapter; they confronted most progressive groups of the period that were not devoted to feminism exclusively. For example, a member of the New York chapter who was both the chapter's only woman and its only technical worker, wrote to the men in the group, "I have a feeling that many of you are comfortable with your sexism, but hip enough to mask its more overt expression, even from yourselves. Intellectually you can accept women as equals, but something inside you hesitates, and we're treated with a bit of deference and a bit of contempt."[101] Feminists in SftP had a more visible presence on the pages of the magazine after 1974. Despite the earlier calls for analyses of sexism, racism, and elitism in the organization, some women felt that there had been very little attention to sexism. They formed two new subgroups: one acted as a support group for women, addressing practical issues of sexism in the workplace and in SftP; the other served as a political base for socialist feminists who were interested in new forms of political action. In 1974, feminists edited a special issue of *Science for the People* on women's concerns.[102] That issue raised questions about the relationship between political ideology and women's biology, the politics of women's sexual response, and the problems that women scientists faced at work, among other topics. Although *Science for the People* published more articles about women and science in the next decade, feminists' concerns were never central to SftP.

In addition to publishing the group's magazine, the Boston chapter played an important role in defining SftP's national identity because it was involved in several highly visible public debates. One was through the actions of Jonathan Beckwith, whose Harvard genetics laboratory isolated and photographed operons, the elements of a gene that control its reproduction in *E. coli*, and Joel Shapiro, a researcher who shared in that discovery. When they announced their findings to the press, the research group used the opportunity to make a political point. They argued that like all scientific research, their work would not necessarily lead to positive benefits. Indeed, there was a strong possibility that their work "might let loose more evil than good."[103] Two months later, Shapiro quit science to undertake political work outside of science. Shapiro cited three reasons for his defection: he believed that the work he did was likely to be put to "evil" use by the government and corporations; he did not want to participate in a system that did not "allow 'the people' to have a say in deciding what work scientists did," and he felt that the most important problems the country faced, including health care and pollution, needed "political solutions rather than scientific ones."[104]

Reactions to Shapiro's decision were varied. The *Science* reporter James K. Glassman commented, "[Even] the older, more conservative scientists will have to agree that Shapiro has made a large sacrifice in an effort to get the word across."[105] Other scientists were outraged over the suggestion that Shapiro's action was a sacrifice. Leaving was a "privilege and an opportunity," and it was a privilege and an opportunity for Shapiro to do what "he most desired."[106] The Nobelist Salvador Luria, one of the initiators of the 1965 *New York Times* statement opposing the use of herbicides in Vietnam and a participant in the MIT March 4 events, however, was highly supportive. He said it was "important that there are scientists like Shapiro who point out the misapplications of science."[107]

Other Chapter Activities

Local chapters in the Midwest and the West developed their own projects. Like the Chicago, New York, and Boston groups, their activities were shaped by local political conditions, other organizations, and the concerns of a few very active people. For example, in 1973 the Madison, Wisconsin, group published *The AMRC Papers: An Indictment of the Army Mathematics Research Center.*[108] In it they documented the types of military research that had been performed at the campus's Army Mathematics Research Center (AMRC) in the 1960s. The document was written in support of sentence mitigation for Karl Armstrong, a Madison man who bombed what he thought was the AMRC in April 1970 because he believed that the center produced weapons for use in Vietnam. The blast killed a young researcher, Robert Fassnacht, and Armstrong was tried and convicted of second degree murder.[109] The Madison group also made a study of new treatments for tuberculosis and began a sustainable forestry project.

The Ann Arbor, Michigan, group was active on farming and farm-worker issues and on more general issues about the politics of biology. It formed Science for Nicaragua (SFN) out of the Ohio SftP working group Farm Labor Organizing Committee (FLOC). In 1982, the biologist John Vandermeer and others from FLOC reorganized as the Alternative Agriculture Group (AAG). With an international group of biologists and agricultural researchers, AAG worked, and continues to work, with third-world farmers and scientists to develop sustainable research programs; the group has been especially active in Mexico and Nicaragua. It sponsored country-specific programs based on the SFV model. In the SFN program, AAG provided information, equipment, training, and personnel to Nicaraguan scientists and farmers. Unlike SFV activists, SFN activists worked closely with the users of scientific ideas and products in order to coproduce practices and knowledge that are scientifically sound, take advantage of indigenous knowledge and materials, and are sustainable over the long term. Like other former SftP scientists involved in science-politics organizations, Vandermeer wrote and spoke about the politics of alternative agriculture, ecology, and genetic reductionism. He also co-edited *The Nicaragua Reader.*[110] Other activities of the Ann Arbor group included the publication of *Biology as a Social Weapon.*[111]

CHALLENGING THE IDEAL OF THE SCIENTIFIC COMMUNITY

In 1975, SftP was a vibrant network of chapters and individuals who were making their mark on local, national, and international debates. They were doing so not in an isolated fashion but, more so than either the SSRS or CNI, through contestations and cooperation with other groups who supported or opposed their perspectives. Their actions were taking place in professional settings, such as professional associations, in factories and industry where some had gone to work, and in academic settings. They were also working with people outside those settings, defending health clinics, organizing workers, and other activities. Like the SSRS, SftP was more a network than a bureaucratic organization, and one of the things it created was activists, who took their views and skills to new settings.

SftP collapsed as a formal organization in the late 1980s. The magazine ceased publication in 1989, and the office was reduced to little more than a closet in a church basement. On the face of it, if we consider organizational survival to be the sine qua non of social movement success, SftP was a failure. But if we think of social movement organizations as means to *generate* more activity, rather than to sustain or reproduce themselves, then SftP's legacy, like that of CNI and the SSRS, is in the networks and ideas that it produced. Following the rivers and brooks that constitute the development and dispersion of the intersections of Marxism, feminism, the New Left, and science reveals that the people who were involved in SftP were less involved in overthrowing capitalism than they were in building relationships with other activists, both scientists and nonscientists. This activity resulted in the simultaneous empowerment of nonscientists and the weakening of the idea that science ought to be segregated from the concerns of ordinary people.[112] The material changes, up through 1975, are difficult to enumerate because of the diversity of people in the group, but I hope that I have identified some of the key activities in which group members were involved (necessarily leaving out many more). Although SftP collapsed in 1989, ceasing to publish *Science for the People*, its former participants did not simply return to their pre-1969 lives. Instead, as I have suggested, they carried out some of the ideals and practices of SftP in new organizational settings. This fluid, changing, project-based work was, in Brian Martin's terms, "science for the people" in the hope of creating "science in a citizen-shaped world."[113] By not constraining scientists to work on one kind of activity in one kind of setting, and by critically engaging their peers, SftP generated new questions and new answers to how science and politics might be linked, and often forcefully made other scientists consider such questions and answers as well.

Two of the distinctive features of SftP were its organizational structure and its emphasis on direct action and intellectual critique. The localism, loose-knit relationships, and project-based work of radicals in science meant that SftP could experiment with new practices and ideas that were generated by the mix of people in the group at the time, the historical moment, and the local context. To a great extent, one could argue that this was the case for scientists in the 1940s and 1950s. Yet very quickly those groups settled into a formal organizational system that did not allow for much exploration of interactions with other groups, other than information provision. SftP's range of interactions were far more varied. Such dynamism is itself, I argue, a force of destabilization of the idea that scientists formed a distinctive political or moral group.

In exposing the rifts among scientists and extending the range of moral and political factors that shaped knowledge production and use, SftP made it difficult to treat scientific knowledge as distinct from the power relations that produced it. SftP never claimed that there could be no such thing as truth about the natural world. It claimed that in the search for understanding the patterns that structure the natural world, power relationships shape the choice of problems and their interpretation. SftP did not suggest that politics ought to substitute for appeal to the material world and the patterns that scientists had identified in it, but rather that the values and beliefs of scientists, their sponsors, and those who used science ought to be included in debates about the veracity and social value of scientific claims.

Science, more than any other profession or way of knowing, is based on the idea that participants may not come up with the same answers, but they operate with common assumptions and techniques. Radicals' ideology focused on the inseparability of science from other parts of social life. The form of their organizations followed. Sustaining ongoing and fluid interactions with other groups and individuals who were not professional scientists became a priority. In this way, the form of the organization marked a more fluid boundary between science and other social practices. More frequent contact with nonscientists on issues of mutual concern resulted in efforts to undermine the expert/nonexpert dichotomy, as well as in a radical break with the idea of science as a community of like-minded, "autonomous" people.

More concretely, what was radical about the form of knowledge production that Marxist and feminist biologists advocated was that they rejected the idea that scientific facts can be separated from the values and beliefs of the *political system* in which they are created. Most scientists were aware of this, but the standard response had always been to argue that one should try to avoid "contamination" by nonscientific values (or run the risk of repeating what the Nazis had done by separating Aryan and Jewish science) and encourage organized skepticism to correct any errors. Radical scientists and feminist scientists took a very different view: explicate the ways in which values and power—especially sources "outside" the scientific community—shape science, and then use organized skepticism to verify findings. This method challenged what many considered the basis of the political strength of science: the production of politically neutral facts by a closed community. Not only could organized skepticism be used to undermine an argument about scientific groups (i.e., logic, mathematical soundness, connections to other credible knowledge, ability to solve important problems), it could be used to expose an opponent's argument as based on political ideas, not technical skill and reason.

SftP's questions and answers were not the only matters of importance: the group's activists, progeny, and networks symbolized and concretized relationships between parties and actions that had previously been separate. By *symbolize* I do not mean symbolic as opposed to real change. The creative labor of reorganizing cultural symbols and, in the process, power constitutes social change. By creating new forms of organization, people join or separate logics of thought and action in new ways, activity that is, in and of itself, socially important (regardless of how well the organizations succeed in accomplishing their explicit goals).

In 1975, the possibilities for scientists' engagement with moral and political issues were vastly more diverse than they had been in 1947. They could choose to operate inside or outside professional associations, at their places of work, in the classroom or outside; to provide information to citizens, the government, interest groups, or other intellectuals and scientists; to challenge the idea that science was separate from politics; to select research projects based on commitments to solving public political problems, such as environmental pollution; and to refuse to participate in some kinds of work. Although some scientists may have wished that these options did not exist, and may have wished to box off scientific practice from other concerns, organizationally, intellectually, politically, and morally, it was less possible to do so. SftP's views and actions were a challenge to the information-provision and moral individualist models, but they did not entirely replace them. Instead, they added means by which scientists could act on their moral and political commitments.

CHAPTER 7

Conclusions: Disrupting the Social and Moral Order of Science

> It is largely due to the assumption that science confirms the distinction between culture and reality, as distinct from orientations guided by myth and fantasy, that scientists could function in our culture as sages and critics. And it is because of the new doubts about these earlier claims that the authority of scientists has been devalued in the course of the twentieth century. . . . [T]he attacks on the unity, universality, and autonomy of the reality of the world have hurt their claims to represent neutral and generalized standards of discourse and action which are sharply distinguishable from the particular domains of myth and ideology.
> —Yaron Ezrahi, *The Descent of Icarus*, 1990

Like many other scholars, Yaron Ezrahi offers a philosophical argument that the decline in the authority of scientists in the twentieth century is owing to the removal of science from sites of public debate and scrutiny.[1] Many other critics have laid blame on the hostility of certain social groups, such as religious conservatives, postmodernists, or an ignorant public, or on general social systems such as capitalism.[2] My argument here is different. It is neither the privatization of scientists' role in political decision making nor the pressures from nonscientists or large-scale social systems alone that have caused this shift in scientists' authority. It is also due to scientists' own efforts to more closely link science with public moral and political concerns. As a result, scientific authority has been "unbound" from scientists themselves, so that claims in the name of and about science are successfully made by many other groups. This has occurred through interactions among the state, scientists, and other activists over scientists' relationships to the military. "Science" shows no signs of disappearing from the public political landscape in America; indeed, it continues to serve, as Jennifer Croissant has argued, as "the legitimating icon that serves as both synecdoche and metonym for all that is rational and good in western society, if not western society itself."[3] What has declined, however, is the authority of *scientists* to serve as unchallenged mediators between nature and the public.

Few of the scientists I have studied in this book could have anticipated that in the early twenty-first century, scientists' authority in public political debate would be so shaken. Many would not have wanted this outcome; others aimed for exactly this state of affairs. It is less important, for the purposes of this book, to understand how each group individually shifted the authority of scientists and science in public life than to understand them as linked efforts. Each subsequent group responded to earlier forms of action and tried to supplant them. Rather than an evolution of forms, however, the story I have told suggests that there were ever more ways of joining political and narrowly professional interests during the mid-twentieth century. The cumulative effect of these actions was that by the early 1970s, there was what many scientific leaders called a "crisis" in science, provoking "antiscience" attitudes that resulted in the public's loss of confidence in science. By drawing attention to the perceived moral failures of individual scientists, to the interests of sponsors and users of science, and to the possibility that scientific ideas could be shaped by political and moral issues, these groups undermined the traditional scope of scientists' authority: a means of informed and rational consensus without resort to coercion.

In this chapter, I return to the three main questions with which I began this book: Why did scientists engage in activism against the relationship between the military and science? Why did the forms of this activism vary? And how did activists' efforts simultaneously contribute to buttressing the power of science in American political life and transforming it? In answering these questions, I develop their broader theoretical implications for the sociology of science, for analysis of social movements, and for the place of science in modern political life.

GETTING INVOLVED: MILITARIZATION, MORALITY, AND MOVEMENTS

To understand scientists' participation in groups that challenged the value of close ties between science and the military, it is useful to consider that their actions were not unprecedented. Before 1945, they participated in public health campaigns, opposed and supported the use of eugenics for political ends, and organized against U.S. participation in the World War I. Others engaged in antiracism and antifascism campaigns during the 1930s.[4] Scientists, like other people, have varied commitments to work, religion, and politics; their tastes for engagement in political debate are no doubt equally diverse. Thus, there is nothing unusual about scientists' participation in American public life, in the most general sense, particularly with the expansion of the state in the twentieth century. But in the postwar period, the most powerful impetuses for scientists' engagement in these activities were the growth of military-science ties, and the development of a wide range of social movements devoted to peace.

The prospect of limitations on scientists' research immediately spawned a political movement among scientists that sought to influence the formation of policies for scientific research and development. But the national security system that developed concurrently raised the stakes of participation in the political process, and most scientists dropped out of public political activity, save for a devoted handful of elite scientists and the largely Quaker scientists whose religious beliefs and social networks prompted them to resist the colonization of scientific research by the military. During the 1950s, scientists had access to nearly unlimited research funds and were increasingly sought after as advisors. These benefits came with a high price: scientists' work and political lives were highly scrutinized, resulting in the arrest and prosecution of hundreds of scientists. As Charles Thorpe has demonstrated, the high-profile prosecution of the government science advisor Robert Oppenheimer revealed that the policing of scientists' behavior included restricting their political advising to the "technical." As Thorpe argues, "the freedom of expert scientific opinion was to be respected, but the scientific opinion was to be regarded as 'free' and 'expert' only so long as it was separate from moral and political concerns."[5]

Yet as a group, scientists were hardly pawns of the state. They willingly developed and tested the weapons that were one of the centerpieces of state power during the cold war. In a material fashion, they helped to create the state, a role they have played for centuries.[6] As levels of surveillance declined, and the presidential campaign of 1956 legitimated debate about atomic testing, there were opportunities for more scientists and other Americans to both develop and express their political views on science-military issues.[7] Although some of the scientists who formed the Society for Social Responsibility in Science (SSRS), the Committee for Nuclear Information (CNI), and Science for the People (SftP) had previously been involved in public political activities, not all had been, and the founders did not consider their actions narrowly political. This suggests that opportunities cannot explain why scientists acted, but neither can transcendent preferences for autonomy, for the redirection of the uses of science, or for some other fixed idea.

The scientists who both formed and participated in activists' views about the military-science relationship were largely emergent, a product of the intersection between nonscientist activists and broader intellectual debates that were often taking place on the college campuses where many worked or attended school. The location of so many scientists and science students on college campuses was itself a product of the federal management of scientific research: federal grants were distributed to a growing number of colleges and universities throughout the United States in the mid-twentieth century, and the science policies implemented after Sputnik meant that there were many more science students in the country than there had been at any time in American history. These encounters between social movement activists and scientists, as well as broader intellectual debates about atomic weapons and war, caused some scientists to oppose some aspects of the military-science tie. Scientists participated at different levels, too, some serving in central roles while others maintained somewhat marginal relationships to the group because of ambivalence or disagreement, or because they had other interests and concerns.

Too strong a distinction between "activists" and "scientists" fails to capture the ways in which scientists themselves served as critics of their peers, the state, and the broader systems that structured scientific knowledge production. Most sociologists of science who study scientists' participation in politics implicitly or explicitly refer to their role as "experts." In this view, scientists act as purveyors of technical information and as legitimators of political projects.[8] To see scientists only as experts, however, neither captures the variety of roles they have played in politics nor enables us to understand how scientists conceive of their own actions. Only in one case did scientists see themselves as "experts." The diversity of their self-conceptions can also be seen in their perceptions of themselves as "activists." SftP participants likely would have perceived themselves as activists, but members of CNI or the SSRS would not. At the same time, it is hard to see CNI's and the SSRS's actions as anything other than deliberate efforts to upset the balance of power in their own field and in American politics. Each in its own way called into question the dominance of one form of scientist engagement in politics in the mid-twentieth century: the scientist as technical advisor to the government. The alternatives that each group provided reflected very different ideas about scientists' appropriate roles. In each case, these groups articulated a different answer to the question "what is political about science?"

What Is Political about Science or Scientists? The Forms of Scientists' Engagement in the Mid-Twentieth-Century Politics of the Military

In our everyday and even scholarly discussions of "science," we often confound the people and institutions that create it, the techniques and technologies used to produce scientific ideas, the ideas themselves, and the ideological bases of science. As Bruno Latour argues, treating "science" as an already-made entity missed the processes by which ideas, people, and practices come to be considered part of—or separated from—the social idea of "science."[9] By drawing attention to the processes of assembly—and disassembly—it is possible to understand science as a dynamic, changing form of action. It also reveals that during the mid-twentieth century, the forms that scientists' political activities took were increasingly diverse.

In their attempts to articulate a place for science in broader public and moral debates, members of the SSRS, CNI, and SftP, as well as members of the Jason Program, the MIT Science Action Coordinating Committee, the Union of Concerned Scientists, and the American Association for the Advancement of Science (AAAS), tried to disentangle which aspects of science were properly subject to political scrutiny and reform. In each case, preexisting models of action, political opportunities, and social movement activism about specific aspects of the military-science relationship shaped their perspectives. Their views were thus powerfully shaped by existing relations, not deduced from transcendent ideologies of science or constructed de novo.[10] As Maren Klawiter argues in her research on breast cancer activism, in any social movement, there are different "cultures of action."[11] Cultures consist of ways of defining a problem, notions of enemies and allies, probable strategies, and ideas of what the future could look like. The scientists who formed the SSRS, CNI, and SftP were members of very different cultures of action. These cultures were shaped by the political issues of the day, the political and intellectual networks in which members were embedded, and the extant solutions to the problem of science and militarism. Central to understanding these differences is that each group defined the problem and developed actions through interaction, not through an a priori idea.[12] Their answers, both individually and as they cumulated over a twenty-five-year period, helped to disrupt the associations among scientists, scientific ideas, and the ideological power of science in liberal democracy.

Scientists as Moral Individuals

The "personal responsibility" model that the SSRS developed was based on the idea that the proper public role for scientists was to take personal responsibility for the uses made of one's work. Following Quaker tradition, SSRS activists argued that such personal choices should be shared with and interrogated by the public and by their peers. They hoped that individual scientists would follow the "way of conscience" that was so familiar to many of the founders of the SSRS. This claim had two important meanings for the political authority of scientists. The first was that scientists were not a community that shared the same norms and values, pursuing truth in the service of progress, as the Western Enlightenment narrative has suggested. Individual scientists made their own choices about what paths of knowledge production to pursue. The SSRS challenged the idea that scientists were mere conduits of truth, purveyors of technical truths with no interest in their use.

The profundity of making this assertion during the early 1950s cannot be missed, and it may account for the lack of popularity of the moral individualist view. By asserting that they were making moral choices, these activists implied that other scientists were not, thereby politicizing not only their own actions, but also the "nonaction" of their peers. Moreover, by drawing attention to what they considered the immoral applications of science, they raised questions about the universalism of the scientific project. The state still depended on scientists to serve as dispassionate advisors and producers, and there were strong formal sanctions for publicly mixing the technical and the political, in the form of a highly developed security system that demanded secrecy and allegiance to the cold war political regime. The personal responsibility model directly challenged one of the key tools needed for state actors to legitimate their actions as universalistic and rational.[13]

Producing a body of scientists who made explicit ethical choices about their work could not occur by force, in the SSRS's view, because its project was based on the idea that people would attend to their inner consciences if given exemplars and social support. The proper public role for the scientist was as a model of action for other scientists. Their role was not to organize or to convince through logic or evidence, but to present themselves as exemplars of the power of individual choice. In some ways, this idea was consistent with the image of the scientist as maverick, unbeholden to group pressures.[14] Scientists have played this role in public debates for centuries, bringing new perspectives into view by refusing to go along with the views of others. In the context of the 1950s, when the notion of a "community of scientists" was high, the eliding of political and intellectual individualism was a challenge to the notion that scientists shared the same norms and values.

In answer to the question of "what is political about science," SSRS members in the 1950s would probably have answered, "Nothing." This is because they understood themselves to be engaged in moral, not political, action. Yet it would be difficult not to see their work as an effort to influence the authoritative distribution of valued goods in a social system—a generic definition of "politics." They did not conceptualize the political world in terms of liberal political theory that differentiated between the public citizenry and rulers. Their actions were more akin to what Michael P. Young has called the "confessionalist" model in American politics, in which groups collectively act to change the moral behavior of individuals in the hope of redeeming the nation. If the notion of "political" can be understood to include the moral meaning of what is right and good, what was "political" about science, for SSRS members, was scientists themselves and the uses to which science was put.

Scientists as Conduits

The information model was in many ways the exact opposite of the individualist model. Whereas the moral individualists exhorted scientists to take personal responsibility, the information-provision model portrayed responsibility as the collective duty of a community of scientists. In classic political theories of liberalism, citizens have unalienable rights in exchange for carrying out duties. One of the rights of citizens is to make public choices about how their society should allocate its resources. The popularity of the idea that citizens need to be specially educated in order to participate in politics has varied historically. In the 1950s it had mixed support. For some, an "educated" public was a bulwark against irrationality. For others, an orderly public did not depend on education. The "public" that scientists wished to serve was the citizenry, writ large, as it is understood in classical liberal theory. It would be possible, even easy, to make a decision for or against an issue without any special technical information at all. Yet CNI scientists and women activists believed that citizens needed facts, separated from their contexts of production, in order to make such decisions wisely.

During the first year that CNI took shape, the group articulated a special role for scientists. They conceived of scientists as experts who were, by virtue of their special access to scientific ideas, neither "citizens" writ large, nor the government. Scientists had the capacity to stand outside of government and the masses, unfettered by ideology. In a framework that has a long history in the West, often fought for by scientists wishing to claim autonomy from other groups and control by the state, scientists were understood as neutral "conduits."[15]

At least in its public presentations, CNI depicted knowledge itself as apolitical, and the scientist, qua scientist, as a machinelike interpreter of facts about nature. What was political, in CNI's view, was the way that knowledge was distributed in American society. CNI members' actions were an indictment of the government for its failure to provide citizens with the facts that they needed to make informed decisions. Yet the government did, and does, provide information to its citizens. By providing additional or different information to the public and challenging government interpretations of scientific claims, CNI exacerbated the growing sense among the public that different groups of scientists interpreted scientific data differently. In doing so, they also undermined the idea of the state as a neutral arbiter of the public good. They invited a closer inspection of the interests of those who produced and distributed knowledge, even as they asserted that they provided "unbiased" information against the "biased" or incomplete information from the government. What was political about science, then, was that the interests of those who produced it—but particularly the state—could shape interpretations of knowledge and their fair distribution to citizens.

Scientists as Advocates

What made SftP distinct from most other postwar models of public political action by scientists was that the group rejected the idea that scientific knowledge and scientists were inherently apolitical. In their view, the content of scientific ideas, the uses to which ideas were put, and the actions of scientists could be shaped by the political systems in which they existed. No SftP scientist would have argued that it was impossible to produce robust understandings of nature, or would have equated nature with politics. But they did assert that the choice of questions, applications, and some processes were shaped by the interests of powerful groups.

Drawing on Marxist, and especially anti-imperialist, political theory, the main line of argumentation in SftP's magazine *Science for the People* and many of their political activities was that scientists had become tools of a capitalist-militarist state, producing knowledge that was useful primarily to a small class of people. This assertion had similarities to the moral exhortations implied in the personal responsibility model of public political action for scientists, because it laid some responsibility on individual scientists to take action to change this arrangement. But it went beyond that model by locating the necessary changes in the structures of social and political life: universities, states, economic systems, and professional organizations.

If the SSRS aimed to change the hearts of scientists, and CNI to enlighten the public, SftP's goals were to change the ways that existing economic and political systems shaped what problems were chosen, and to encourage the development of scientific ideas that were more useful to a broad range of people. They called for closer alliances with groups struggling for social change. Rather than calling for an apolitical science, the radical model called for more, not less, discussion of the political implications of scientific claims and the motivations of those who made them. SftP was less interested in serving the generic "public"—for that was the terminology of the liberal that masked the class differences among the ruled—than in addressing the problems of "the people," conceived of as the working class and the disempowered. Although some members produced analyses that examined the relationship between the content of knowledge and political systems, in the main, SftP members were focused on why some kinds of knowledge were produced and benefited a small class of people, while "science for the people" was left "undone."[16]

Yet the radical model never came up with a role for scientists upon which its members agreed. To some extent, this was because the group was chapter-based, with no central authority. Other professionals struggled with the same issue during this era. Different chapters developed their own ideas about how members could go about doing "science for the people," based on their own political interests and skills. This was, in itself, a challenge to the then dominant framework of using technical skill rather than values as the major basis of political intervention among scientists.

SftP's use of disruption of meetings and public challenges to the motivations of other scientists publicly revealed that not all scientists acted purely on the basis of fact and reason; some acted on emotion and political values. These tactics also made it difficult for scientists to present a unified political face, just as the disputes about the interpretation of evidence about the health effects of atomic testing had done in the 1950s. What was political about science thus included the social relationships among scientists, in addition to its content, the subjects that were chosen, and the way that science was used.

Each of these models of action implied a particular way of organizing social life more broadly. Because science and liberal democracy have developed together and are interdependent, shifts in how science is organized also represent shifts in the foundations of public political life.[17] The scope and importance of such challenges are not equal, nor do all groups have the same capacity to challenge knowledge systems and the power that is distributed through them.

Separating Scientists from Authority over Scientific Knowledge

Taken together, the individual and cumulative effects of these methods of reorganizing the moral economy of science in American life have had two overarching effects. The first is that they demonstrate the capacity—and interest—of scientists to respond to the moral and political problems of their day, not only as "experts" whose authority is based on technical knowledge, but as moral individuals and/or movement activists. As Steven Shapin argues, the perceived moral qualities of scientists shape the extent to which their claims are credible.[18] If one of the bases of the authority of scientists is the public demonstration of their commitment to work toward a greater social good rather than for their own benefit, then we can judge the actions of mid-twentieth-century scientists to be successful. This form of success came at a price, however, for it ironically undermined some of the other key bases of the authority of *scientists:* that they form a community that can and does share a set of rules that separate the moral and the technical in judging the veracity of claims; that they are, as a group, autonomous from the influences of other groups; and that science inevitably leads to moral progress through material well-being.

The questions scientist activists raised about the morality of their peers, the uses of science, and the relationships between the concerns of citizens and those of scientists challenged three key assumptions undergirding the authority of scientists and scientific knowledge: (1) that scientific knowledge inevitably led to material and moral progress; (2) that scientists were an autonomous group that operated according to their own internal rules and norms; and (3) that scientists formed a special community that agreed on the basic rules for judging the validity of scientific ideas and the credibility of its practitioners. Reason and logic, not emotion or value, were supposed to be the main forms of exchange among scientists. Their public displays of disunity about the extent to which morality and emotion could and should be used to assess the credibility of other scientists helped to weaken this basis of authority.

Progress

At the most obvious level, scientists' public political actions have challenged the idea that science is inevitably linked to generalized human progress. Since the seventeenth century, Europeans and Americans have had great faith in the idea that science, as a system of thought and practices and their outcomes, will ultimately improve human existence. The scientists whose stories I tell in this book took great pains to ensure that the public understood that scientific products could be a destructive force. To be sure, Americans were aware of the obvious dangers of weapons, but scientists, individually and in groups, called into question the idea that all material goods made life easier and healthier. From Albert Einstein's calls for disarmament in the early 1950s, to the scientists who raised the alarm over the toxicity of human and natural chemicals in the late 1960s, this was an era in which the destructive possibilities of technology were brought to the fore by scientists.

The notion of progress in Western thought is associated with more than the material benefits brought about by technological advance. It is also associated with overcoming superstition and myth by replacing it with reason and rational decision making. Scientists have often been treated as the representatives of progress through reason. By drawing attention to the preferences and beliefs that shaped their own choices about the kind of work they did and what they thought others should do, the scientists portrayed here made it more difficult to see reason as the guiding force underlying the political and economic systems of the United States.

Unity

Part of the political power of science has been based on the idea that "science" constitutes a unified field of action, in the sense that its practitioners largely agree on the standards for judging ideas. As scholars such as Galison and Stump and Knorr Cetina have argued, science, as a set of ideas and practices, is not as unified as philosophers of science and other researchers have considered it to be.[19] Among these standards are reason and evidence, not values or emotion. Scientists' specialized training is supposed to ensure that they share certain ways of thinking, certain goals, and certain practices. If science refers to heterogeneous practices, strong disagreements on the nature and purpose of scientific investigations, and the rules by which claims and claimants were judged, then it slides into the messy world of value or, at worst, mere opinion. Thus, to the extent that science is understood to refer to a distinct set of practices and rules for judgment, it is more politically powerful.

When scientists publicly reveal not merely minor disagreements about procedures, but serious conflicts about how to judge the credibility of other scientists and the rules by which scientific claims can be made, they disrupt the idea that "science" is a homogenous field of action. Scientists disagree frequently, of course, but their power in public life is based on the idea that these differences can be adjudicated by reference to shared procedures. This idea is the essence of science: it is supposed to be a way of collectively seeing and judging that is based on universalistic rather than personalistic principles. Clearly, scientists disagree frequently about how to judge evidence, and there are always conflicts in any given scientific field.[20] Their authority, though, is based in part on the idea that these disagreements can be resolved by experiment and reason, without resort to extrascientific factors.

The varied views of scientists about the health effects of fallout placed disagreements about how to decide whether entities were harmful at the forefront of the public imagination. Given the dire warnings that some scientists put forth about the genetic effects of atomic fallout, their disagreements were especially worrisome to the public—and to the scientists who worried about their credibility with the public. As more models for public political action developed over time, and radicals and feminists raised more questions about the extent to which personal characteristics could shape scientists' interpretations, they demonstrated a public face of disunity over how to judge some scientific claims. In arguing that science has become "disunified," I do not mean the sense of unity that the sociologist Talcott Parsons or other structural functionalist analysts assumed. I mean here the ideological notion that "science" refers to a bounded set of practices and shared ideals.

The public perceptions of disunity, the front stage display of disagreements that are normally kept backstage, as Stephen Hilgartner has shown, weaken scientists' authority for other, more general reasons.[21] The historian Brian Balogh argues that multiplying forms of expertise have led to a lack of public faith in expertise.[22] I argue that scientists and others' participation in social movements raised questions about the extent to which experts were politically neutral and had the best interests of stakeholders at heart. Groups that are divided have a more difficult time fending off challengers, as analysts of organizations, social movements, and revolutions have shown. Divisions offer opportunities for new alliances, but they very often weaken existing regimes by destabilizing resource flows and organizational systems, making the routine reproduction of these systems more difficult.[23]

Autonomy

Scientists' capacity to universalize political decision making is based on the idea that they belong to a self-regulating system. Daniel Sarewitz has argued that one of the myths of science is the myth of unfettered research: that scientists are detached from the concerns of daily life, but operate in a special community with its own rules and norms.[24] The relationship between scientists and the military clearly threatened that notion, as scientists quickly recognized. Despite the attempts of the government and scientists themselves to assert, and police, the strict image of the scientist as autonomous, rational conduits whose ideas developed independent of preferences and values, other scientists refused to accede to this view.

In their own way, each of the models of action I describe challenged the image of an autonomous community—and the notion of autonomous individuals in that community. Laboratories could not be considered segregated technical spaces, where one's values were left at the door, if the SSRS, Albert Einstein, and Norbert Wiener were right. The values that scientists had by virtue of their connections to other moral communities ought to and often did affect what scientists studied. Individual scientists could thus not be considered "autonomous" in their profession, in any meaningful sense of the term. The idea that the *community* of science operated with its own norms and values was also unsustainable over time. CNI decided to provide information directly to the public, while other scientists who worked for the government did not. The organizational distribution of the disagreements about fallout (government scientists said it was harmless; many critics of the government's perspective worked at universities) further called into question whether the imagined community transcended the locations where scientists worked. The radical position provided the most thoroughgoing critique of the ideology of autonomy. Scientists, in this view, were working for the benefit of the state and capital. Nor did radicals insist that "autonomy" was itself an important goal, but advocated closer connections to "the people."

Scientists alone are not responsible for the changes in the authority of scientists. But the depth and breadth of scientists' own arguments about the proper relationship between science and politics disrupted images of the objectivity of scientists and scientific knowledge, and the assumption that science and progress were synonymous. By drawing public attention to the idea that scientists were morally implicated in the development and use of their ideas, they made it difficult for scientists to distance themselves from the negative effects of scientific products. Their public self-criticisms—which were influenced by the dissatisfaction of nonscientists with scientists—opened the door for more critiques of science. The depth, breadth, and constancy of their challenges played a critical role in weakening these bases of the authority of scientists. To more clearly understand the transformed place of scientists in public political life, I turn next to a brief overview of three post-1975 changes that have intersected with scientists' public political activities: health-based social movements, the growing regulation of the production processes and uses of scientific knowledge, and the market.

Unbinding in Theory

Understanding how scientists maintain authority over claims making has produced a rich array of work in the sociology of science. Much of this work has come from the "anthropological turn" in the sociology of science, in which researchers follow scientists and engineers in their everyday work to understand their language, machines, and cultures, and the social networks in which they are embedded. The "black box" of scientific fact making that Bruno Latour and Steve Woolgar wrote about in their 1977 book *Laboratory Life* has been opened.[25] Some scientists' responses to the "science wars" of the late 1990s were based on what they believed was a misunderstanding of how scientists work. The vehemence of their responses suggests that the influx of new commentators on scientists' work—especially researchers' refusal to treat science as a priori distinct from other kinds of social action—made scientists uneasy about who had the authority to make claims about the scientific process.

Given the reaction of scientists to the "science wars" of the 1990s, and twenty-five years of understanding confrontations over scientists' political authority as "boundary disputes," one might have predicted that scientists, faced with a public uneasy with the arms race, skeptical of whether atomic testing was safe, and under fire for their contributions to the war in Vietnam, would have engaged in "boundary work" to reestablish their moral authority. To be sure, some scientists did that. But many of the scientists in this book were among the dissenting voices, rather than among those who reacted defensively by trying to shore up the "boundaries" of science.

First introduced into the sociology of science by Thomas F. Gieryn in 1983, the analysis of boundaries is concerned with what kinds of languages, organizations, objects, and people serve to mediate between "science" and "nonscience."[26] Although few writers in 1983 would likely have thought of scientists as warriors for truth, Gieryn's argument made explicit the ways that scientists sought power and status for themselves instead of automatically garnering it. Since 1983, the analysis of how scientists maintain boundaries between their work and that of contenders has illuminated how boundaries are defended and shifted through language, via objects ranging from molecules, to books, to machines, through organizations, and through rules. In many of these analyses of collective boundary work, researchers have found that scientists often reassert their role as legitimate authorities.[27]

The imagery of boundaries that has guided so much of the analysis of scientific authority in sociology has been complicated in recent years by the proliferation of analyses of "in-betweenness" as an outcome of authority struggles. "Institutional hybrids" of various sorts, "catchment areas," "interdisciplines," "cyborgs," and other concepts help to illuminate the fundamental entwining of scientific knowledge production and other systems.[28] These "in-between" places are not challenges to the idea that scientists may maintain their authority by defeating challengers, as analysts of boundary work have concluded. They are reminders that the authority of science is not entirely derived from its segregation from other practices or from the existence of clearly defined notions of who is or is not scientific in making claims.

Although much of this work has been concerned with the "how" of boundary work, as Kleinman and Kinchy note, it does not tell us much about why scientific authority is fixed or changes.[29] The analysis that I have put forth is an effort to address that problem. Far from defending their domains of action, scientists have sometimes sought to engage wider audiences as participants in and critics of science as usual. I hope to have shown that they did so in response to concrete, material changes in their work lives, and in the intellectual and political debates about the role of science in public life.

It would be easy to argue that this work was purely defensive, a response to psychological feelings of guilt or fears about the loss of social honor, but that is a distortion of how scientists came to participate in organizations that invited peers and the public to consider the moral imperatives of science more carefully. I do not wish to imply that scientists brought people "inside" the institution, or that they went "outside" its edges, for that would perpetuate an almost Parsonian image of a building or other entity that scientists inhabit. I am more concerned here with how scientists changed the possible cultural configurations of what science, and what politics, mean. I hope to have shown that the activities of scientists did not explode the "boundaries" of science so much as they exposed the ways in which science was implicated in broader political projects, and how it might be used for other ones. The imagery I have in mind here is a network, rather than a series of fixed institutions that only occasionally encounter one another or change. In the process of engaging public political issues, scientists made it difficult to return to the ideology of science as an ideology of purity and truth that could easily be used to justify public political decisions.

By using a method of comparative case study over time, I hope to have shown how these different configurations overlapped, challenged one another, and unfolded in an uneven fashion. There was no linear movement from one form to another or complete replacement of one form by another, but rather, increasing contestation over forms of action. The shift in scientists' authority, thus, was not an event but rather the result of the slow accumulation of increasingly broad and contentious examination of the ways in which values shaped scientists' work and their interpretations of nature. I have chosen to compare three cases in which the same overarching issue—the influence of the military on science, and scientists' influence on the military—was the cause that generated action. But what profoundly shaped the forms that scientists' claims took were the political mobilizations of nonscientist activists, or "social movements" with whom scientists were in contact. Many of the scientists who came to form the new organizations that I profile here had some experience in public politics, held religious or political beliefs that supported the idea that they ought to take action, or were convinced of this idea through their contact with other activists.

Each iteration of scientists' invention of new forms of collective engagement with issues of how scientific knowledge should be produced and used was in some way a response to the methods that came before it. As these forms proliferated, it became increasingly difficult to see scientists as a group with a unified sense of its own moral obligations. Thus, to understand scientists' involvement, it is necessary to "unbind" them from the straightjacket of monodimensional identities as professionals or experts. In a historical period when scientists' own capacity to freely express their own views was under surveillance, and when other activists were mobilizing against military-related issues, scientists were also moved and motivated to engage in public debates. Feminists have long examined when and why scientists' gender may have shaped their choices of questions and their answers.[30] My suggestion is that we extend our analysis to other features of scientists' identities. This is not a call to read scientists' actions and political viewpoints from social categories but rather to examine how and why scientists self-consciously engage in public political debates, and how these actions may lead to the re- or unbinding of scientists' authority over "scientific" claims.

It is not that scientists' authority has become unbound from them only as a result of their participation in new forms of political action. The new influences of law and markets, and the rise of health-based social movements after 1975, helped to shift the authority over decisions about the veracity of scientific claims to judges and lawyers, patent offices, corporations, and patients. The unbinding process, therefore, is a more general phenomenon that has spread over the past thirty years.

OTHER SOURCES OF THE REDISTRIBUTION OF SCIENTIFIC AUTHORITY, 1975–2006

Social Movements

In the early 1970s, community-based groups in the United States and abroad began to organize around the broad issue of the politics of health. Women and African Americans were especially important in asking why patients' perspectives were not included in studies or discussions of health.[31] Women were especially important in challenging how the Food and Drug Administration assessed the safety of contraceptives. Women distributed health information and offered a critical perspective on medicine via organizations such as the National Women's Health Network and in books such as the 1969 *Our Bodies, Ourselves*. Gay activists, mental health activists, and the disabled followed suit, rejecting scientifically prescribed treatments for mental health and categorizations of disability and homosexuality as medical problems. Activists were not limited to nonscientists, however. Clinical researchers, scholars, and other health care professionals worked with these groups to protest their treatment, and importantly, to find new ways to address their concerns.

"Embodied" social movements, in Rachel Morello-Frosch and her coauthors' analysis, are especially important in challenging the authority of scientists. Activists involved in toxic waste, disability-based movements, and illness-based movements have been able to contest the expertise of scientists by relying on collective action based on their common bodily experiences. As Morello-Frosch and her coauthors argue, "The body, combined with people's embodied experience, thus becomes a counter-authority in which movement groups can base their critique of medical science."[32] Nick Crossley's analysis of antipsychiatry movements similarly illustrates the ways in which the "mentally ill" resist treatments assigned to them by researchers and clinicians on the basis of patient's claims to the moral autonomy of the body.[33] One of the most successful challenges to the authority of scientists in the past three decades comes from the AIDS movement. AIDS activists, most of them white and middle-class, learned the science of AIDS and its treatment, and, using disruptive political actions and guerilla drug trials, convinced the National Institutes of Health to revise its treatment policies.[34] Although health activists have challenged scientific evidence, one of the ways that they have done so is by raising questions about how scientists' assumptions or their interests, by virtue of their association with a specific employer or their financial sponsorship, shape the claims they make. In some cases, such challenges have led to cooperative, participatory forms of scientific research with activists.[35] In others, they have led to the abandonment of research projects, such as the Human Genome Diversity Project.[36]

Law

Legal and bureaucratic regulations have intruded on the practices of scientists over the past fifty years. As a result of more elaborated human and animal subjects regulations than in the past, scientists have slowly lost the authority to treat research subjects as they please. The use of humans as subjects used to be based on the supposition that humans were objects whose individual rights were overridden by the larger public good presumed to come from scientific study. The deliberate exposure of American soldiers to radiation, the tragedy of the Tuskegee syphilis experiments, plutonium experiments, birth control experiments on Puerto Rican women, and other studies in which harm was done to human subjects led scientists, politicians, and lay people to develop protocols for informed consent. Moving humans out of the category of objects that were equivalent to nonliving things means that individuals can make greater legal and moral claims on scientists and those who employ them.[37] Regulation implicitly asserts that scientists themselves are not capable of adequately protecting subjects. Informed consent rules and laws protecting laboratory animals mean that legal authorities, people in research studies, and concerned organized constituencies now have the power to challenge scientists' authority over the morality and ethics of research.[38]

Other forms of regulation guide what kinds of scientific knowledge are admissible in court, and credible speakers about science have also affected scientists' authority. The 1980 *Daubert v. Dow Pharmaceutical* decision gave judges the power to differentiate between "junk" and credible science. So did the subsequent *Kuhmo Tire v. Carmichael.* Both *Daubert* and *Kuhmo* elaborated judges' capacity to make decisions about what constitutes "good" science in the courtroom.[39] As Sheila Jasanoff has argued, it is not that scientists have no power in the courtroom, but that what they do and say is more highly scrutinized.[40] Similarly, as Michael

Lynch has demonstrated, as scientists' testimony about the veracity of fingerprinting (or DNA) became subject to close scrutiny, the authority of this knowledge has been placed in machinery, not people. In the process of walling off the work of humans from the machines that calculate and evaluate DNA and fingerprint evidence, scientists themselves have become less important as verifiers of the knowledge.[41]

Markets

The market has also played a key role in the "unbinding" of scientific authority from scientists. Whether something is profitable increasingly defines whether an invention "works," as Daniel L. Kleinman and Jennifer Fishman have shown. The dramatic rise in patenting in the United States means that ownership, not the credibility of a scientist, is the basis for controlling knowledge. This new focus on patenting has given rise to new collaborations between scientists in universities and corporations that have distributed knowledge production and control over a much wider array of actors than in the past.[42]

This redistribution has had another effect: consumers have become more aware of cases in which scientists have knowingly or unknowingly skewed results in ways that are favorable to the profit-making firms that sponsor their research. In the past four years, the editors of the *New England Journal of Medicine* have had to retract studies of the painkillers Vioxx and Celebrex because of data omissions or errors that misrepresented the extent and severity of side effects of these drugs.[43] The fact that these were high-profile cases in which articles were retracted might reinforce the idea that most scientists do not skew their results to please sponsors, but it could also indicate a more general problem with the credibility of scientists. The point here is that the intersection between scientists and markets has distributed authority over what works away from scientists' testimonials and toward the owners and sponsors of the research itself.

SOCIAL MOVEMENTS AS SOURCES OF NEW IDENTITIES

Organizing Identities

One of the most important sets of findings about social movements over the past thirty years is that they face the same kinds of problems that other kinds of organizations do, including oligarchization, recruiting and retaining members, and acquiring resources.[44] Another set of analyses has examined the processes by which groups form organizations, and, in the process, new political identities, the ways in which social movement organizations serve as expressions of meaning, and their role as crucibles of creativity. As they struggled to reconcile the promise of science to improve life with science's relationship to social problems, scientists formed organizations whose goals were often vague. Scientists knew they wanted to create something new, but they were not entirely sure what the new organizations might look like, nor were they certain of the practices in which they would engage or the claims they would make.

By focusing on the process of organizing rather than on ready-made organizations, I have analyzed a series of what Erving Goffman calls "encounters"—moments when an individual's or a group's attention is focused on one or just a few significant actors. Most of the time, Goffman argued, our social lives do not require us to be conscious of or rework moral or normative action. But in "encounters," such moral and normative rules are rethought. Gamson argues that during encounters, social movements may engage in "rim talk," in which they implicitly or explicitly question authorities' conduct.[45] This talk is prior to the act of "framing," or the delimiting of how the problem at hand is understood.[46]

When organizing the SSRS, CNI, and SftP, participants tried out a number of different ways of articulating who they were as scientists and as political actors, and who they were not. The debates about how to name themselves are indicative of the debates about "who we are." In no case would it be possible to understand their answers only with reference to a priori interests; the shifting political climate, their awareness of their opponents, and their own uncertainty led them to try out different identities. The notion of a "rim," of course, has similarities to the concept of a "boundary." The utility of Goffman's term is that it moves attention away from strategy and interest as a basis for action, and toward an image of social life that is constituted more by attempts at sense-making. The sense-making is about what kinds of relationships people want to have, and how they enact them.

Sociologists of social movements and community organizing have paid close attention to these processes among political activists. In *Freedom Is an Endless Meeting*, for example, Francesca Polletta closely examines how participatory democracy groups used different models of social relationships, such as a family or friendship network, as a basis for their action. Similarly, Paul Lichterman has shown that different groups of activists rely on very different models of "community" when they organize for social change. For Lichterman, the answer to "what is a community?" has consequences for who can be involved and what actions they undertake. The scientists that I have studied in this book were not always sure that they were "activists" or even "political," but the same processes can be observed among them.[47]

The importance of attention to "rim talk" and self-identity is that it is at the foundation of what collective contentious political action tries to do: formulate new social relationships. Joseph Gusfield makes a broader argument about the significance of examining this kind of organizing. He argues that one of the ways that social movements are significant is that they provide "a vocabulary and an opening of ideas and actions which in the past was either unknown or unthinkable."[48] Some ideas become more widespread than others, of course, and most analyses of social movement outcomes have tried to understand how these ideas and actions become institutionalized in laws, rules, material changes, or new identities.

By attending to the process of organizing, I have also shown that the social networks in which scientists were involved were critical to their formulations of the proper role of science in public life. The debates they had with opponents and with prospective participants helped to generate debate in new settings. Throughout the mid-twentieth century, for example, scientists attempted to use the AAAS as a vehicle for their expression of their political views, with greater or less success, depending on their perspective and the historical moment in which they were acting. No matter what the AAAS's initial response was, the continued efforts of scientists to engage their peers resulted, over time, in an association that was more democratic and more open to engaging public political issues.

New political identities can be sustained through networks, laws and rules, material constructions, symbolic systems, and economic exchanges. I draw attention to how organizations can stabilize identities. As I have argued elsewhere, organizations matter for identity construction and preservation because they have two important qualities. The first quality is that they are formally recognized systems of exchange. Rather than acting as boundaries, organizations, at least in a pluralist society, permit the flow of people, ideas, and material objects. The second quality is that they are symbols of specific kinds of relationships and commitments.[49]

Organizations through which scientists can engage public political issues grew significantly in the 1970s, just as they did for other professionals. Before turning to the more general expansion of such organizations, I first turn to an overview of the organizational offshoots of the SSRS, CNI, and SftP. Although the SSRS did not survive, the model of action that it espoused can now be found in the form of "pledges" by scientists not to work on projects they find morally unacceptable. The most recent example of this is the pledge circulated by scientists in the 1980s stating that they would not accept research monies to work on the missile system colloquially known as "Star Wars."[50] Others have proposed a Hippocratic oath for scientists. Sir Joseph Rotblat, one of the founders of the Pugwash Conferences and a member of the SSRS in 1956, worked with the Student Pugwash organization in 1996 and 1997 to develop such an oath.[51] Although the number of scientists who have signed these pledges and oaths is probably less than twenty thousand worldwide, these scientists are evidence of the continuity of the idea that individual scientists ought to be concerned with the uses of their products.[52] Similarly, the movement among pharmacists who subscribe to a "conscience clause" and refuse to prescribe emergency contraception uses the same principle.[53]

CNI, SftP, and the Scientists' Institute for Public Information, the aggregate organization made up of the local information groups, have direct organizational "descendants" that were created by former members of these organizations. Among them are the Council for Responsible Genetics (CRG), a watchdog group that distributes information about the political implications of biotechnology, genetic research, and the employment and political uses of genetic research. CRG was originally formed by a core group of scientists from SftP. CRG continues to play an important role in publicizing the dangers of genetic research on humans, animals, and plants.

In 1966, Barry Commoner formed the Center for the Biology of Natural Systems (CBNS) at Washington University in order to engage in research and provide the public with information that it could use to participate in political decision making about biological issues such as pollution, energy, and habitat destruction. Commoner and the CBNS moved to Queens College in 1981, where it remains an important source of regionally used information about biological issues, with a strong track record of engaging the public in the collection and interpretation of biological information about pollution, genetics, energy issues, and agriculture.[54]

This model of action is widespread today. One can find organizations that employ scientists to create or distribute scientific information about human health issues, animal issues, climate change, nanotechnology, and thousands of other scientific subjects.[55] Prominent examples include the Union of Concerned Scientists, National Women's Health Network, and the Center for Science in the Public Interest. This model of the scientist as advocate is institutionalized in the thousands of nonprofit organizations in the United States that distribute scientific research to their audiences *and* recommend specific kinds of political action based on this evidence. The Union of Concerned Scientists, formed at MIT in 1969, is perhaps the best-known example, but there are thousands of others organized around issues ranging from environmental health to earthquake preparedness.

One of the most important organizational results of midcentury political organizing among scientists has been the formation of new subunits of existing professional associations. As I have argued in earlier chapters, the AAAS's mission beginning in 1951 was to engage public issues. For several decades, this mission was somewhat stalled, bogged down in debates about whether public engagement constituted partisan politics, fears of red-baiting, and similar issues. Between 1969 and 1973, the AAAS organized a series of ad hoc and permanent committees to find ways to increase the numbers of women and minorities in the sciences, and to develop means for scientists to address public political concerns. Between 1969 and 1975, nearly every science organization in the United States organized a new subunit or committee charged with providing the public with more information about the association's mission, membership, and relationship to issues of public concern. These subunits sometimes developed into separate groups that held conferences, sponsored talks by scientists, and investigated popular political concerns relating to a specific subfield. The names of these committees and subunits have changed over time, but this task and the organizational resources that are put into it remain an important part of the work that professional science associations do.

To be sure, the availability of these organizational outlets for public service work by scientists means that other parts of professional organizations remain formally devoted just to the promotion of scientists. Yet considering that in 1957 Barry Commoner could not convince a group of scientists at the AAAS to form a political committee dealing with public issues, and scientists expressed fears in the 1950s and 1960s about being seen as "too political," the routine presence of these organizations and subunits is indicative of a larger acceptance of participation in public life among scientists.

These specific shifts, however, should not be treated as the only important outcome of scientists' multiple forms of mobilization. Scientists' attempts to refashion their public identities—regardless of the substantive, long-lasting effects—should be seen as an end itself. As scientists develop multiple modes of action that have brought them into relationships with the public, with states, with markets, and with law, they have challenged the notion that "science" and scientists are monolithic. It is beyond the scope of this work to delineate the contemporary forms of political action among scientists, but I hope, at least, to conceptualize the methodological issues involved in such a delineation.

The "Social Movementization" of Professions

In their 1973 book *The Trend of Social Movements in America: Professionalization and Resource Mobilization*, John D. McCarthy and Mayer N. Zald observed that social movements were becoming more like interest groups, with formal credentials as the means for employment and participation. They also noted that there were high levels of education among those with power in the organizations. At the same time, McCarthy and Zald argued, these new kinds of organizations spent much of their time seeking resources to further their causes.[56] Whereas subsequent scholarship on social movements has focused on the resource mobilization aspect of their argument, I want to return to the relationship between professions and social movements. In addition to the professionalization of social movements, I argue that the organizations I have analyzed in this book represent forms of the "social movementization" of professions.

From the late 1960s, a large cohort of young professionals looked for ways to make their political commitments fit more clearly with their professional lives. Leaving the universities in which they were trained, they sought to inject their desire to ameliorate racism, sexism, class-based inequality, and environmental degradation into their work. Rather than forming new organizations, some worked within existing organizations (firms, nonprofits, universities, and government agencies) to investigate and try to ameliorate social problems by using their skills as professionals. Although some found that attempting to maintain a "radical" focus was difficult, many others found ways to integrate their political and professional concerns.[57] As the promise of "radicalizing" professions declined, so, too, did the focus on scholarship on social movements and professions, save for the observation that many social movement organizations were now staffed with professionals. In this book I have studied the groups that first undertook this "social movementization," before it became widespread as mass-based political movements declined in the mid-1970s.

To understand the intersection of professions and politics today, looking for "radicals" in social movements, or seeking out separate organizations that explicitly join a profession with an avowedly "political" goal is one possibility. A more robust method is to examine how political, educational, humanitarian, and technical concerns are joined together in educational systems and professional practices. In professional education, for example, especially in law, human services of all kinds, and medicine, students are now trained in understanding the problems of disadvantaged groups and the solutions to resolving them. Examples include the "Innocence Projects" at law schools, or medical schools' programs to involve students in projects aimed to expose them to ways of ameliorating the problems of the poor or those living with minimal access to health care. Once trained and involved in networks of other professionals with similar interests and skills, professionals form movementlike communities of like-minded people.[58] "Activism" is not the province only of the left; many conservatives treat their professions as means for making broader social change as well.

NEW OPPORTUNITIES FOR SCIENTISTS

This book has focused on the ways in which scientific authority has become unbounded. Although the authority of scientists to assert that they are neutral conduits for facts has weakened, public enthusiasm for "scientific" claims more generally has not. Assertions that products ranging from food to automobiles have been "scientifically formulated" show no signs of slowing down, and Americans use scientific ideas in much the same way that they use other consumer items. They are offered an endless array of information about health dangers and scientific sources of solving them, the state of the planet, the habits and tendencies of males and females, and the age of the Earth, to name just a few topics. They consume this information in ways that are in accordance with their own worldviews. Thus, when the *New York Times* reports that scientists have found a gene for "creative dancing," those Americans concerned about their prosaic dancing style and inclined toward individualist explanations might conclude that they probably do not have the creative dance gene.[59] Similarly, Christian fundamentalists who believe in the literal interpretation of the Bible are more inclined to "see" the evidence that creationists present as credible; those in favor of unfettering business from the constraints of government regulations may be less likely to believe that global warming is caused at least in part by human activity. Thus, "scientism" has not lost its power in public life, in the sense that actors continue to claim that their ideas are scientific in order to gain credibility and power: even creationists assert that their claims are scientific in order to win converts. Yet few Americans understand how scientists or their employers and sponsors choose research questions, how they generate evidence, and how they assess the veracity of claims.[60]

Scientists have expressed great dismay about this situation since the 1950s. It is not possible or desirable, however, to return to an era in which it was difficult for the public to acquire scientific information, much less knowledge of how science works. As Brian Wynne writes, "There is a crucial public dimension of science, which has always existed, but whose definitions and significance have increased enormously . . . since the post-1950s."[61] David Hess concurs, arguing that in the present, "we have an emerging system of the 'public shaping of science,' in which there is both greater agency of social movement/lay advocacy organizations and greater recognition of the legitimacy of that agency."[62]

Scientists have responded to this situation by connecting public political issues and the specialized knowledge and techniques that they use in their work. The attention to the influence of the market on professions may have inhibited us from seeing this shift, and few scholars of social movements have studied this process. These scientists may not call their activities "activism" or say that they are part of a "social movement," but scientists in fields ranging from wildlife ecology to "green chemistry" to immunology organize their professional lives around finding solutions to problems that do more than simply make greater profits or bring them professional glory. Some do so through "participatory" research with groups who are or could be affected by their work; others hope that their work will find its way to those who can best implement it.[63] The debates about how scientists ought to link their work to public political questions have thus served as an opportunity for some scientists to develop new and intellectually engaging research projects. In doing so, they are participating in raising another source of their social status: resolving problems in order to make life happier, healthier, and more sustainable.

There are legitimate reasons to worry that as the values of nonscientists are taken into account, scientific ideas may be distorted, even willfully, in the service of political argument. There is no way to prevent people from using ideas in whichever way they choose. But it is possible to help nonscientists to understand that even though values and politics shape choices of ideas and methods, and can shape interpretations, not all scientific claims are equally plausible and reliable. Without ways to understand the power and limitations of the techniques that scientists use—to understand that science is both fallible and shaped by politics *and* produces reliable truths about nature—there is no way for nonscientists to assess the validity of scientific information that flows like a river into people's lives. At the same time, inventing ways to educate the public or have them "appreciate" science will do no good unless we encourage scientists to continue along the path many other scientists have taken and are taking: to treat seriously the moral and political dimensions of science and to learn from nonscientists. As the scientists I have studied in this book, and the varied ways that scientists engage public political issues today, have shown, there is no one way to do this. The time has perhaps passed, too, when any single method would be convincing or useful. What we need to encourage is the proliferation of ways that scientists and nonscientists can learn from each other.

Notes

CHAPTER 1: INTRODUCTION

1. "Men of the Year," *Time*, January 22, 1960, p. 40.
2. "The Scientists' Dilemma," *The Nation*, January 18, 1971, p. 69.
3. Philip Abelson, "Science and Immediate Social Goals," *Science*, August 21, 1970, p. 1.
4. Max Weber, "Science as a Vocation," in *From Max Weber: Essays in Sociology*, ed. Hans Gerth and C. Wright Mills (New York: Oxford University Press, 1946), p. 140.
5. The subject of this book is the contestation over what science is. Science is considered to be simultaneously a body of knowledge, a group of people, and the means by which knowledge is acquired and disseminated. The body of knowledge includes conventions for how to find out about nature, and statements and theories about nature. The means include instrumentation, conventions for sharing knowledge with scientists and nonscientists, processes of inclusion and exclusion of those who make claims in the name of science, and meetings, among many other things. Scientists are those people who are deemed credible speakers about knowledge by scientists who are already members of the scientific community. None of these features of science was spared scrutiny by scientists or other critics during the period I consider in this book. Central to my argument is the idea that the meaning of science is far more contested today than it was in 1945. At the most general level, in 1945 science could be considered systematized knowledge about nature derived from observation or theory in order to understand the nature and principles of what is being studied. Although that definition provides the bare bones of science, it does not include several other characteristics attributed to science in 1945, including autonomy from nonscientific values and interests, and a source of benefits for all people. These should not be considered distinct from the process of knowing about nature. By politics, I mean the set of formal and informal relationships by which power to rule or dominate others is distributed in a society.
6. Kristen Luker, *Abortion and the Politics of Motherhood* (Berkeley and Los Angeles: University of California Press, 1984).
7. Lily Hoffman, *The Politics of Knowledge: Activist Movements in Medicine and Planning* (Albany: State University of New York Press, 1989); Scott Frickel, *Chemical Consequences: Environmental Mutagens, Scientist Activism, and the Rise of Genetic Toxicology* (New Brunswick, NJ: Rutgers University Press, 2004).
8. Christian Smith, *The Emergence of Liberation Theology: Radical Religion and Social Movement Theory* (Chicago: University of Chicago Press, 1991).
9. Andrew Abbott, *The System of Professions* (Chicago: University of Chicago Press, 1982); Thomas F. Gieryn, George M. Bevins, and Stephen C. Zehr, "Professionalization of American Scientists: Public Science in Creation/Evolution Trials,"

American Sociological Review 50 (1985): 392–409; Magali Sarfatti Larson, *The Rise of Professionalism* (Berkeley and Los Angeles: University of California Press, 1977); Eliot Freidson, *Professionalism Reborn: Theory, Prophecy, and Policy* (Chicago: University of Chicago Press, 1994); Pierre Bourdieu, "The Specificity of the Scientific Field and the Social Conditions of the Progress of Reason," *Social Science Information* 14 (1975): 19–47; Steven G. Brint, *In an Age of Experts: The Changing Role of Professionals in Politics* (Princeton, NJ: Princeton University Press, 1994).

10. Elizabeth Hodes, "Precedents for Social Responsibility among Scientists: The American Association of Scientific Workers and the Federation of American Scientists, 1938–1948" (Ph.D. dissertation, University of California–Santa Barbara, 1982); Peter J. Kuznick, *Beyond the Laboratory: Scientists as Political Activists in 1930s America* (Chicago: University of Chicago Press, 1987); Edwin T. Layton Jr., *The Revolt of the Engineers: Social Responsibility and the American Engineering Profession* (Cleveland: Press of Case Western Reserve University, 1971).

11. Daniel Lee Kleinman, *Politics on the Endless Frontier: Postwar Research Policy in the United States* (Durham, NC: Duke University Press, 1995); Daniel J. Kevles, *The Physicists: The History of a Scientific Community in Modern America* (Cambridge, MA: Harvard University Press, 1995), pp. 324–366; Jessica Wang, *American Scientists in an Age of Anxiety: Scientists, Anticommunism, and the Cold War* (Chapel Hill: University of North Carolina Press, 1999); Peter Galison and Bruce Hevly, eds., *Big Science: The Growth of Large-Scale Research* (Stanford, CA: Stanford University Press, 1992); Richard G. Hewlett and Oscar E. Anderson, *A History of the United States Atomic Energy Commission* (University Park: Pennsylvania State University Press, 1962–1969); Alice Kimball Smith, *A Peril and a Hope: The Scientists' Movement in America, 1945–1947* (Chicago: University of Chicago Press, 1965); Daniel Lee Kleinman and Mark Solovey, "Hot Science/Cold War: The National Science Foundation after World War II," *Radical History Review* 63 (1995): 110–139; Jessica Wang, "Liberals, the Progressive Left, and the Political Economy of Postwar American Science: The National Science Foundation Debate Revisited," *Historical Studies in the Physical and Biological Sciences* 26 (1995): 139–166.

12. Kuznick, *Beyond the Laboratory*, p. 87; 1970: U.S. Bureau of the Census, "Federal Funds for Research and Development, by Agency: 1947–1970," series W126–143, in *Historical Statistics of the United States, Colonial Times to 1970, Bicentennial Edition* (Washington, DC: GPO, 1975), 2:966. In 1970 dollars.

13. Bruce L. R. Smith, *American Science Policy since World War II* (Washington, DC: Brookings Institution Press, 1990); J. Stefan Dupré and Sanford A. Lakoff, *Science and the Nation: Policy and Politics* (Englewood Cliffs, NJ: Prentice-Hall, 1962), pp. 11–15 and chapter 2; Robert Gilpin and Christopher Wright, eds., *Scientists and National Policy-Making* (New York: Columbia University Press, 1964); Richard C. Atkinson, "Science Advice at the Cabinet Level," in *Science and Technology Advice to the President, Congress, and the Judiciary,* ed. William T. Golden (New York: Pergamon Books, 1993), pp. 11–15; Spencer Klaw, *New Brahmins: Scientific Life in America* (New York: William Morrow, 1968); Michael D. Reagan, *Science and the Federal Patron* (New York: Oxford Univer-

sity Press, 1969); Don K. Price, *The Scientific Estate* (Cambridge, MA: Harvard University Press, 1965).

14. Jessica Wang, "Scientists and the Problem of the Public in Cold War America, 1945–1960," *Osiris* 17 (2002): 323–347; David A. Hollinger, "The Defense of Democracy and Robert K. Merton's Formulation of the Scientific Ethos," in *Science, Jews, and Secular Culture: Studies in Mid-Twentieth-Century Intellectual History*, ed. David A. Hollinger (Princeton, NJ: Princeton University Press, 1996), pp. 80–96; Paul Boyer, *By the Bomb's Early Light: American Thought and Culture at the Dawn of the Atomic Age* (Chapel Hill: University of North Carolina Press, 1994); Jessica Wang, "Merton's Shadow: Perspectives on Science and Democracy since 1940," *Historical Studies in the Physical and Biological Sciences* 30 (1999): 279–306; Joseph Ben-David, *The Scientist's Role in Society: A Comparative Study* (Englewood Cliffs, NJ: Prentice-Hall, 1971); Edward Shils, "The Autonomy of Science," in *The Torment of Secrecy: The Background and Consequences of American Security Policies* (Chicago: Ivan R. Dee, 1956), chapter 7.

15. Bruce V. Lewenstein, "The Meaning of 'Public Understanding of Science' in the United States after World War II," *Public Understanding of Science* 1 (1992): 47–48; Piel and Flanagan quoted in Bruce V. Lewenstein, "Magazine Publishing and Popular Science after World War II," *American Journalism* 6 (1989): 220.

16. Wang, *American Scientists in an Age of Anxiety;* Walter Gellhorn, *Security, Loyalty, Science* (Ithaca, NY: Cornell University Press, 1950).

17. Michael P. Young, *Bearing Witness against Sin: The Evangelical Birth of the American Social Movement* (Chicago: University of Chicago Press, 2006).

18. These three perspectives do not exhaust the traditions and ideas on which criticisms of science and scientists were based. Romanticism, for example, played a critical role not only in challenging the uses of science, scientists' responsibility, and issues concerning democracy, but in critiquing instrumental rationality itself. Western culture, argued some intellectuals and members of the counterculture in the 1950s and 1960s, was based on rationalist forms of thinking that inhibited people's ability to undertake a full range of sensual experiences and had nearly eliminated an attitude of awe and enchantment toward nature and the self. The result was the wholesale destruction of nature and of human experience. The application of instrumental rationality to political debate and its use as a means of organizing experience led to what many people in the romantic tradition, particularly members of the counterculture, considered to be utterly insane practices of mutually assured destruction and a refusal to consider the moral bases of political arguments and claims.

19. I have wrestled with the terms *unbinding* and *unbounding*—when to use them and their differing implications, despite a basic similarity. *Unbinding* suggests that the tie binding science and authority is loosened. In the story I tell here, this is an important shift: scientists' capacity to legitimate the decisions of government and to represent a system of universalism and ability to generate "progress" became weaker throughout the mid-twentieth century. *Unbounding* indicates that scientific authority is transgressing boundaries beyond scientists. Scientists were never in an isolated network, but the concept of "unbounding" is consistent with the scholarly literature on "boundary work" and "boundaries," which emphasizes the ways in which systems of power are demarcated. Unbounding, as an action,

suggests that unintended things happen and surprise people—a sense that is not normally captured in the analysis of boundaries, but which is essential to understanding the fate of the authority of scientists. Both terms and their implications have, I believe, relevance to the story I tell here. I have tried to use one or the other of the terms to be consistent with the analytic point I am making at any given moment. That social life is a complex interplay of active processes, unintended consequences, and structuring forces is nicely revealed in this ambiguity.

20. Joseph R. Gusfield, *Symbolic Crusade: Status Politics and the American Temperance Movement* (Urbana: University of Illinois Press, 1963); James M. Jasper, *The Art of Moral Protest: Culture, Biography, and Creativity in Social Movements* (Chicago: University of Chicago Press, 1997); Luker, *Abortion and the Politics of Motherhood;* Nicola Beisel, "Constructing a Shifting Moral Boundary: Literature and Obscenity in Nineteenth-Century America," in *Cultivating Differences: Symbolic Boundaries and the Making of Inequality,* ed. Michèle Lamont and Marcel Fournier (Chicago: University of Chicago Press, 1992), pp. 104–128; Sharon Erikson Nepstad, *Convictions of the Soul: Religion, Culture, and Agency in the U.S. Central America Solidarity Movement* (New York: Oxford University Press, 2004).

21. Ben-David, *The Scientist's Role in Society;* Ralph S. Bates, *Scientific Societies in the United States* (Cambridge: Cambridge University Press, 1965); George H. Daniels, ed., *Nineteenth-Century American Science* (Evanston, IL: Northwestern University Press, 1972).

22. Thomas F. Gieryn, "Boundary-Work and the Demarcation of Science from Non-Science: Strains and Interests in Professional Ideologies of Scientists," *American Sociological Review* 48 (1983): 781–795.

23. Daniel Lee Kleinman and Abby J. Kinchy, "Boundaries in Science Policy Making: Bovine Growth Hormone in the European Union," *Sociological Quarterly* 44 (2003): 577–595; Emanuel Gaziano, "Ecological Metaphors as Scientific Boundary Work," *American Journal of Sociology* 101 (1996): 874–907; David Indyk and David Rier, "Grassroots AIDS Knowledge: Implications for the Boundaries of Science and Collective Action," *Knowledge* 15 (1993): 3–43; David H. Guston, "Stabilizing the Boundary between US Politics and Science: The Role of the Office of Technology Transfer as a Boundary Organization," *Social Studies of Science* 29 (1999): 87–111; Anne Kerr, Sarah Cunningham-Burley, and Amanda Amos, "The New Genetics: Professionals' Discursive Boundaries." *Sociological Review* 45 (1997): 279–303; Edmund Ramsden, "Carving Up Population Science: Eugenics, Demography and the Controversy over the 'Biological Law' of Population Growth," *Social Studies of Science* 32 (2002): 857–899.

24. Thomas F. Gieryn, *Cultural Boundaries of Science: Credibility on the Line* (Chicago: University of Chicago Press, 1999), p. 24.

25. Bourdieu, "The Specificity of the Scientific Field and the Social Conditions of the Progress of Reason"; Bruno Latour and Steve Woolgar, *Laboratory Life: The Construction of Scientific Facts* (Princeton, NJ: Princeton University Press, 1986), pp. 197–198.

26. Scott Frickel, "Building an Interdiscipline: Collective Action Framing and the Rise of Genetic Toxicology," *Social Problems* 51 (2004): 269–287; Edward J. Woodhouse, "Change of State? The Greening of Chemistry," in *Synthetic Planet:*

Chemical Products in Modern Life, ed. Monica J. Casper (New York: Routledge, 2003), pp. 177–193; Steven Epstein, *Impure Science: AIDS, Activism, and the Politics of Knowledge* (Berkeley and Los Angeles: University of California Press, 1996). Although I coined the concept of activist scientist in 1996 and continue to believe it to be analytically useful, it is also the case that many scientists whose work is explicitly concerned with changing power relations, whether in terms of what does and does not get made or in terms of solving a specific problem, choose to see themselves as "apolitical." There are advantages to raising questions about why.

27. This perspective owes much to the institutionalist perspective in the sociology of organizations, in which choices are made based on the limited range of norms, stories, and rules that are available to people (John Meyer and Brian Rowan, "Institutionalized Organizations: Formal Structure as Myth and Ceremony," *American Journal of Sociology* 83 (1977): 340–363; Paul J. DiMaggio and Walter W. Powell, "The Iron Cage Revisited: Institutional Isomorphism and Collective Rationality in Organizational Fields," *American Sociological Review* 48 (1983): 147–160). At the same time, it differs from them in emphasizing change, not stasis, in organizations

28. Marianne Gosztonyi Ainley, "The Contribution of the Amateur to North American Ornithology: A Historical Perspective," *Living Bird* 18 (1979/80): 161–177; Phil Brown, "Popular Epidemiology Revisited," *Current Sociology* 45 (1997): 137–156; Richard Sclove, "Research by the People, for the People," *Futures* 29 (1997): 541–551; Frank Fischer, *Citizens, Experts, and the Environment: The Politics of Local Knowledge* (Durham, NC: Duke University Press, 2000); Robert Futrell, "Technical Adversarialism and Participatory Collaboration in the U.S. Chemical Weapons Disposal Program," *Science, Technology, and Human Values* 28 (2003): 451–482; Meredith Minkler, Angela Glover Blackwell, Mildred Thompson, and Heather Tamir, "Community-Based Participatory Research: Implications for Public Health Funding," *American Journal of Public Health* 8 (2003): 1210–1214; Frank N. Laird, "Participatory Analysis, Democracy, and Technical Decision Making," *Science, Technology, and Human Values* 18 (1993): 341–361; David H. Guston and Daniel Sarewitz, "Real-Time Technology Assessment," *Technology in Society* 24 (2002): 93–109; Alan Irwin, *Citizen Science: A Study of People, Expertise, and Sustainable Development* (New York: Routledge, 1995).

29. Young, *Bearing Witness against Sin;* Barbara Epstein, *Political Protest and Cultural Revolution: Nonviolent Direct Action in the 1970s and 1980s* (Berkeley and Los Angeles: University of California Press, 1991); Christian Smith, ed., *Disruptive Religion: The Force of Faith in Social Movement Activism* (New York: Routledge, 1999); Doug Rossinow, *The Politics of Authenticity: Liberalism, Christianity, and the New Left in America* (New York: Columbia University Press, 1998); James J. Farrell, *The Spirit of the Sixties: The Making of Postwar Radicalism* (New York: Routledge, 1997).

30. Scott Frickel and Kelly Moore, "Prospects and Challenges for a New Political Sociology of Science," in *The New Political Sociology of Science: Institutions, Networks, and Power*, ed. Scott Frickel and Kelly Moore (Madison: University of Wisconsin Press, 2006), pp. 3–34.

31. Most leading scholars of states and social movements, including Charles Tilly, Doug McAdam, Sidney Tarrow, Elisabeth S. Clemens, Edwin Amenta, Theda Skocpol, and Jeff Goodwin, assume that states are the targets of social movements. They also share a concern with the effects of state centralization and capacities for repression in explaining types and frequency of protest and its timing. Yet even if states are understood by actors to be the source of grievances, activists may not necessarily target the state, but instead seek to influence other actors whose behavior could affect the state. Moreover, most of these writers assume that states are well-bounded entities. In the post–World War II American state, such boundaries were not entirely rigid. Scientists and other professionals who worked under contract with the state were simultaneously part of the state and also independent of it. This gray area makes it more difficult to conceive of the state as an entity entirely separate from challengers. See Edwin Amenta et al., "Challengers and States: Toward a Political Sociology of Social Movements," *Research in Political Sociology* 10 (2002): 47–83; Elisabeth S. Clemens, *The People's Lobby: Organizational Innovation and the Rise of Interest Group Politics in the United States, 1890–1925* (Chicago: University of Chicago Press, 1997); Jeff Goodwin, *No Other Way Out: States and Social Revolutions* (New York: Cambridge University Press, 2001); Doug McAdam, Sidney Tarrow, and Charles Tilly, *Dynamics of Contention* (New York: Cambridge University Press, 2001); Theda Skocpol, *Protecting Soldiers and Mothers: The Political Origins of Social Policy in the United States* (Cambridge, MA: Harvard University Press, 1992).

32. Patrick Carroll, *Science, Culture, and Modern State Formation* (Berkeley and Los Angeles: University of California Press, 2006).

33. Wang, *American Scientists in an Age of Anxiety;* Jennifer Earl, "Repression and the Control of Protest," *Mobilization* 11, no. 2 (2006): 129–143; Jennifer Earl, "Controlling Protest: New Directions for Research on the Social Control of Protest," *Research in Social Movements, Conflict and Change* 25 (2004): 55–83; Jennifer Earl, "Tanks, Tear Gas, and Taxes: Toward a Theory of Movement Repression," *Sociological Theory* 21 (2003): 44–68. Earl's typology of repression suggests that the visibility of the repression matters in its effects on movements. Elaborating on her claim, in this book I demonstrate that the red-baiting that was commonplace in the 1950s, expressed by state and nonstate groups and individuals in both visible and "invisible" ways, had a clear effect of "depoliticizing" the political claims that actors made. See also Doug McAdam, *Political Process and the Development of Black Insurgency, 1930–1970* (Chicago: University of Chicago Press, 1982).

34. Christian Davenport, "Introduction: Repression and Mobilization: Insights from Political Science and Sociology," in *Repression and Mobilization*, ed. Christian Davenport, Hank Johnston, and Carol Mueller (Minneapolis: University of Minnesota Press, 2004), pp. vii–xli; Gilda Zwerman and Patricia Steinhoff, "When Activists Ask for Trouble: State-Dissident Interactions and the New Left Cycle of Resistance in the United States and Japan," in *Repression and Mobilization*, pp. 87–105.

35. Elisabeth S. Clemens and Debra C. Minkoff, "Beyond the Iron Law: Rethinking the Place of Organizations in Social Movement Research," in *The Blackwell Companion to Social Movements*, ed. David A. Snow, Sarah A. Soule, and

Hanspeter Kriesi (Malden, MA: Blackwell Publishing, 2004), pp. 155–168. For critiques of this approach, see Jeff Goodwin and James M. Jasper, "Caught in a Winding, Snarling Vine: The Structural Bias of Political Process Theory," *Sociological Forum* 14 (1999): 27–54; and David A. Snow, "Social Movements as Challenges to Authority: Resistance to an Emerging Conceptual Hegemony," *Research in Social Movements, Conflict and Change* 25 (2004): 3–25.

36. Kelly Moore and Nicole Hala, "Organizing Identity: The Creation of Science for the People," *Research in the Sociology of Organizations* 19 (2002): 309–355.

37. Elisabeth S. Clemens, "Organizational Repertoires and Institutional Change: Women's Groups and the Transformation of U.S. Politics, 1890–1920," *American Journal of Sociology* 98 (1993): 755–798.

38. David A. Snow, "Social Movements," in *Handbook of Symbolic Interactionism*, ed. Larry T. Reynolds and Nancy J. Herman-Kinney (Walnut Creek, CA: Altamira Press, 2003), pp. 811–834; Francesca Polletta, *It Was Like a Fever: Storytelling in Protest and Politics* (Chicago: University of Chicago Press, 2006); Maren Klawiter, "Racing for the Cure, Walking Women, and Toxic Touring: Mapping Cultures of Action within the Bay Area Terrain of Breast Cancer," *Social Problems* 46 (1999): 104–126; Ann Mische, *Partisan Publics: Activist Trajectories and Communicative Styles in Brazilian Youth Politics, 1977–1997* (Princeton, NJ: Princeton University Press, 2006); Ralph H. Turner and Lewis M. Killian, *Collective Behavior*, 3rd edition (Englewood Cliffs, NJ: Prentice-Hall, 1987), pp. 7–8. The analysis here is based on a psychology of *bounded rationality*, a term coined by Herbert Simon to refer to the inability of any individual or group to make the optimal decision, given that it does not have access to perfect information; see Herbert A. Simon, "Theories of Decision-Making in Economics and Behavioral Science," *American Economic Review* 49 (1959): 253–283. Most analysts of organizing and social movements are concerned with labor or community organizing, which are characterized by the persuasive work designed to encourage others to join an existing group or union. This activity comes after the organizing process to which I refer.

39. William A. Gamson, "Goffman's Legacy to Political Sociology," *Theory and Society* 14 (1985): 605–622; Erving Goffman, *Encounters: Two Studies in the Sociology of Interaction* (1961; reprint, London: Allen Lane, 1972).

40. Jasper, *The Art of Moral Protest*, p. 65.

41. Paul J. DiMaggio and Walter W. Powell, eds., *The New Institutionalism in Organizational Analysis* (Chicago: University of Chicago Press, 1991); Elisabeth S. Clemens and James M. Cook, "Politics and Institutionalism: Explaining Durability and Change," *Annual Review of Sociology* 25 (1999): 441–466; Clemens, *The People's Lobby*; Paul M. Hirsch and Michael Lounsbury, "Ending the Family Quarrel: Towards a Reconciliation of 'Old' and 'New' Institutionalism," *American Behavioral Scientist* 40 (1997): 406–418; Michael Lounsbury and Marc J. Ventresca, "'Social Structure and Organizations' Revisited," *Research in the Sociology of Organizations* 19 (2002): 3–38; Michael Lounsbury, "Institutional Sources of Practice Variation: Staffing College and University Recycling Programs," *Administrative Science Quarterly* 46 (2001): 29–56.

42. On the issue of "unbounding," see Peter Galison and David J. Stump, eds., *The Disunity of Science: Boundaries, Contexts, and Power* (Stanford, CA: Stanford University Press, 1996).

43. Smith, *A Peril and a Hope;* Donald A. Strickland, *Scientists in Politics: The Atomic Scientists Movement, 1945–46* (Lafayette, IN: Purdue University Studies, 1968).

CHAPTER 2: THE EXPANSION AND CRITIQUES
OF SCIENCE-MILITARY TIES, 1945–1970

1. Daniel J. Kevles, *The Physicists: The History of a Scientific Community in Modern America* (Cambridge, MA: Harvard University Press, 1995), pp. 324–348 ($500M on p. 341); Lawrence Badash, *Scientists and the Development of Nuclear Weapons: From Fission to the Limited Test Ban Treaty, 1939–1963* (Atlantic Highlands, NJ: Humanities Press, 1993).

2. Steven Epstein, "Sexualizing Governance and Medicalizing Identities: The Emergence of 'State-Centered' LGBT Health Politics in the United States," *Sexualities* 6 (2003): 131–171; Patrick Carroll, "Science, Power, Bodies: The Mobilization of Nature as State Formation," *Journal of Historical Sociology* 9 (1996): 139–167; Chandra Mukerji, *A Fragile Power: Scientists and the State* (Princeton, NJ: Princeton University Press, 1989); Adele E. Clarke et al., "Biomedicalization: Technoscientific Transformations of Health, Illness, and U.S. Biomedicine," *American Sociological Review* 68 (2003): 161–194; A. Hunter Dupree, *Science in the Federal Government: A History of Policies and Activities to 1940* (Cambridge, MA: Belknap Press of Harvard University Press, 1957); Douglas R. Weiner, *Models of Nature: Ecology, Conservation, and Cultural Revolution in Soviet Russia* (Pittsburgh, PA: University of Pittsburgh Press, 2000); Chandra Mukerji, *Territorial Ambitions and the Gardens of Versailles* (New York: Cambridge University Press, 1997).

3. Daniel Lee Kleinman, *Politics on the Endless Frontier: Postwar Research Policy in the United States* (Durham, NC: Duke University Press, 1995); Daniel J. Kevles, "Scientists, the Military, and the Control of Postwar Defense Research: The Case of the Research Board for National Security, 1944–46," *Technology and Culture* 16 (1975): 20–47.

4. Alice Kimball Smith, *A Peril and a Hope: The Scientists' Movement in America, 1945–1947* (Chicago: University of Chicago Press, 1965); Paul Boyer, *By the Bomb's Early Light: American Thought and Culture at the Dawn of the Atomic Age* (Chapel Hill: University of North Carolina Press, 1994), pp. 33–46.

5. Nathan Reingold, "Vannevar Bush's New Deal for Research: Or the Triumph of the Old Order," *Historical Studies in the Physical and Biological Sciences* 17 (1987): 299–344; Michael Aaron Dennis, "Reconstructing Sociotechnical Order: Vannevar Bush and U.S. Science Policy," in *States of Knowledge: The Coproduction of Science and Social Order*, ed. Sheila Jasanoff (London: Routledge, 2004), pp. 225–253; Kleinman, *Politics on the Endless Frontier.*

6. Kevles, *The Physicists*, p. 352.

7. U.S. Bureau of the Census, "Federal Funds for Research and Development, by Agency: 1947–1970," series W126–143, in *Historical Statistics of the United States, Colonial Times to 1970, Bicentennial Edition* (Washington, DC: GPO, 1975), 2:966.

8. J. Stefan Dupré and Sanford A. Lakoff, *Science and the Nation: Policy and Politics* (Englewood Cliffs, NJ: Prentice-Hall, 1962), p. 34.

9. House Committee on Science and Technology, *A History of Science Policy in the United States, 1940–1985*, Science Study Background Report no. 1, 99th Congress, 2nd session (Washington, DC: GPO, 1986), p. 38.

10. *Federal Funds for Research and Development: Detailed Historical Tables* (Bethesda, MD: Prepared for the National Science Foundation, Division of Science Resource Studies, by Quantum Research Associates, 1993); National Science Foundation, *Federal Funds for Research, Development and Other Scientific Activities*, vols. 1–14 (Washington, DC: National Science Foundation, 1952/1953–1963/64).

11. National Science Foundation, Table B-1, "Transfers of Funds Expended Annually for Performance of Research and Development by Sector, Distributed by Source, 1953–1977," in *National Patterns of R&D Resources: Funds and Manpower in the United States, 1953–1977* (Washington, DC: GPO, 1977), p. 22.

12. Roger Geiger, *Research and Relevant Knowledge: American Research Universities since World War II* (New York: Oxford University Press, 1993).

13. National Science Foundation, "Federal Contract Research Centers," in *Federal Funds for Science* (Washington, DC: GPO, 1962), pp. 71–72; Office of Technology Assessment, Congress of the United States, *A History of the Department of Defense Federally Funded Research and Development Centers* (Washington, DC: GPO, 1995).

14. National Science Foundation, *Federal Support for Academic Science and Other Educational Activities in Universities and Colleges, Fiscal Year 1963–66* (Washington, DC: National Science Foundation, 1967), pp. 36–42.

15. Quoted in Roger D. Launius, "Sputnik and Its Repercussions: A Historical Catalyst," *Aerospace Historian* 17 (1970): 89.

16. Joseph Turner, "Meeting the Challenge," *Science*, October 25, 1957, p. 22.

17. National Science Foundation, *Federal Organization for Scientific Activities* (Washington, DC: GPO, 1962), p. 6.

18. Asif A. Siddiqi, "Korolev, Sputnik, and the International Geophysical Year," in *Reconsidering Sputnik: Forty Years since the Soviet Satellite*, ed. Roger D. Launius, John M. Logsdon, and Robert W. Smith (London: Routledge, 2002), pp. 43–72; Dwayne A. Day, "Cover Stories and Hidden Agendas: Early American Space and National Security Policy," in *Reconsidering Sputnik*, pp. 161–196.

19. U.S. Department of Defense, *The Advanced Research Projects Agency, 1958–1974* (Washington, DC: Richard J. Barber Associates, December 1975).

20. Kenneth J. Heineman, *Campus Wars: The Peace Movement at American State Universities in the Vietnam Era* (New York: New York University Press, 1993), p. 17.

21. Heineman, *Campus Wars*.

22. Ann Finkbinder, *The Jasons: The Secret History of Science's Postwar Elite* (New York: Viking, 2006), p. 23. Finkbinder's book provides an excellent account of the clubbiness of the extraordinarily elite group of scientists who were members of the first Jason group, and the nearly unlimited power they had to decide which weapons would be developed.

23. Ibid., pp. 23, 27.

24. "Findings and Declaration of Policy," in *National Defense Education Act*, 85th Congress, 2nd session (Washington, DC: GPO, 1958), p. 260.

25. House Committee on Armed Services, *Hearings on Investigation of National Defense Establishment, Report on the National Defense Education Act, Fiscal Year Ending June 30, 1960*, 86th Congress, 2nd session (Washington, DC: GPO, 1960), p. 16.

26. Barbara Barksdale Clowse, "Education as an Instrument of National Security: The Cold War Campaign to 'Beat the Russians' from Sputnik to the National Defense Education Act of 1958" (Ph.D. dissertation, University of North Carolina–Chapel Hill, 1977).

27. House Committee on Armed Services, *Hearings on Investigation of National Defense Establishment, Report on the National Defense Education Act, Fiscal Year Ending June 30, 1960*, p. 16.

28. James L. Weston, Testimony before the U.S. House of Representatives, April 24, 1958, in *Hearings on the National Defense Education Act*, 85th Congress, 2nd session (Washington, DC: GPO, 1958), p. 23.

29. David Kaiser, "Cold War Requisitions, Scientific Manpower, and the Production of American Physicists after World War II," *Historical Studies in the Physical and Biological Sciences* 33 (2002): 133; see also Mukerji, *A Fragile Power*.

30. ROTC was part of the National Defense Act of 1916, which established reserve officer training in colleges and universities, and military academies such as the Citadel and Norwich. Eugene M. Lyons and John W. Masland, *Education and Military Leadership: A Study of the R.O.T.C.* (Princeton, NJ: Princeton University Press, 1959).

31. Quoted in Finkbinder, *The Jasons*, p. 25.

32. Stuart W. Leslie, *The Cold War and American Science: The Military-Industrial-Academic Complex at MIT and Stanford* (New York: Columbia University Press, 1993), p. 9.

33. Richard T. Sylves, *Nuclear Oracles: A Political History of the General Advisory Committee of the Atomic Energy Commission, 1947–1977* (Ames: University of Iowa Press, 1987); George T. Mazuzan and J. Samuel Walker, *Controlling the Atom: The Beginnings of Nuclear Regulation, 1946–1962* (Berkeley and Los Angeles: University of California Press, 1985); Richard G. Hewlett and Oscar E. Anderson, *A History of the United States Atomic Energy Commission* (University Park: Pennsylvania State University Press, 1963–1969).

34. Bruce L. R. Smith, *The Advisers: Scientists in the Policy Process* (Washington, DC: Brookings Institution Press, 1992), pp. 48–54.

35. Rexmond C. Cochrane, *The National Academy of Sciences: The First Hundred Years 1863–1963* (Washington, DC: National Academy of Sciences, 1978), pp. 531–558.

36. Kevles, *The Physicists*, p. 377.

37. Smith, *The Advisers*, pp. 48–52.

38. Arthur S. Fleming, "The Philosophy and Objectives of the National Defense Education Act," *Annals of the American Academy of Political and Social Science* 327 (1962): 132–138.

39. Richard C. Atkinson, "Science Advice at the Cabinet Level," in *Science and Technology Advice to the President, Congress, and Judiciary*, ed. William T. Golden (New Brunswick, NJ: Transaction Publishers, 1995), pp. 11–15.

40. Meg Greenfield, "Science Goes to Washington," *Science*, October 18, 1963, pp. 363–364.

41. Kevles, *The Physicists*, p. 394.

42. Joseph Turner, "An Academic Question," *Science*, March 8, 1957, p. 425.

43. Thomas F. Gieryn, *Cultural Boundaries of Science: Credibility on the Line* (Chicago: University of Chicago Press, 1999), pp. 37–64.

44. Bruce V. Lewenstein, "The Meaning of 'Public Understanding of Science' in the United States after World War II," *Public Understanding of Science* 1 (1992): 48.

45. In his study of the role of scientists in the development of science curricula in the 1950s, John L. Rudolph makes a related point. He demonstrates that scientists separated "pure science" from technology and from the humanities. Their goal, among other things, was to inculcate in students an appreciation of rational empiricism, in the hope that this would lead them to support scientific endeavors in the future. John L. Rudolph, *Scientists in the Classroom: The Cold War Reconstruction of American Science Education* (New York: Palgrave, 2002).

46. James C. Wood, "Scientists and Politics: The Rise of an Apolitical Elite," in *Scientists and National Policy-Making*, ed. Robert Gilpin and Christopher Wright (New York: Columbia University Press, 1964), pp. 41–72.

47. Kevles, *The Physicists*, p. 386.

48. House Committee on Science and Technology, *A History of Science Policy in the United States, 1940–1985*, p. 50.

49. Bruce L. R. Smith, *American Science Policy since World War II* (Washington, DC: Brookings Institution Press, 1990), p. 74.

50. U.S. Department of Defense, *Project THEMIS* (Washington, DC: Office of the Director of Defense Research and Engineering, 1967).

51. Lee A. Du Bridge, "Twenty-Five Years of the National Science Foundation," *Proceedings of the American Philosophical Society* 121 (1977): 191–194.

52. Smith, *American Science Policy since World War II*, p. 80.

53. Kevles, *The Physicists*, p. 387.

54. House Committee on Science and Technology, *A History of Science Policy in the United States, 1940–1985*, p. 42.

55. *National Aeronautics and Space Act of 1958*, Public Law #85–568, 72 Stat., 426, p. 1. Signed by the president on July 29, 1958.

56. Kevles, *The Physicists*, p. 385.

57. National Science Foundation, *Federal Funds for Research, Development and Other Scientific Activities* (Washington, DC: GPO, 1969), p. 12.

58. Cochrane, *The National Academy of Sciences*, pp. 573–575.

59. House Committee on Science and Technology, *A History of Science Policy in the United States, 1940–1985*, p. 65.

60. Sam Archibald, "The Early Years of the Freedom of Information Act, 1955–1975," *PS: Political Science and Politics* 26 (1993): 726–731.

61. Robert N. Mayer, *The Consumer Movement: Guardians of the Marketplace* (Boston: Twayne, 1989); Paul J. Culhane, "Federal Organizational Change in Response to Environmentalism," *Humboldt Journal of Social Relations* 2 (1974): 31–44.

62. National Science Foundation, *Science and Engineering in Higher Education* (Washington, DC: GPO, 1978), p. 28.

63. National Research Council, *Survey of Earned Doctorates, 1950–1970* (Washington, DC: National Research Council, 1950–1970).

64. A. J. Muste, "Conscience against the Bomb," *Fellowship*, December 1945, p. 209.

65. Michael P. Young, *Bearing Witness against Sin: The Evangelical Birth of the American Social Movement* (Chicago: University of Chicago Press, 2006).

66. Peter Brock and Nigel Young, *Pacifism in the Twentieth Century* (Syracuse, NY: Syracuse University Press, 1999), pp. 246–250; Scott H. Bennett, *Radical Pacifism: The War Resisters League and Gandhian Nonviolence* (Syracuse, NY: Syracuse University Press, 2003), pp. 98–134; Allen Smith, "The Renewal Movement: The Peace Testimony and Modern Quakerism," *Quaker History* 85 (1996): 1–23.

67. Reinhold Niebuhr, "The Christian Faith and World Crisis," *Christianity and Crisis* 1, no. 1 (1941): 4.

68. H. Larry Ingle, "The American Friends Service Committee, 1947–1949: The Cold War's Effect," *Peace and Change* 23 (1998): 27–48.

69. R. Allen Smith, "Mass Society and the Bomb: The Discourse of Pacifism in the 1950s," *Peace and Change* 18 (1993): 347–372.

70. War Resisters League, *History of the War Resisters League* (New York: War Resisters League, 1980); see also James J. Farrell, *The Spirit of the Sixties: The Making of Postwar Radicalism* (New York: Routledge, 1997), p. 22.

71. Sam Gottlieb, "The F.B.I. as Big Brother: Filming Polaris," *Liberation* 7 (June 1962): 9. "Yale Alumni against Polaris," 1957, "Peace Groups, Misc., 1957" box 2, folder "Speeches and Press Releases," Swarthmore College Peace Collection.

72. New England Committee for Nonviolent Action, series III, Printed Releases box 2, folder "Flyers and Bulletins (1958–62)," Swarthmore College Peace Collection.

73. Dick Bruner, "Goofmanship," *Liberation* 3 (May 1958): 9.

74. John C. Bennett, "Christian Ethics and International Affairs," *Christianity and Crisis*, August 5, 1963, reprinted in *Witness to a Generation: Significant Writings from Christianity and Crisis, 1941–1966*, ed. Wayne H. Cowan (New York: Bobbs-Merrill, 1966), p. 97.

75. Rabbi Abraham Heschel, "In Whose Name?" (1968), in "Addresses etc. by Clergy and Laity Concerned at the Feb 5 and 6 '68 Mobilization against the War in Vietnam, Washington, D.C.: Meditations, Addresses, and Prayers Offered at the Ecumenical Worship Service the Evening of Feb 5," CALC Papers, series IV, Swarthmore College Peace Collection.

76. Martin Luther King Jr., "Addresses Delivered at the Closing Session of the Mobilization Conference, Washington, D.C., February 6," February 6, 1968, CALC Papers, series IV, box 2, folder "Speeches and Press Releases," Swarthmore College Peace Collection.

77. Alan Brinkley, *Liberalism and Its Discontents* (Cambridge, MA: Harvard University Press, 1998), pp. 96–98.

78. As Shapin argues, "the public" is as much a social construct as the meaning of "science." Indeed, constructing a "public" that is not conversant with all of the languages and practices of experts is part of the process of differentiating—and connecting—science and "the public" (Steven Shapin, "Science and the Public," in *Companion to the History of Modern Science*, ed. R. C. Olby, G. N. Cantor, J.R.R. Christie, and M.J.S. Hodge [London: Routledge, 1990], pp. 990–1007). Witness, for example, the recent debates about creationism, in which "public" accounts were put on trial and found wanting. As I will argue in chapter 4, scientists who provided citizens with information about atomic weapons testing had a particular kind of public in mind, whereas the scientists I examine in chapters 3, 5, and 6 were less concerned with the notion of the "public" in their political actions.

79. Daniel Bell, *The End of Ideology: On the Exhaustion of Political Ideas in the Fifties* (New York: Free Press, 1962).

80. Jessica Wang, "Scientists and the Problem of the Public in Cold War America, 1945–1960," *Osiris* 17 (2002): 342–347.

81. David A. Hollinger, "Free Enterprise and Free Inquiry: The Emergence of Laissez-Faire Communitarianism in the Ideology of Science in the United States," in *Science, Jews, and Secular Culture: Studies in Mid-Twentieth-Century Intellectual History*, ed. David A. Hollinger (Princeton, NJ: Princeton University Press, 1996), p. 112.

82. David Hawkins, "Should the Scientist Take Part in Politics?" *New York Times Magazine*, June 16, 1946, pp. 13–46.

83. Sidney Hook, "The Scientist in Politics," *New York Times Magazine*, April 9, 1950, p. 25.

84. Herman Kahn, *On Thermonuclear War* (New York: MacMillan, 1962).

85. Walter Goldstein and S. M. Miller, "Herman Kahn: Ideologist of Military Strategy," *Dissent* 10, no. 1 (1963): 85.

86. Hans Morgenthau, "The Intellectual and Moral Dilemmas of History," *New York Review of Books* 8 (1963): 4.

87. Dwight D. Eisenhower, "Farewell Radio and Television Address to the American People, January 17, 1961," posted on www.eisenhower.archives.gov/farewell.htm (accessed March 17, 2007).

88. C. P. Snow, *The Two Cultures and the Scientific Revolution* (Cambridge: Cambridge University Press, 1959), p. 11.

89. Rachel Carson, *Silent Spring* (Boston: Houghton Mifflin, 1962), p. 6.

90. Wood, "Scientists and Politics."

91. Don K. Price, *The Scientific Estate* (Cambridge, MA: Harvard University Press, 1965).

92. Greenfield, "Science Goes to Washington," p. 363.

93. Everett Mendelsohn. "The Politics of Pessimism: Science and Technology Circa 1968," in *Technology, Pessimism, and Postmodernism*, ed. Yaron Ezrahi, Everett Mendelsohn, and Howard Segal (Amherst: University of Massachusetts Press, 1994), pp. 151–274.

94. Maurice Isserman, *If I Had a Hammer: The Death of the Old Left and the Birth of the New Left* (New York: Basic Books, 1987); Kirkpatrick Sale, *SDS* (New York: Random House, 1973).

95. C. Wright Mills, *The Power Elite* (Oxford: Oxford University Press, 1956).

96. C. Wright Mills, *The Causes of World War Three* (New York: Ballantine Books, 1958), p. 26.

97. Ibid., p. 109.

98. For an analysis of Arendt's critique of instrumental rationality, see Wang, "Scientists and the Problem of the Public in Cold War America."

99. Herbert Marcuse, *One-Dimensional Man: Studies in the Ideology of Advanced Industrial Society* (Boston: Beacon Press, 1964), p. 3.

100. Ibid., p. 251.

101. Ibid., p. 233.

102. Erich Fromm, "The Pathology of the Cold War," *Liberation* 6 (October 1, 1961): 11.

103. Boyer, *By the Bomb's Early Light*, pp. 353–367.

104. Anonymous, "Witness at the Conspiracy Trial, 1969," in *The Sixties Papers: Documents of a Rebellious Decade*, ed. Judith Clavir Albert and Stuart Albert (New York: Praeger, 1984), p. 391.

105. Quoted in Sol Stern, "A Deeper Disenchantment," *Liberation* 9 (February 1965): 25.

106. Charles DeBenedetti with Charles Chatfield, *An American Ordeal: The Antiwar Movement of the Vietnam Era* (Syracuse, NY: Syracuse University Press, 1990), pp. 95–105.

107. Barbara Ehrenreich and John Ehrenreich, *Long March, Short Spring: The Student Uprisings at Home and Abroad* (New York: Monthly Review Press, 1969), chapters 2, 4; Kevles, *The Physicists*, p. 355; Leslie, *The Cold War and American Science*, chapters 4, 5; Pennsylvania State, SUNY–Buffalo, Kent State University, Georgetown University: Heineman, *Campus Wars*, pp. 196, 214–217, 228; Columbia University: Jerry Avorn, *Up against the Ivy Wall: A History of the Columbia University Crisis*, ed. Andrew Crane, with the editors of the *Columbia University Spectator* (New York: New Atheneum, 1969); MIT and Stanford: Leslie, *The Cold War and American Science*, chapter 4; University of Chicago: Bradford Lyttle, *The Chicago Anti–Vietnam War Movement* (Chicago: Midwest Pacifist Center, 1988); University of Wisconsin: Tom Bates, *Rads: The 1970 Bombing of the Army Math Research Center and Its Aftermath* (New York: Harper Collins, 1992); Northwestern: Jack Nusan Porter, *Student Protest and the Technocratic Society: The Case of ROTC* (Chicago: Adams Press, 1973).

108. Leslie, *The Cold War and American Science*, p. 244.

109. Ibid.

110. Compiled and annotated by Sol Stern, "War Catalog of the University of Pennsylvania," *Ramparts* 5, no. 3 (1965): 32–40; the institute's brochure, 1962,

p. 34; "Campus Peace Activism, 1962–1965" file, Swarthmore College Peace Collection.

111. Letter to Grayson Kirk from Mark Rudd and Nicolas Freudenberg, March 27, 1968, Columbiana Archives, Columbia University.

112. "IDA Must Go," flyer, "Student Groups: SDS" Collection, Columbiana Archives.

113. Columbia Student Strike Committee, "Why We Strike," 1968, Columbiana Archives.

114. Bates, *Rads*, p. 139.

115. Todd Gitlin, *The Sixties: Years of Hope, Days of Rage* (New York: Bantam, 1987), p. 254; Michael Miles, "Tactics of Disruption," *New Republic*, November 1967, p. 10.

116. Sale, *SDS*, pp. 380–381.

CHAPTER 3: SCIENTISTS AS MORAL INDIVIDUALS

1. Alice K. Smith, *A Peril and a Hope: The Scientists' Movement in America, 1945–1947* (Chicago: University of Chicago Press, 1965); Daniel J. Kevles, *The Physicists: The History of a Scientific Community in Modern America* (Cambridge, MA: Harvard University Press, 1995), pp. 351–357; Paul Boyer, *By the Bomb's Early Light: American Thought and Culture at the Dawn of the Atomic Age* (Chapel Hill: University of North Carolina Press, 1994), pp. 49–64, 93–106; Donald Strickland, *Scientists in Politics: The Atomic Scientists Movement, 1945–1946* (Lafayette, IN: Purdue University Studies, 1968); Lawrence S. Wittner, *One World or None: A History of the World Nuclear Disarmament Movement through 1953* (Stanford, CA: Stanford University Press, 1997), pp. 59–66. Jessica Wang's *American Scientists in an Age of Anxiety: Scientists, Anticommunism, and the Cold War* (Chapel Hill: University of North Carolina Press, 1999) covers the movement, with an emphasis on how security and loyalty programs affected scientists' political, work, and personal lives.

2. John A. Simpson, "The Scientist as Public Educator: A Two-Year Summary," *Bulletin of the Atomic Scientists* (June 1947): 243–246; "Pittsburgh Did It—Your City Can, Too!" *Atomic Information*, March 4, 1946, p. 3; "How Can I Help to Prevent an Atomic War?" *Atomic Information*, April 22, 1946, p. 1; "Report on the Institute," *Atomic Information*, August 19, 1946, pp. 10–11; "What the League of Women Voters Has Done," *Atomic Information*, October 26, 1946, p. 9; "What AAUW Has Done," *Atomic Information*, January 20, 1947, p. 10; Atomic Scientists of Chicago, n.d., "Summary of Principal Points on Atomic Bomb," University of Chicago Archives and Special Collections, Atomic Scientists Collection, box "Atomic Scientists—Misc.," folder "Atomic Scientists of Chicago." See also mastheads in *Atomic Information*, volumes 1, 2, and 3, for listings of affiliated organizations.

3. Peter J. Kuznick, *Beyond the Laboratory: Scientists as Political Activists in 1930s America* (Chicago: University of Chicago Press, 1987); Elizabeth Hodes, "Precedents for Social Responsibility among Scientists: The American Association of Scientific Workers and the Federation of American Scientists, 1938–1948"

(Ph.D. dissertation, University of California–Santa Barbara, 1982); Stephen Duggen and Betty Drury, *The Rescue of Science and Learning* (New York: Macmillan, 1948). On mathematicians, see Nathan Reingold, "Refugee Mathematicians in the United States of America, 1933–1941: Reception and Reaction," *Annals of Science* 38 (1981): 313–338; on physicists, see Charles Weiner, "A New Site for the Seminar: The Refugees and American Physics in the Thirties," in *The Intellectual Migration: Europe and America, 1930–1960*, ed. Donald Fleming and Bernard Bailyn (Cambridge, MA: Belknap Press of Harvard University Press, 1969), pp. 190–234.

4. Philip Morrison and Robert R. Wilson, in *Bulletin of the Atomic Scientists* (March 1947): 18.

5. For analyses of the development and effects of these programs and policies, see David Caute, *The Great Fear: The Anti-Communist Purge under Truman and Eisenhower* (New York: Simon and Schuster, 1978); Athan Theoharis, *Seeds of Repression: Harry S. Truman and the Origins of McCarthyism* (New York: Quadrangle, 1971); George R. Stewart, *The Year of the Oath: The Fight for Academic Freedom at the University of California* (Garden City, NY: Doubleday, 1950); and Sandra Weinstein, *Personnel Security Programs of the Federal Government* (New York: Fund for the Republic, 1954).

6. Kevles, *The Physicists*, p. 378.

7. Wang, *American Scientists in an Age of Anxiety*; Charles Thorpe, *Oppenheimer: The Tragic Intellect* (Chicago: University of Chicago Press, 2006); Jessica Wang, "Science, Security, and the Cold War: The Case of E. U. Condon," *Isis* 83 (June 1992): 238–269; Thomas Hagen, *Force of Nature: The Life of Linus Pauling* (New York: Simon and Schuster, 1995); Martin D. Kamen, *Radiant Science, Dark Politics: A Memoir of the Nuclear Age* (Berkeley and Los Angeles: University of California Press, 1985); Harlow Shapley, *Through Rugged Ways to the Stars* (New York: Charles Scribner and Sons, 1969).

8. Wang, *American Scientists in an Age of Anxiety*.

9. A. G. Mezerik, "Scientists and the Great Debate," *New Republic*, February 2, 1951, p. 12.

10. Jo Ann Robinson, *Abraham Went Out: A Biography of A. J. Muste* (Philadelphia: Temple University Press, 1981), p. 143.

11. A. J. Muste, "An Open Letter to Dr. Einstein," *Fellowship* 12, no. 7 (May 1946): 121.

12. Boyer, *By the Bomb's Early Light*, pp. 220–222.

13. Robinson, *Abraham Went Out*, p. 143.

14. Victor Paschkis, "Double Standards," *Friends Intelligencer*, January 1947, p. 463. George W. Hartmann warned, in a similar article, "Pacifists should not repeat the error of Veblen who thought that engineers' professional interest in production was a sufficient psychological force to make them opponents of a profit-dominated system." George W. Hartmann, "Beware the Scientists Bearing Olive Branches," *Fellowship*, January 1947, p. 9.

15. Hartmann, "Beware the Scientists Bearing Olive Branches," p. 9.

16. Victor Paschkis, "Victor Starts the Society for Social Responsibility in Science," n.d., box 1, Papers of Victor Paschkis, Swarthmore College Peace Collec-

tion; Victor Paschkis, "10 Years—SSRS," presented at the 1959 annual meeting of the SSRS, Edward G. Ramberg (ER) Papers, "Correspondence," American Philosophical Society. Muste's encouragement of Paschkis was not unique. Muste worked actively to transform the religious and intellectual, as well as the scientific elite. His political strategy, as Jo Ann Robinson argues, was to organize "cells" of individuals who were "affiliated with a key social institution and committed to the way of love; then mobilizing the institutions, in turn, to convert the power centers to which they related." Jo Ann Robinson, "A. J. Muste and Ways to Peace," in *Peace Movements in America*, ed. Charles Chatfield (New York: Schocken, 1973), p. 88.

17. The participants were Henry C. Babcock, Hidden Springs, NJ; John E. Baer, Chemistry, Carleton College; Otto T. Benfey; Irving S. Bengelsdorf, Chicago; Andrew A. Benson, Berkeley, Edward K. Blum, Brooklyn; William E. Cadbury, Haverford College; R. J. Cox, Baltimore; Loring P. Crosman, Darien, CT; Nancy Cross, Staten Island; S. Leonard Dart, Swarthmore, PA; Don and Jeanette De-Vault, Chicago; Francis Evants, Ann Arbor; Marcus A. Frieder, Brooklyn; Adoph Furth, Mickelton, NJ; Nelson Fuson, Johns Hopkins; Robert L. Graham, New York; Theodore B. Hetzel, Haverford; Marion Hollingsworth, Columbus, OH; Herbert Jehle, Philadelphia; Robert Martin, Physics, University of Michigan; S. G. Mathews III, New York; A. Melvin, East Sound, WA; Franklin Miller, New Brunswick, NJ; Victor and Susanne Paschkis; Edward Ramberg, Feasterville, PA; Le Roy Schulz, Chicago; William T. Scott, Northampton, MA; Arthur M. Squires, Brooklyn; Albert B. Stewart, Antioch; Richard M. Sutton, Haverford; C. Wilbur Ufford, Haverford; Jerome Weiner, Brooklyn; Gilbert F. White, Haverford; Michael Yanowitch, Brooklyn, "Participants in Conference of Scientists at Haverford College, June 5–6, 1948," Otto T. Benfey (OTB) Papers, Haverford College.

18. Norbert Wiener, "A Scientist Rebels," *The Atlantic*, January 1947, p. 46.

19. Leo Szilard, "The Physicist Invades Politics," *Saturday Review of Literature*, May 3, 1947.

20. Quoted in Boyer, *By the Bomb's Early Light*, p. 60.

21. By the middle of the 1950s, pacifists were debating the meanings of *nonviolence*, *peace*, and *passive resistance*. For the most radical, *nonviolence* meant not only eschewing violence, but finding positive means of creating a just society. This new meaning influenced some of the younger, later members of the SSRS, as well as other younger activists in the anti–Vietnam War movement.

22. Others elected include William Scott, physics, Smith College/Brookhaven National Laboratory (membership chairman); Theodore Hetzel, engineering, Haverford (chairman of the Occupational Division); Franklin Miller Jr., physics, Kenyon College (chairman of the Educational Division); S. Leonard Dart, physics, American Viscose Corporation, and James E. Vail, former president, American Institute of Chemical Engineers (member of the council at large); Marion Hollingsworth Sr., Howard Alexander, Andrew Benson, Robert Warner, and Wesley Pendleton (Board of Appeals); and Toshiyuki Fukushima, Marion Hollinsgworth Jr., and Robert Martin (Nominating Committee), *SSRS Newsletter*, December 1949, p. 3.

23. As Elisabeth S. Clemens and Francesca Polletta have both shown, the forms that organizations take are highly dependent on the ways in which founders envision their identities. They do not in some rational-choice sense "choose" these forms from a menu, but rather create or use existing forms that are commensurate with their moral and social identities. Elisabeth S. Clemens, "Organizational Form as Frame: Collective Identity and Political Strategy in the American Labor Movement, 1880–1920," in *Comparative Perspectives on Social Movements: Political Opportunities, Mobilizing Structures, and Cultural Framings*, ed. Doug McAdam, John D. McCarthy, and Mayer N. Zald (Cambridge: Cambridge University Press, 1996), pp. 205–226; Francesca Polletta, *Freedom Is an Endless Meeting: Democracy in American Social Movements* (Chicago: University of Chicago Press, 2004).

24. Letter from Victor Paschkis to Dear Colleague, September 9, 1949, OTB Papers, Haverford College. The group's name did become a liability. The SSRS was frequently assumed to be communist because its initials were thought to be similar to U.S.S.R.

25. Letter from Victor Paschkis to Dear Colleague, September, 20, 1948, OTB Papers, Haverford College.

26. Albert Einstein, Letter to the Editor, *Science*, December 22, 1950, pp. 760–761.

27. Religious News Service, "Scientists to Bar 'Destructive' Aid,'" *New York Times*, September 18, 1949, p. 71; O. T. Benfey, "The Scientists' Conscience," *Bulletin of the Atomic Scientists* 12, no. 5 (May 1956): 177–178.

28. Arthur E. Morgan, *Social Responsibility in Science*, 1949, Pamphlet no. 1 by the SSRS, p. 11, OTB Papers, Haverford College.

29. Boyer, *By the Bomb's Early Light*, p. 349.

30. Wittner, *One World or None*, p. 42.

31. Allen Smith, "Converting America: Three Community Efforts to End the Cold War, 1956–1973" (Ph.D. dissertation, American University, 1995), p. 26.

32. SSRS Constitution, September 1949, SSRS Collection, Swarthmore College Peace Collection.

33. Victor Paschkis to Colleagues, June 20, 1949, OTB Papers, Haverford College.

34. Minutes of Council Meeting, January 30, 1955, OTB Papers, Haverford.

35. *SSRS Newsletter*, January 1953, p. 6.

36. Franklin Miller, telephone interview by the author, July 7, 2004.

37. O. Theodore Benfey, telephone interview by the author, June 11, 2004.

38. Bruce V. Lewenstein, "Shifting Science from People to Programs: AAAS in the Postwar Years," in *The Establishment of Science in America: 150 Years of the American Association for the Advancement of Science*, ed. Sally Gregory Kohlstedt, Michael M. Sokal, and Bruce V. Lewenstein (New Brunswick, NJ: Rutgers University Press, 1999), p. 112.

39. "SSRS at AAAS, Dec. 28," *SSRS Newsletter*, December 1951, p. 1.

40. Letter quoted in "AAAS Turns Down SSRS," *SSRS Newsletter*, December 1954, p. 1.

41. "Science Association Bars Pacifist Group," *New York Times*, December 14, 1954, p. 41; "'Constructive' Science Unit Is Barred from A.A.A.S.," *New York Herald Tribune*, December 14, 1954, p. 16.

42. Some of the core members of the SSRS, including Edward Ramberg, Paschkis, Herbert Jehle, and Ted Benfey, came from German Jewish families that converted to Christianity between the late nineteenth century and the 1930s.

43. *SSRS Newsletter*, November 1950, p. 2.

44. *SSRS Newsletter*, June 1952, p. 1.

45. Peter Brock and Nigel Young, *Pacifism in the Twentieth Century* (Syracuse, NY: Syracuse University Press, 1999), p. 159.

46. Kathleen Lonsdale, "The Ethical Problems of Scientists," *Bulletin of the Atomic Scientists* 7, no. 7 (1951): 204.

47. David A. Hollinger, "Jewish Intellectuals and the De-Christianization of American Public Culture in the Twentieth Century," in *Science, Jews, and Secular Culture: Studies in Mid-Twentieth-Century American Intellectual History*, ed. David A. Hollinger (Princeton, NJ: Princeton University Press, 1996), pp. 27–28.

48. Robert K. Merton, "The Normative Structure of Science," in *The Sociology of Science*, ed. Norman W. Storer (Chicago: University of Chicago Press, 1973), pp. 267–278.

49. Edward Shils, "The Autonomy of Science," in *The Torment of Secrecy: The Background and Consequences of American Security Policies* (Chicago: Ivan R. Dee, 1956), p. 177.

50. Michael Polanyi, *Science, Faith, and Society* (London: Oxford University Press, 1946).

51. David A. Hollinger, "The Defense of Democracy and Robert K. Merton's Formulation of the Scientific Ethos," in *Science, Jews, and Secular Culture: Studies in Mid-Twentieth-Century American Intellectual History*, ed. David A. Hollinger (Princeton, NJ: Princeton University Press, 1996), pp. 80–96.

52. Wang, *American Scientists in an Age of Anxiety*.

53. Max Weber, "The Social Psychology of World Religions," in *From Max Weber: Essays in Sociology*, ed. Hans Gerth and C. Wright Mills (New York: Oxford University Press, 1946), pp. 267–301.

54. William Scott, "President Scott Describes Purposes and Aims of SSRS," *SSRS Newsletter*, May 1956, pp. 2–3.

55. "Harvard Discusses Science and Society," *SSRS Newsletter*, October 1954, pp. 1–8.

56. "Council Meeting March 29, 1952." *SSRS Newsletter*, October 1952, p. 5.

57. Ibid.

58. Victor Paschkis, Letter to the Editor, *Science*, December 2, 1960; William F. Hewitt Jr., Letter to the Editor, *Science*, December 8, 1950, p. 760; O. T. Benfey, Book Review of *Science and the Social Order*, *Science*, February 13, 1952, p. 996; and O. T. Benfey, "Nuclear Tests and Ethics," *Science*, March 14, 1952, are typical of the types of letters that SSRS members wrote to science publications.

59. *SSRS Newsletter*, June 1953, pp. 1, 4.

60. *SSRS Newsletter*, January 1953, pp. 5–6; Victor Paschkis, "Approaching the Annual Meeting," *SSRS Newsletter*, September 1952, pp. 1–2.

61. *SSRS Newsletter*, June 1952, p. 1.

62. William Parke, "Tribute to Professor Emeritus Herbert Jehle," www.gwu.edu/~physics/jehle.htm (accessed July 6, 2006).

63. *SSRS Newsletter*, December 1951, p. 6.

64. Charles A. Coulson, *Science, Technology, and the Christian* (London: Abington Press, 1960).

65. "Council Briefing," *SSRS Newsletter*, September 12, 1955, p. 3.

66. Bertrand Russell, "Man's Peril," in *Portraits from Memory* (New York: Simon and Schuster, 1951), pp. 223–228.

67. It was signed by Percy W. Bridgman, Albert Einstein, Leopold Infeld, Frédéric Joliot-Curie, Herman J. Muller, Linus Pauling, Cecil F. Powell, Joseph Rotblat, Bertrand Russell, and Hideki Yukawa. Sandra Ionno Butcher, "The Origins of the Russell-Einstein Manifesto" (manuscript, Pugwash Councils on Science and World Affairs, May 2005).

68. Letter from Franklin J. Miller to SSRS Council, May 22, 1955, OTB Papers, Haverford College. On Einstein's support of the personal responsibility model, see Albert Einstein, "The Real Problem Is in the Hearts of Men," interview by Michael Amrine, *New York Times Magazine*, June 23, 1946, p. 44.

69. A. G. Mezerik, *The Pursuit of Plenty: The Story of Man's Expanding Domain* (New York: Harper, 1950).

70. Council Briefing, September 9, 1953, OTB Papers, Haverford College.

71. Council Briefing, September 12, 1955, OTB Papers, Haverford College.

72. "Another Nobel Winner Joins the SSRS," *SSRS Newsletter*, April 1956, p. 1; "SSRS Activities to Be Expanded: Council Announces Wider Program," *SSRS Newsletter*, May 1956, p. 1.

73. "Candidates Propose SSRS Tasks," *SSRS Newsletter*, July 1956, p. 1.

74. Dave Dellinger, "The Here and Now Revolution," *Liberation* 1 (1956): 7–18, quoted in Charles DeBenedetti with Charles Chatfield, *An American Ordeal: The Antiwar Movement of the Vietnam Era* (Syracuse, NY: Syracuse University Press, 1990), p. 25.

75. "SSRS Leaders Consider Effect of AAAS Stand on Science, Society," *SSRS Newsletter*, February 1957, p. 1.

76. Wittner, *One World or None*, p. 35; Joseph Rotblat, *History of the Pugwash Conferences* (London: Taylor and Francis, 1962); Joseph Rotblat, *Pugwash, the First Ten Years: History of the Conferences of Science and World Affairs* (London: Heinemann, 1967); Herbert F. York, *Making Weapons, Talking Peace: A Physicist's Odyssey from Hiroshima to Geneva* (New York: Basic Books, 1987), pp. 246–255.

77. "Are Scientists Morally Responsible for Applications of Science? CBS-TV Debates the Issue; SSRS View Gets No Support," *SSRS Newsletter*, June 1959, p. 1.

78. It is likely that the SSRS wished to avoid the appropriation of its name by communist or communist-influenced organizations. "Legion vs. SSRS," *SSRS Newsletter*, October 1957, p. 1.

79. Ibid.

80. Richard Gid Powers, *Not without Honor: The History of American Anti-communism* (New Haven, CT: Yale University Press, 1995), pp. 246–249.

81. "Legion vs. SSRS," *SSRS Newsletter*, October 1957, p. 1.

82. John D'Emilio, *Sexual Politics, Sexual Communities: The Making of a Homosexual Minority, 1940–1970* (Chicago: University of Chicago Press, 1983).

83. Victor Paschkis, "European Organizations and Communities," *SSRS Newsletter*, March 1964, pp. 1–2; April 1964, p. 2; May 1964, pp. 2–3.

84. Victor Paschkis, Report of the President [of SSRS], July 28, 1959.

85. E. Ackerman, "Social Responsibility of Scientists," *SSRS Newsletter*, July 1962, pp. 1–3.

86. Herbert Jehle, "Testimony on Fallout," *SSRS Newsletter*, October 1962, p. 3.

87. Victor Paschkis, "SSRS—10 Years," 1959, ER Papers, folder "SSRS+."

88. Ibid.

89. "SSRS Educational Division Report: 1949–1953," by Franklin Miller Jr., Annual Meeting Minutes, Antioch, OH, August 28–29, 1954; Annual Meeting Minutes, September 19–20 1964. All in OTB Papers, no folder.

90. "Dr. Bauer Addresses SSRS Annual Meeting," *SSRS Newsletter*, November 1959, p. 1.

91. J. H. Rothschild, "Germs and Gas," *Harper's*, June 1959, p. 29.

92. Editorial, *Chemical and Engineering News*, June 1959.

93. Victor Paschkis, "Foes of Germ Warfare Picket Army Test Base in Maryland," *New York Times*, January 12, 1960; Don DeVault, "Council of SSRS Opposes Germ Warfare," *SSRS Newsletter*, July 1959, p. 2.

94. Theodore Rosebury, "Science and Biological Warfare," *SSRS Newsletter*, October 1962, pp. 1, 3–4.

95. DeVault, "Council of SSRS Opposes Germ Warfare," pp. 1–2.

96. Theodore Roszak, "Scientists for Peace," *The Nation*, September 30, 1961, pp. 205–206.

97. "Membership Ballot on Admission of Social Scientists as Members of the SSRS," *SSRS Newsletter*, June 1952, p. 4.

98. Don DeVault, in *SSRS Newsletter*, October 1962, p. 3.

99. Ibid.

100. "Report of Greater Boston Chapter," *SSRS Newsletter*, September 1964, p. 3.

101. Steven H. Kaiser, "Students in the SSRS," *SSRS Newsletter*, December 1964, p. 3.

102. Herbert Meyer, "Annual Report from Boston," *SSRS Newsletter*, July 1965, p. 1

103. Stuart A. Leslie, *The Cold War and American Science: The Military-Industrial-Academic Conflict at MIT and Stanford* (New York: Columbia University Press, 1993).

104. A. L. Allen, "New England Meeting Supports Challenge to Loyalty Oath," *SSRS Newsletter*, December 1965, p. 1.

105. Stephen H. Kaiser, "Fear, Apathy, and Security Clearance," *SSRS Newsletter*, July 1965, p. 1.

106. "Report of the MIT-SSRS," *SSRS Newsletter*, May 1966, p. 1.

107. Ibid.

108. Stephen H. Kaiser, "MIT-SSRS Annual Report," *SSRS Newsletter*, December 1966, p. 7.

109. "Minutes of SSRS Council Meeting" (April 23, 1960), ER Papers, folder "SSRS+"; "Boston Group Plans CBR Meeting," *SSRS Newsletter*, October 1960, p. 2.

110. "Minutes of the SSRS Council Meeting" (November 8, 1959), ER Papers, folder "SSRS+."

111. Minutes of the Chicago Regional Meeting of the Society for Social Responsibility in Science, March 12, 1963, Papers of the Atomic Scientists, box "Atomic Scientists—Misc.," folder, "SSRS," University of Chicago Archives and Special Collections.

112. Herbert Meyer, "Does Man Have a Novel Opportunity Today?" *SSRS Newsletter*, March 1966, p. 1.

113. Stephen H. Kaiser, "Vietnam and the SSRS," *SSRS Newsletter*, May 1966, p. 2.

114. Peter Ralph, Letter to the Editor, *SSRS Newsletter*, October 1966, p. 8.

115. Barry Blesser, "Identity Crisis," *SSRS Newsletter*, May 1966, p. 165.

116. DeBenedetti with Chatfield, *An American Ordeal*, p. 155.

117. William C. Davidon, "SSRS: Bystander or Participant?" *SSRS Newsletter*, April 1966, p. 1.

118. Klaus Arons, "Whither the SSRS: The Dialogue Continues," *SSRS Newsletter*, June 1966, p. 1.

119. Peter J. Kuznick, "Scientists on the Stump," *Bulletin of the Atomic Scientists* 60, no. 6 (2004): 28–35.

120. Salvador Luria and Albert Szent-Györgyi, Letter to the Editor, *Science*, October 5, 1967, p. 47.

121. Stanley Buckster, Letter to the Editor, *Science*, December 15, 1967, p. 1393.

122. John M. Pfeiffer, Letter to the Editor, *Science*, December 15, 1967, p. 1393.

123. John Reiner, Letter to the Editor, *Science*, December 15, 1967, p. 1393.

124. Evelyn Fox Keller, *Secrets of Life, Secrets of Death: Essays on Gender, Language, and Science* (Cambridge, MA: MIT Press, 2000), p. 97.

125. I do not mean to suggest that physics and biology were entirely separate fields; on the contrary, physics played an important role in the development of ideas about genetics. See ibid., pp. 93–110.

126. Quoted in Peter J. Kuznick, "The Ethical and Political Crisis of Science: The AAAS Confronts the War in Vietnam" (paper presented at the Annual Meeting of the History of Science Society, New Orleans, October 1994), p. 4.

127. Ibid.

128. "Highlights of the Annual Meeting," *SSRS Newsletter*, October 1966, p. 3.

129. Letter from Victor Paschkis to Edward U. Condon, March 20, 1967, box 112, folder 4, ER Papers.

130. Letter from Franklin Miller to Edward U. Condon, June 19, 1967, box 112, folder 4, ER Papers. There is an interesting parallel here with the story that E. Digby Baltzell tells in *Puritan Boston and Quaker Philadelphia* (New Brunswick, NJ: Transaction Publishers, 1996). Baltzell claims that Puritan religious cul-

ture was corporatist and focused on a collective response to reforming the world, whereas the Quaker heritage of Philadelphia encouraged individualized "right action" over institution building. These two approaches seem mirrored in the tensions between SSRS's Boston chapter and its Philadelphia-based founders.

131. Letter from Edward G. Ramberg to Professor Harald Wergerland, Seminar of Theoretical Physics, Norwegian Institute of Technology, Trondheim, Norway, December 2, 1967, ER Papers, box 112, folder 4.

132. This organization was a Marxist group, not a group devoted to the kinds of individual-level responsibility that the SSRS advocated.

133. David Nichols, "The Associational Interest Groups of American Science," in *Scientists in Public Affairs*, ed. Albert H. Teich (Cambridge, MA: MIT Press, 1974), p. 167.

CHAPTER 4: INFORMATION AND POLITICAL NEUTRALITY

1. Lily E. Kay, *Who Wrote the Book of Life? A History of the Genetic Code* (Stanford, CA: Stanford University Press, 2000).

2. Charles Bazerman, "Nuclear Information: One Rhetorical Moment in the Construction of the Information Age," *Written Communication* 18, no. 3 (2001): 263.

3. Thomas Jefferson is often invoked as the source of the idea that an "informed" public is necessary for democracy. Research has shown, however, that far from being a populist, Jefferson wanted the public to be more informed about his political policies rather than those of his adversaries. His exhortation to inform the public was not the result of an abstract devotion to democracy. Jeffrey L. Pasley, *"The Tyranny of Printers": Newspaper Politics in the Early American Republic* (Richmond: University of Virginia Press, 2001).

4. Robert A. Divine, *Blowing on the Wind: The Nuclear Test Ban Debate, 1954–1960* (New York: Oxford University Press, 1978), pp. 1–30; Lawrence S. Wittner, *Resisting the Bomb: A History of the World Nuclear Disarmament Movement, 1954–1970* (Stanford, CA: Stanford University Press, 1997), pp. 1–9.

5. Harry Schwartz, "Long Deadly Part Found in Atom Ash," *New York Times*, March 26, 1954, p. 12.

6. Statement by Lewis L. Strauss, Chairman, United States Atomic Energy Commission, USAEC Release, March 31, 1957. Quoted in Carolyn Kopp, "The Origins of the American Scientific Debate over Fallout Hazards," *Social Studies of Science* 9 (1979): 405.

7. Edward Folliard, "H-Bomb Can Wreck N.Y.," *Washington Post and Times Herald*, April 1, 1954, p. 1; "H-Bomb Can Destroy Any City: Strauss," *Chicago Daily Tribune*, April 1, 1954, p. 1; "H-Bomb Can Wipe Out Any City, Strauss Reports after Test," *Los Angeles Times*, April 1, 1954: p. 1; Wittner, *Resisting the Bomb*, pp. 2–3; Divine, *Blowing on the Wind*, pp. 1–30.

8. Wittner, *Resisting the Bomb*, pp. 10–11.

9. A. H. Sturtevant, "Social Implications of the Genetics of Man," *Science*, September 10, 1954, pp. 405–407.

10. Divine, *Blowing on the Wind*, pp. 33–35.

11. Kopp, "The Origins of the American Scientific Debate over Fallout Hazards," pp. 411–413.

12. Nate Haseltine, "H-Bomb Tests Called Danger to Heredity," *New York Times*, April 26, 1955, p. 3.

13. Kopp, "The Origins of the American Scientific Debate over Fallout Hazards," pp. 411–413.

14. Warren Unna, "Lower Limit Is Urged on Safe A-Exposure," *Washington Post and Times Herald*, June 2, 1956, p. 1.

15. "Text of Genetics Committee Report Concerning Effects of Radiation on Heredity," *New York Times*, June 13, 1956, pp. 18–20.

16. Lloyd Norman, "Guards Urged against Rise in A-Deformities," *Chicago Daily Tribune*, June 13, 1956, p. B9; Nate Haseltine, "Nation's Top Scientists Term Radiation a Peril to Future of Man," *Washington Post and Times-Herald*, June 13, 1956, p. 1.

17. William Cuyler Sullivan Jr., *Nuclear Democracy: A History of the Greater St. Louis Citizens Committee on Nuclear Information*, University College Occasional Papers, no. 1. (St. Louis: Washington University, 1982), p. 10.

18. "Text of Stevenson Speech at Meeting Here," *Washington Post and Times Herald*, April 22, 1956, p. A9; "Text of Stevenson Speech on A-Bomb," *Los Angeles Times*, October 16, 1956, p. 15; Divine, *Blowing on the Wind*, pp. 92–98.

19. Lloyd Norman, "A-Tests Pollute Midwest," *Chicago Tribune*, October 13, 1956, p. 7; "AEC Reveals New Data on Atomic Testing," *Los Angeles Times*, October 13, 1956, p. 3.

20. "Grave Peril Seen in Strontium 90," *New York Times*, October 24, 1956.

21. "Atom Experts Urge Bomb Study," *New York Times*, October 20, 1956, p. 1; "G.O.P. Gives Nation a 'New Glow,' Says Ike," *Chicago Tribune*, October 21, 1956, p. 4.

22. E. B. Lewis, "Leukemia and Ionizing Radiation," *Science*, May 17, 1957, p. 965.

23. Ernest C. Anderson, Robert L. Schuch, William R. Fisher, and Wright Langham, "Radioactivity of People and Foods," *Science*, June 28, 1957, p. 1273.

24. Wittner, *Resisting the Bomb*, pp. 148–156.

25. Louis E. Carlat, "Partners with God: Arthur Holly Compton and the Moral Leadership of Science" (A.B. thesis, Washington University, 1986); Thomas Charles Lassman, "Compton's Effect on Washington University" (A.B. thesis, Washington University, 1991).

26. Sullivan, *Nuclear Democracy*, pp. 2–3.

27. Ibid., p. 3

28. Martin D. Kamen, *Radiant Science, Dark Politics: A Memoir of the Nuclear Age* (Berkeley and Los Angeles: University of California Press, 1985), p. 187.

29. Ibid.

30. Allen Smith, "Democracy and the Politics of Information: The St. Louis Committee for Nuclear Information," *Gateway Heritage* 17 (1996): 6.

31. Sullivan, *Nuclear Democracy*, p. 6.

32. Ibid., p. 10.

33. Jessica Wang, "Science, Security and the Cold War: The Case of E. U. Condon," *Isis* 83 (1992): 246–248.

34. Sullivan, *Nuclear Democracy*, p. 9.

35. "Hydrogen Bomb!" advertisement, *New York Times*, October 30, 1956, p. 20.

36. "Medical Men Form Group to Back Stevenson," *New York Times*, October 4, 1956, p. 22.

37. "Radiation Food Inquiry Hidden, Stevenson Says," *Los Angeles Times*, November 3, 1956, p. 6; "Stevenson Sees Coverup on Bomb," *New York Times*, November 3, 1956, p. 20.

38. Divine, *Blowing on the Wind*, p. 104.

39. Ibid., p. 98.

40. Edna Gellhorn Papers finding aid, John M. Olin Library, Washington University.

41. Allen Smith, "Converting America: Three Community Efforts to End the Cold War, 1956–1973" (Ph.D. dissertation, American University, 1995), pp. 142–144.

42. Virginia Brodine, interview by D. Scott Peterson, May 12, 1972, Center for the Study of History and Memory, Indiana University, Bloomington; Smith, "Converting America," pp. 143–144.

43. Sullivan, *Nuclear Democracy*, p. 14.

44. Ibid., p. 7; John M. Fowler, "A CNI History," 1961, p. 2, Committee for Environmental Information Records (CEIR), Western Historical Manuscript Collection, University of Missouri–St. Louis.

45. Mrs. George Gellhorn, Statement before the Sub-Committee on Disarmament of the Senate Foreign Relations Committee, December 12, 1956, CEIR, folder 5; Gertrude B. Faust, Doris B. Wheeler, Marcelle Malamas, Virginia Brodine, Edna Gellhorn, and Elizabeth G. [illegible], "In October, 1956 . . . ," May 1957, CEIR, folder 5.

46. Smith, "Converting America," p. 142.

47. Ibid., pp. 142–143.

48. Sullivan, *Nuclear Democracy*, p. 9.

49. Ibid., p. 6.

50. Smith, "Converting America," p. 138.

51. Faust et al., "In October, 1956 . . . "

52. Smith, "Converting America," p. 137.

53. Sullivan, *Nuclear Democracy*, pp. 30–38.

54. Joseph Klarman, Dan Bolef, and Gertrude Faust, interview by William Sullivan, February 4, 1982, CEIR.

55. Mrs. Frederick Faust, Statement before the Sub-Committee on Disarmament of the Senate Foreign Relations Committee, December 12, 1956; Gellhorn, Statement before the Sub-Committee on Disarmament of the Senate Foreign Relations Committee, both in CEIR, folder 5.

56. Smith, "Converting America," pp. 140–141; Alice Kimball Smith, *A Peril and A Hope: The Scientists Movement, 1945–1947* (Chicago: University of Chicago Press, 1965), p. 151; Barry Commoner, tape recorded interview by D. Scott

Peterson, April 24, 1973, tape 1, pp. 1–3, Center for History and Memory, Indiana University.

57. Barry Commoner, interview by D. Scott Peterson, April 24, 1973, p. 20.

58. "Text of Convention Report on the Impact of Science on Social Forces," *New York Times*, December 31, 1956. Committee member Chauncey D. Leake recalled in 1957 that the committee had considered taking political positions on issues. Chauncey D. Leake, Letter to the Editor, *New York Times*, November 16, 1957, p. 18; Bruce V. Lewenstein, "Shifting Science from People to Programs: AAAS in the Postwar Years," in *The Establishment of Science in America: 150 Years of the American Association for the Advancement of Science*, ed. Sally Gregory Kohlstedt, Michael M. Sokal, and Bruce V. Lewenstein (New Brunswick, NJ: Rutgers University Press, 1999), p. 124.

59. Dael Wolfle, "Social Responsibility of Science," *Science*, January 25, 1957, p. 1; Lewenstein, "Shifting Science from People to Programs," pp. 122–124; Dael Wolfle, *Renewing a Scientific Society: The American Association for the Advancement of Science from World War II to 1970* (Washington, DC: American Association for the Advancement of Science, 1989).

60. Federation of American Scientists, quoted in Lewenstein, "Shifting Science from People to Programs," p. 123.

61. "Biggest Story of the Year," *The Nation*, January 5, 1957, pp. 6–7.

62. "Science and Society," *New York Times*, January 1, 1957, p. 20.

63. "Against the Clear Call," *Wall Street Journal*, December 31, 1956, p. 6.

64. Joseph Turner, "An Academic Question," *Science*, March 8, 1957, p. 1.

65. M. W. Thistle, "Popularizing Science: Can It Be Done?" *Science*, April 25, 1958, pp. 951–955.

66. "Social Aspects of Science: Preliminary Report of the AAAS Interim Committee," *Science*, January 25, 1957, p. 143; Dael Wolfle, "AAAS Council Meeting, 1956," *Science*, February 15, 1957, p. 280. Commoner also contacted Gerald Wendt, a prominent chemical researcher devoted to the popularization of science and the editor of the journal *Impact of Science on Society*. Commoner proposed the possibility of publishing debates among scientists about scientists' role in public life. Like the AAAS, *Impact* was a dead end. Wendt told Commoner that "*Impact* has had very few American authors . . . because of the State Department regulations that require 'clearance' of an author before he is invited to write a manuscript. In practice it is most uncomfortable to have a man investigated by the F.B.I. when he has no idea what it is all about, and to invite him to write an article only after he has been cleared." Letter from Gerald Wendt to Barry Commoner, February 13, 1957, Barry Commoner Collection (BCC), Library of Congress, box 435; see also Bruce V. Lewenstein, "Magazine Publishing and Popular Science after World War II," *American Journalism* 6 (1989): 220, on Wendt as a science popularizer.

67. Divine, *Blowing on the Wind*, pp. 161–163.

68. Wittner, *Resisting the Bomb*, pp. 52–53; Divine, *Blowing on the Wind*, p. 120.

69. Sullivan, *Nuclear Democracy*, p. 31–34.

70. Linus Pauling, *No More War!* (New York: Dodd, Mead, 1958), pp. 170–171.

71. Milton Viorist, "25 Scientists Petition for Halting of Atom Tests," *Washington Post and Times Herald*, May 27, 1957, p. A1; "2000 Scientists Appeal for Ban on A-Tests," *Washington Post and Times Herald*, June 4, 1957, p. A10; "Pauling Scores Libby, Teller," *Washington Post and Times Herald*, May 12, 1958, A2.

72. Pauling, *No More War!*, p. 162.

73. Ibid., pp. 161–162.

74. For a discussion of the fate of the Deweyian project and scientists in the decade and a half after World War II, see Jessica Wang, "Scientists and the Problem of the Public in Cold War America, 1945–1960," *Osiris* 17 (2002): 323–347.

75. "President Defends Nuclear Tests," *Los Angeles Times*, June 7, 1957, p. B4; Paul Jacobs, in *Washington Post and Times Herald*, June 6, 1957, p. A18; "UC Expert Questions H-Foes Qualifications," *Los Angeles Times*, June 5, 1957, p. 13; Warren Unna, "Libby Defends, Lapp Criticizes AEC's Opinion of 'Small Risk,'" *Washington Post and Times Herald*, June 6, 1957, p. A1.

76. Divine, *Blowing on the Wind*, p. 129.

77. Milton Katz, *Ban the Bomb: A History of SANE, the Committee for a Sane Nuclear Policy, 1957–1985* (Westport, CT: Greenwood, 1987); Wittner, *Resisting the Bomb*, pp. 55–82.

78. Jonathan Turner, "When Scientists Disagree," *Science*, September 13, 1957, p. 483.

79. Graham DuShane, "Scientists and Legislation," *Science*, October 4, 1957, p. 635.

80. Ralph E. Lapp, Letter to the Editor, *New York Times*, November 2, 1958, p. E8.

81. Divine, *Blowing on the Wind*, p. 183; "Scientists in Debate," *New York Times*, February 21, 1958, p. 10.

82. Barry Commoner, "The Fallout Problem," *Science*, May 2, 1958, p. 1025.

83. Ibid., pp. 1023–1026.

84. Kay, *Who Wrote the Book of Life?*, pp. 14, 20.

85. Letter from Marcelle Malamas to Mrs. Bolling, October 24, 1958, CEIR, folder 5.

86. Smith, "Converting America," p. 146.

87. Sullivan, *Nuclear Democracy*, p. 22–23.

88. Barry Commoner, Memo to Steering Committee, April 14, 1958, CEIR, folder 56.

89. Gertrude Faust, in Joseph Klarman, Dan Bolef, and Gertrude Faust, interview by William Sullivan, February 4, 1982, CEIR.

90. Walter Bauer, interview by D. Scott Peterson, April 27, 1973, p. 9, Center for History and Memory, Indiana University.

91. Virginia Brodine, interview by D. Scott Peterson, May 12, 1972, p. 18.

92. Greater St. Louis Citizens Committee for Nuclear Information, "Proposed Policies and Programs," April 14, 1958, CEIR, folder 56.

93. Greater St. Louis Citizens Committee for Nuclear Information, "Resolution of March 23 Meeting," April 16, 1958, CEIR, folder 56.

94. Sullivan, *Nuclear Democracy*, p. 19.

95. Sullivan, *Nuclear Democracy*, p. 22; Letter to Barry Commoner from Halsted Holman, June 28, 1958, BCP, box 3, folder H; Letter to Barry Commoner from E. Richard Weinerman, M.D., April 29, 1959, BCP, box 3, folder W.

96. Sullivan, *Nuclear Democracy*, p. 27.

97. Smith, "Converting America," p. 154. Even though Commoner and Arthur Compton shared an interest in the intersections of science and public politics, Compton was suspicious of Commoner's political loyalties and considered him a communist.

98. Letter from Leslie C. Drews, M.D., to John M. Fowler, June 5, 1958, CEIR, folder 7.

99. Letter from Wendell G. Scott, M.D., to John Fowler, May 27, 1958, CEIR, folder 7. Virginia Brodine recalled in a 1972 interview that radiologists from the medical school were particularly opposed to CNI: "Some of the radiologists considered it sort of their professional business and it wasn't something that we should talk about to the man and the woman on the street, because they would get all worried and they wouldn't be able to evaluate it and so forth." Virginia Brodine, interview with D. Scott Peterson, May 12, 1972, p. 13.

100. H. Bentley Glass, letter in *Science*, May 2, 1958, pp. 1349–1350.

101. Minutes of Emergency Executive Board Meeting, February 23, 1959, CEIR, folder 7.

102. John M. Fowler, interview by William Sullivan, January 15, 1982, interview no. T-674, untranscribed, CEIR.

103. Minutes of Emergency Executive Board Meeting, February 23, 1959, CEIR, folder 7.

104. Letter from the Greater St. Louis Citizens' Committee for Nuclear Information, February 23, 1959, CEIR, folder 5.

105. Sullivan, *Nuclear Democracy*, p. 29.

106. Florence Moog and the Committee for Nuclear Information, *Nuclear Disaster* (St. Louis: Committee for Nuclear Information, 1959).

107. Sullivan, *Nuclear Democracy*, p. 29; Minutes of the Executive Board Meeting, April 6, 1959, CEIR, folder 7.

108. Herman M. Kalckar, "A Radiation Census," *Nature*, August 1, 1958, p. 283; CNI, "Summary of Activities, January 1960–May 1961," CEIR, folder 65; "Fallout: The Moment of Tooth," *Newsweek*, April 25, 1960, p. 70; E. S. Khalifa, "Teeth and Strontium 90: The Story behind the Baby Tooth Survey," *Journal of the Missouri State Dental Association* 39 (February 1959): 28–31.

109. Amy Swerdlow, *Women Strike for Peace: Traditional Motherhood and Radical Politics in the 1960s* (Chicago: University of Chicago Press, 1993), pp. 193–196.

110. Wittner, *Resisting the Bomb*, p. 249.

111. Ibid., pp. 325–330.

112. Richard Gid Powers, *Not without Honor: The History of American Anticommunism* (New Haven, CT: Yale University Press, 1995), p. 87.

113. Edward U. Condon, "Copy of a Letter to Mr. [Richard H.] Amberg," publisher of the *St. Louis Globe-Democrat*, October 11, 1960, BCP, box 435.

114. The accusation came as a result of Condon's and Ralph Abele's membership in the local chapter of SANE, which had members and past members who were members of the communist party. Smith, "Converting America," p. 159.

115. Minutes of Joint Meeting of the CNI Community Relations and Executive Committees of the Committee for Nuclear Information, January 3, 1959, BCC, box 427.

116. "Senators Say Reds Used Pauling Plea," *Washington Post and Times Herald*, March 18, 1961, p. 5.

117. Gene Bylinsky, "Fight on Fallout: Red Blasts Spur Hunt for Ways to Protect Humans, Crops, Milk from Nuclear Fallout," *Wall Street Journal*, September 8, 1961, p. 1; "Strontium-90 Level in British Milk Falls," *Washington Post and Times Herald*, April 14, 1961, p. A2; "Beltsville Clears Strontium from Milk," *Washington Post and Times Herald*, September 10, 1961, p. A15.

118. Sullivan, *Nuclear Democracy*, p. 38.

119. Ibid.

120. Walter Bauer, interview by D. Scott Peterson, April 27, 1973, pp. 12–13.

121. Sullivan, *Nuclear Democracy*, pp. 62–63.

122. "Caribou May Bar Alaska A-Blasts," *New York Times*, June 4, 1961, p. 63. Native Alaskans also challenged the project, citing aboriginal hunting and fishing rights that they were guaranteed when Alaska became a state in 1958. "Eskimos Demand Hunting Privilege," *New York Times*, December 3, 1961, p. 148; Raymond J. Crowley. "Eskimos Win Fight with Science; AEC Drops Plan to Blast Harbor," *Washington Post and Times Herald*, August 25, 1962, p. D18; Virginia Brodine, interview by D. Scott Peterson, May 12, 1972.

123. Letter from Margaret Mead to Mr. J. M. Kaplan, Thanksgiving Day, 1959, BCC, box 438, folder CEI.

124. Members of the AAAS Committee on Science in the Promotion of Human Welfare (Barry Commoner, chair; Robert B. Brode; Harrison Brown; T. C. Byerly; Laurence K. Frank; Jack Geiger; Frank W. Notestein; Margaret Mead; and Dael Wolfle, ex officio), "Science and Human Welfare," *Science*, July 8, 1960, pp. 68–73.

125. Margaret Mead, "Opening Address," National Conference for Scientific Information, February 16, 1963, BCC, box 438, folder CEI.

126. Wolfle, *Renewing a Scientific Society*, pp. 232–233.

127. Lewenstein, "Shifting Science from People to Programs," p. 124.

128. AAAS Committee on Science in the Promotion of Human Welfare, "Air Conservation," *Science*, July 6, 1962, p. 9; AAAS Committee on Science in the Promotion of Human Welfare, "Science and the Race Problem," *Science*, November 1, 1963, pp. 556–558; Lewenstein, "Shifting Science from People to Programs."

129. H. Bentley Glass, "Scientists and Their 'True and Lawful Goals,'" *Science*, July 20, 1962, p. 217.

130. St. Louis Committee for Nuclear Information, "Summary of Activities, January 1960–May 1961," BCC, box 427, folder "CNI Minutes."

131. Scientists' Institute for Public Information, "Founding Resolutions and Bylaws of the Scientists' Institute for Public Information," 1963, BCC, box 444, folder "Statements, 1963–1965."

132. Scientists' Institute for Public Information, "National Conference for Scientific Information," February 16, 1963, BCC, box 444, folder "Statements, 1963–1965."

133. Sullivan, *Nuclear Democracy*, pp. 71–72.

134. "Minutes of the Executive Committee Meeting," March 3, 1963, CEIR, folder 38; "Minutes of the Executive Committee Meeting," September 9, 1963, CEIR, folder 38.

135. Barry Commoner, *Science and Survival* (New York: Viking, 1966); see also Barry Commoner, *Scientific Statesmanship in Air Pollution Control* (Washington, DC: United States Public Health Service, 1964), a speech he delivered at the National Conference on Air Pollution in December 1962; Lewenstein, "Shifting Science from People to Programs," p. 124.

136. Sullivan, *Nuclear Democracy*, pp. 71–73.

137. CNI, "List of Government Agency Subscribers," CEIR, Scrapbook #4; CNI, Circulation Statistics," CEIR, Scrapbook #4.

138. "Resolutions for the CNI Annual Meeting," May 17, 1967, CEIR, folder 85.

139. U.S. Bureau of the Census, *Statistical Abstract of the United States, 2007*, table 214, "Educational Attainment by Race, and Hispanic Origin, 1960–2005," posted online at www.census.gov/compendia/statab/education (accessed March 30, 2007).

CHAPTER 5: CONFRONTING LIBERALISM

1. Jon Beckwith, "The Origins of the Radical Science Movement," *Monthly Review* 3 (1986): 120.

2. Donna Haraway, "The Transformation of the Left in Science: Radical Associations in Britain in the 30's and the U.S.A. in the 60's's," *Soundings* 58 (1973): 441.

3. Charles DeBenedetti with Charles Chatfield, *An American Ordeal: The Antiwar Movement of the Vietnam Era* (Syracuse, NY: Syracuse University Press, 1990), p. 125.

4. Ibid., pp. 153–154.

5. Mark Kurlansky, *1968: The Year That Rocked the World* (New York: Ballantine Books, 2004), pp. 192–208, 308–375, 409–427; Todd Gitlin, *The Sixties: Years of Hope, Days of Rage* (New York: Bantam, 1987); DeBenedetti, *An American Ordeal*, pp. 33–52, 165–184, 215–237.

6. Peter J. Kuznick, "Scientists on the Stump," *Bulletin of the Atomic Scientists* 60 (2004): 28–35. The Boston area was also home to the Boston Area Faculty Group on Public Issues, one of the groups that was organized in the information-provision movement spawned by the St. Louis Committee for Nuclear Information (CNI). Its special focus was opposing the construction of fallout shelters. Members publicized their views by providing speakers, testifying before legislative and federal committees, and publishing anti–fallout shelter literature advertisements in the *New York Times*. Murray Eden, "Historical Introduction," in *March*

4: *Scientists, Students and Society*, ed. Jonathan Allen (Cambridge, MA: MIT Press, 1970), p. ix.

7. "29 Scientists Score Use of Chemicals on Vietcong Crops," *New York Times*, January 16, 1966, p. A1. The signers were John Edsall, Bernard Davis, Keith R. Porter, George Gaylord Simpson, Matthew S. Meselson, George Wald, Stephen Kuffler, Mahlon B. Hoagland, Eugene P. Kennedy, David H. Hubel, Warren Gold, Sanford Gifford, Peter Reich, Robert Goldwyn, Jack Clark, and Bernard Lown (Harvard University); Victor W. Sidel, Stanley Cobb, and Herbert M. Kalckar (Massachusetts General Hospital); Alexander Rich, Patrick D. Wall, and Charles Coryell (Brandeis); Nathan O. Kapland and William P. Jencks (MIT); Henry T. Yost (Brown); Peter H. Von Hippel (Amherst); Charles E. Magraw (Tufts); Albert Szent-Györgyi (Institute for Muscle Research, Woods Hole, MA); and Hudson Hoagland (Worcester Institute of Experimental Biology). John Edsall had already taken unorthodox political positions. In 1962, he wrote a long letter to the editor of *Science* arguing that the conventional wisdom that science and democracy were "natural" allies was at best an unproven assumption, based on his examination of the wide variety of political systems under which science had flourished in the past. Letter to the Editor, *Science*, August 10, 1962, pp. 456–458.

8. Eden, "Historical Introduction," pp. ix–x.

9. Thomas A. Halsted, "Lobbying against the ABM: 1967–1970," *Bulletin of the Atomic Scientists* 27, no. 4 (1971): 23–28.

10. Jerry E. Bishops, "Controversy Increases over Whether America Needs Nuclear Defense," *Wall Street Journal*, August 15, 1968, p. 1.

11. Anne Hessing Cahn, *Eggheads and Warheads: Scientists and the ABM* (Cambridge, MA: MIT Science and Public Policy Program, Department of Political Science and Center for International Studies, 1971), pp. 54–55.

12. "Experts See 'Thin' ABM Vulnerable," *Washington Post*, March 3, 1968, p. F12.

13. Merry Selk, "Sentinel in the Backyard: The Transitional Reaction," *Bulletin of the Atomic Scientists* 25, no. 1 (1969): 7; George C. Wilson, "Foes Picture ABM Risks to City Defended," *Washington Post*, November 22, 1968, p. A8.

14. Norman C. Miller, "Army's Choice of Sites for Antimissile Bases Stirs Opposition," *Wall Street Journal*, January 28, 1969, p. 1.

15. Dorothy Nelkin, *The University and Military Research: Moral Politics at MIT* (Ithaca, NY: Cornell University Press, 1972), pp. 42–53.

16. "How Decisions Are Made," *The Nation*, March 24, 1969, p. 355.

17. War Research Subcommittee, MIT Students for a Democratic Society, *M.I.T. and the Warfare State*, n.d., author's files.

18. Ibid., p. 10.

19. Noam Chomsky, "The Cold War and the University," in *The Cold War and the University: Toward an Intellectual History of the Postwar Years*, ed. André Schiffrin (New York: New Press, 1997), p. 180.

20. Joel Feigenbaum, tape-recorded interview by the author, West Barnstable, MA, March 19, 1991; Alan Chodos, tape-recorded interview by the author, New Haven, CT, March 18, 1991; Anonymous member of SACC, tape-recorded interview by the author, Greenfield, MA, March 20, 1991.

21. Alan Chodos, interview by the author, March 18, 1991; Eden, "Historical Introduction," p. xii.

22. SACC, "Draft Statement of Principles," December 19, 1969, author's files.

23. Alan Chodos, Joel Feigenbaum, Jonathan Kabat, and Ira Rubenzahl, for the SACC, with the names of 182 MIT graduate students and faculty attached, Letter to Lee A. DuBridge, Special Assistant to the President, January 17, 1969, author's files; "MIT Group Urges New Science Goals," *New York Times*, January 21, 1969, p. 94.

24. Anonymous member of SACC, interview by the author, March 20, 1991.

25. Alan Chodos, interview by the author, March 18, 1991.

26. "Scientists Plan One-Day Strike," *New York Times*, January 24, 1969, p. 73.

27. SACC, "The March Fourth Research Stoppage—A Statement of Policy," n.d.; SACC, "March Fourth Student Statement"; SACC, "Information Sheet," February 2, 1969, all in Massachusetts Institute of Technology Archives and Special Collections, Cambridge, MA, Science Action Coordinating Committee (SACC) Papers, "SACC History, 1965–1976."

28. SACC, "SACC's Demands Have Not Been Met," April 15, 1969, author's files.

29. David E. Rosenbaum, "R.O.T.C. Being Challenged on Campuses across the Country," *New York Times*, January 5, 1969, p. 64; "Faculty at Harvard Overturns Decision to Oust Protesters," *New York Times*, January 16, 1969, p. 38; "Harvard Faculty to Open Meetings," *New York Times*, January 22, 1969, p. 22; Robert Reinhold, "Harvard Report Calls for Degree in Negro Studies," *New York Times*, January 22, 1969; "Yale Faculty Votes to End Credit for R.O.T.C. Work," *New York Times*, January 31, 1969. p. 1.

30. Gitlin, *The Sixties*, p. 343.

31. Anonymous MIT professor of physics, tape-recorded interview by the author, Cambridge, MA, June 13, 1991.

32. "Science Stoppage Pushed at MIT," *New York Times*, February 23, 1969, p. 72.

33. Alan Chodos, interview by the author, March 18, 1991.

34. Bryce Nelson, "Scientists Plan Research Strike at M.I.T. on 4 March," *Science*, January 24, 1969, p. 1373.

35. Elinor Langer, "A West Coast Version of the March 4 Protest," *Science*, March 14, 1969, p. 117; Crocker Snow Jr., "MIT's 'Day of Reflection' Protests Defense Research," *Boston Globe*, March 4, 1969, p. A17; Walter Sullivan, "Strike to Protest 'Misuse' of Science," *New York Times*, February 6, 1969, p. 1.

36. Nancy Zaroulis and Gerald Sullivan, *Who Spoke Up? American Protest against the War in Vietnam, 1963–1975* (Garden City, NY: Doubleday, 1984), p. 167.

37. Nelkin, *The University and Military Research*, pp. 55–60.

38. Among them were Bernard T. Feld, Kurt Gottfried, Lee Grodzins, Salvador Luria, Philip Morison, and Victor Weisskopf.

39. Francis Low, quoted in "Faculty, Students Demand Research Options," *Scientific Research*, March 3, 1969, p. 11. Gary L. Downey has argued that the formation of UCS was largely based on the faculty's concern to protect their status

as senior faculty members. This interpretation is supported by the evidence, but it does not adequately appreciate the extent to which the model of action that UCS scientists had in mind was that of an earlier generation. Many of these faculty members' mentors were involved in the Manhattan Project and the later system of government advising that was built up between 1946 and 1965. This liberal model treated facts and the political actions undertaken in light of them as separate, and saw scientists not as a group that ought to be ashamed of their role in military advising, as SACC activists perceived them, but as public servants. See Gary L. Downey, "Reproducing Cultural Identity in Negotiating Nuclear Power: The Union of Concerned Scientists and Emergency Core Cooling," *Social Studies of Science* 18 (1988): 231–264.

40. SACC, "The March Fourth Research Stoppage—A Statement of Policy."

41. Snow, "MIT's 'Day of Reflection' Protests Defense Research."

42. Chodos, "The Protest Movement in the United States in Recent Years," n.d., author's files.

43. David Baltimore, Ira Gerstein, Ethan Signer, Vigdor Teplitz, and Kosta Tsippis, "To Our Colleagues of the Science and Engineering Teaching Staff," n.d., SACC Papers, "SACC History, 1965–1976."

44. Joel Feigenbaum and Ira Rubenzahl, Letter to the Editor, *New York Times*, February 27, 1969, p. 40.

45. SACC, "March Fourth Student Statement."

46. UCS, "Beyond March 4," n.d., author's files.

47. In my previous work I interpreted the activities of UCS, SftP, and SIPI in a functionalist fashion, arguing that the formation of each organization offered a "solution" for how to keep science and politics apart while joining them together. Gary Downey's interpretation of the formation of UCS makes a similar claim. Downey, "Reproducing Cultural Identity in Negotiating Nuclear Power."

48. Downey, for example, sees UCS's later use of expertise in a controversy over nuclear power as an attempt to preserve its members' professional status. Ibid.

49. UCS, "Program for March 4 and March 8," February 21, 1969, Massachusetts Institute of Technology Special Collections and Archives, Union of Concerned Scientists (UCS) Papers.

50. Joel Feigenbaum, quoted in *March 4: Scientists, Students and Society*, ed. Jonathan Allen (Cambridge, MA: MIT Press, 1970), pp. 4–7. The MIRVs to which he refers are offensive missile systems that use one rocket to launch multiple warheads aimed at independent targets.

51. Anonymous MIT professor of physics, interview by the author, June 13, 1991.

52. Thomas C. Schelling, "The Future of the Academic Community," in *March 4: Scientists, Students and Society*, ed. Jonathan Allen (Cambridge, MA: MIT Press, 1970), pp. 83–87.

53. Lee A. DuBridge, "The Social Control of Science," *Bulletin of the Atomic Scientists* 25, no. 5 (May 1969): 26–28, 35.

54. Other New York area schools that held March 4 events include Columbia University, Fordham University, New York University, Brooklyn Polytechnic Institute, and the State University of New York–Stony Brook. Nationally, students

and faculty at Cornell University, the University of California–Los Angeles, the University of California–Irvine, and the University of California–San Francisco held March 4 events. Murray Ilson, "Hundreds at Columbia Join," *New York Times*, March 5, 1969, p. 14; Robert Reinhold, "Scientists Halt Research for a Day; Troubled over Role in Research," *New York Times*, March 5, 1969, p. A1; Walter Sullivan, "Strike to Protest 'Misuse' of Science," *New York Times*, February 6, 1969, p. A1; "Science Stoppage Pushed at M.I.T.," *New York Times*, February 3, 1969, p. 72; "Announcing a Day-Long Symposium on the Use and Misuse of Science and Technology," *Daily Californian*, March 3, 1969, p. 9.

55. "Yale Prof. Palmer Defeats Lynd for Historian Group Presidency," *Washington Post*, December 29, 1969, p. A2.

56. Paul L. Montgomery, "Vietnam Debated by Philosophers," *New York Times*, December 29, 1969, p. A24.

57. "Language Unit Radicals Rap Pentagon," *Washington Post*, December 29, 1969, p. A7.

58. Ruth M. Oltman, "Women in the Professional Caucuses," *American Behavioral Scientist* 18 (1972): 281–302.

59. Ted Steege, "Introduction," in *Radicals in Professions: Selected Papers* (Ann Arbor, MI: Radical Education Project, 1967), p. 7.

60. Barbara Haber and Al Haber, "Getting By with a Little Help from Our Friends," in *Radicals in Professions*, p. 47.

61. AAAS Board of Directors, "War, Militarism, and Violence," Minutes of the Meeting of the Board of Directors of the AAAS, June 27, 28, 29, 1969, folder "Minutes and Agenda, May–September, 1969," American Association for the Advancement of Science (AAAS) Archives, Washington, DC.

62. "AAAS 1969 Report of the Chairman to the Board of Directors," folder "Minutes and Agenda, November–December 1969," AAAS Archives.

63. Charles Schwartz, oral history by Finn Aaserud, May 15, 1987, American Institute of Physics, College Park, MD.

64. Lawrence E. Davies, "Students Occupy Stanford Electronics Laboratory," *New York Times*, April 11, 1969, p. 24; Peter Braestrup, "Researchers Aid Thai Rebel Fight," *New York Times*, March 20, 1967, p. 7; see also John Maffre, "Think Tank Put on the Defensive," *Washington Post*, March 30, 1968, p. A10.

65. "APS to Consider Whether It Should Discuss Public Issues," *Physics Today*, November, 18, 1967, p. 81.

66. "Special Announcement: Should the American Physical Society Broaden Its Purpose and Aims to Include Discussion of Public Issues?" *Bulletin of the American Physical Society* 12 (1971): 1028.

67. Charles Schwartz, "Should the APS Discuss Public Issues? For the Schwartz Amendment," *Physics Today* 19 (1967): 9–10.

68. R. Hobart Ellis, "What We Are Not Against," *Physics Today* 19 (August 1968): 104.

69. See *Physics Today*, vol. 19, September, October, and November 1968 issues.

70. Charles Schwartz, tape-recorded interview by the author, Berkeley, CA, August 13, 1998.

71. Ibid.

72. University of Michigan Teach-In Program, "America in Crisis," 1965, p. 2, author's files.

73. Goldhaber et al., "Announcing the Formation of a New Organization Dedicated to Vigorous Social and Political Action"; see also "New Radical Science Group Is Formed," *Chicago Tribune*, February 9, 1969, p. B13.

74. Goldhaber et al., "Announcing the Formation of a New Organization Dedicated to Vigorous Social and Political Action."

75. "New Radical Science Group Is Formed."

76. Scientists for Social and Political Action, *Scientists for Social and Political Action Newsletter*, February 23, 1969, p. 1. Those who volunteered to help organize local chapters of the SSPA were mainly located at research institutions.

77. Walter Sullivan, "Physicists Enter Debate over ABM," *New York Times*, April 30, 1969, p. 13.

78. Martin L. Perl and Michael H. Goldhaber, "A New Organization of Scientists Concerned with All the Problems of Today's World and Seeking a Radical Redirection in the Control of Modern Science and Technology," February 1969, author's files.

79. SSPA, *Scientists for Social and Political Action Newsletter*, February 23, 1969.

80. Walter Sullivan, "Fighting 'The Misuse of Knowledge,'" *New York Times*, February 9, 1969, p. E7.

81. Michael Goldhaber, tape-recorded interview by the author, San Francisco, February 2, 1992.

82. Ibid.

83. Anonymous biologist, tape-recorded interview by the author, Cambridge, MA, December 4, 1991.

84. Britta Fischer, tape-recorded interview by the author, Brookline, MA, December 5, 1991.

85. The APS declined to create a division, and instead formed a committee called the Committee on Problems of Physics and Society. Council Minutes, American Physical Society, November 1969, American Physical Society (APS) Papers, American Institute of Physics; SSPA, *Scientists for Social and Political Action Newsletter*, February 23, 1969.

86. Herb Fox, "SftP: A History," *Science for the People*, December 1970, p. 2.

87. Charles Schwartz, oral history by Finn Aaserud, May 15, 1987. Indeed, FAS did become a liberal lobbying group by 1971.

88. Deborah Shapley, "FAS: Reviving Lobby Battles ABM, Scientists' Apathy," *Science*, March 26, 1971, p. 1227.

89. The rapid development of public interest groups during this period was caused by three major changes: foundations and individual donors increased their contributions to such groups, partly in response to new tax laws; new federal regulation of industrial production and its effects on humans and the environment drove demand for professional assessments of pollution; and many young professionals, including scientists, hoped to use their skills to address specific issues such as nutrition and health, pollution, and environmental issues. John D. McCarthy and Mayer N. Zald, *The Trend of Social Movements in America: Professionaliza-*

tion and Resource Mobilization (Morristown, NJ: General Learning Press, 1973); Kelly Moore, "Political Protest and Institutional Change: The Anti–Viet Nam War Movement and American Science," in *How Social Movements Matter*, ed. Marco Guigni, Doug McAdam, and Charles Tilly (Minneapolis: University of Minnesota Press, 1999), pp. 97–118.

90. Max Elbaum, *Revolution in the Air: Sixties Radicals Turn to Lenin, Mao, and Che* (New York: Verso, 2002).

91. Dan Georgakas and Marvin Surkin, *Detroit, I Do Mind Dying: A Study in Urban Revolution* (Boston: South End Press, 1998).

92. Elbaum, *Revolution in the Air*, p. 26.

93. Allen J. Matusow, *The Unraveling of America: A History of Liberalism in the 1960s* (New York: Harper & Row, 1984), pp. 337, 342–344; John Morton Blum, *Years of Discord: American Politics and Society, 1961–1974* (New York: W. W. Norton, 1991), pp. 356–364; Gitlin, *The Sixties*, pp. 391, 399–419.

94. John Kifner, "Police in Chicago Slay Two Panthers," *New York Times*, December 5, 1969, p. 1; Doug Rossinow, *The Politics of Authenticity: Liberalism, Christianity, and the New Left in America* (New York: Columbia University Press, 1998), p. 202.

95. Myra Marx Ferree and Patricia Yancey Martin, *Feminist Organizations: Harvest of the New Women's Movement* (Philadelphia: Temple University Press, 1991); Sandra Morgen, *Into Our Own Hands: The Women's Health Movement in the United States, 1969–1990* (New Brunswick, NJ: Rutgers University Press, 2002); Londa Schiebinger, *Has Feminism Changed Science?* (Cambridge, MA: Harvard University Press, 1999); Evelyn Fox Keller, "Gender and Science: Origin, History, and Politics," *Osiris* 10 (1995): 27–38.

CHAPTER 6: DOING "SCIENCE FOR THE PEOPLE"

1. Brian Martin, "Science in a Citizen-Shaped World," in *The New Political Sociology of Science: Institutions, Networks, and Power*, ed. Scott Frickel and Kelly Moore (Madison: University of Wisconsin Press, 2006), pp. 272–298.

2. The following analysis of the 1969 and 1970 AAAS meetings is based on Kelly Moore and Nicole Hala, "Organizing Identity: The Creation of Science for the People," *Research in the Sociology of Organizations* 19 (2002): 309–355.

3. Herb Fox, Science for the People listserv (accessed February 1, 2006).

4. Science for the People, Press Release, December 20, 1969, Science for the People (SftP) Papers, 1969–1992, folder "AAAS Boston 1969," Massachusetts Institute of Technology Archives and Special Collections, Cambridge, MA. The main student organizers were David J. Jhirad, an instructor in astronomy; Donael E. MacKenzie, a social relations graduate student; Richard J. Paul, a medical student at Harvard University; and Mark S. Tuttle of the Division of Engineering and Applied Physics at Harvard. Richard Knox, "Science Group Taken to Task," *Boston Globe*, December 20, 1969, p. 26; "Sessions May Be Open to Public and Free," *Boston Globe*, December 21, 1969, p. A25.

5. AAAS Board of Directors, Meeting Minutes, June 27, 28, 29, 1969, in folder "Minutes and Agenda, May–September, 1969," American Association for the Advancement of Science (AAAS) Archives, Washington, DC.

6. Meeting Agenda, December 9, 1969; Minutes of the Meeting of the Committee on Council Affairs, Boston, Massachusetts, December 27 and 28, 1969, both in folder "Minutes and Agenda, November–December, 1969," AAAS Archives.

7. Programs of the Annual Meetings of the American Association for the Advancement of Science, 1967 and 1968, folder "Meeting Programs," AAAS Archives.

8. Robert J. Sales, "Student Criticizes 'Misuses of Technology,'" *Boston Globe*, December 29, 1969, p. 20; "Science Criticized for Serving Industry's Planned Obsolescence," *Chicago Tribune*, December 29, 1969, p. 1.

9. Victor K. McElheny, "AAAS Mulls Steps to Stop Disrupter," *Boston Globe*, December 28, 1969, p. 1.

10. Andrew Blake, "Line Forms Here for Tissue Bath," *Boston Globe*, December 28, 1969, p. 2.

11. Janet Christiansen, "12 Intruders Protest Women's 'Inferior Role,'" *Herald Traveler*, December 29, 1969, p. 1; Robert J. Sales, "A Woman's Place in Science," *Boston Globe*, December 29, 1969, p. A28.

12. Richard Knox, "Students to Examine 'Sorry State of Science,'" *Boston Globe*, December 27, 1969, p. A5.

13. Women's International Terrorist Conspiracy from Hell, "Witches' Hex for Phi Beta Kappa–Sigma Xi," 1969, SftP Papers, box 1.

14. Bearded, tam-o-shanter: Victor K. McElheny, "AAAS Mulls Steps to Stop Disrupter," *Boston Globe*, December 28, 1969, p. 1; clean-cut, shoeless: Robert Reinhold, "4 Scientists Urge World Disarmament," *New York Times*, December 27, 1969, p. 1; wool shirt and work pants: Robert Reinhold, "Scientists Wary on Use of Tear Gas in Vietnam," *New York Times*, December 28, 1969, p. 1.

15. Peter J. Kuznick, "The Ethical and Political Crisis of Science: The AAAS Confronts the War in Vietnam" (paper presented at the Annual Meeting of the History of Science Society, New Orleans, October 1994).

16. McMahon quoted in Richard Knox, "AAAS Council Hauls Science Out of the White Tower," *Boston Globe*, December 31, 1969, p. A8.

17. "Three Resolutions . . . ," Press Release, December 26, 1969, SftP Papers, box 1.

18. "Minutes of the Meeting of the Committee on Council Affairs, Boston, Mass., December 27 and 28, 1969," folder "Minutes and Agendas, Nov.–Dec. 1969," AAAS Archives.

19. Doug McAdam, "Tactical Innovation and the Pace of Insurgency," *American Sociological Review* 48 (1983): 735–754.

20. National Strike Information Committee, *Newsletter*, no. 4, May 6, 1970, p. 1, SftP Papers, folder 2.

21. AAAS Board of Directors, "Proposed Procedures in the Event of Serious Disruption at the Chicago Meeting," Minutes of the Meeting of the Board of Directors, December 12 and 13, 1970, Executive Office Files, Board of Directors, Council, and Committee Minutes, box 7, AAAS Archives.

22. John Noble Wilford, "Knitting Needle Thrust Interrupts One Dissident," *New York Times*, December 29, 1970, p. 8.

23. Ibid.

24. Philip M. Boffey, "AAAS Conference: Radicals Harass the Establishment," *Science*, January 8, 1971, p. 48.

25. Wilford, "Knitting Needle Thrust Interrupts One Dissident"; John Noble Wilford, "Teller Deplores Secret Research," *New York Times*, December 30, 1970, p. 1; Boffey, "AAAS Conference," p. 47.

26. Boffey, "AAAS Conference," p. 46; Mrs. Garrett Hardin, quoted in Ronald Yates, "Women Fed Up with Heckler at Parley Gives Him 'Needle,'" *Chicago Tribune*, December 30, 1970, p. A1; Stuart Auerbach, "Irate Knitter Needles Heckler," *Washington Post*, December 29, 1970, p. A1.

27. Boffey, "AAAS Conference," p. 46; Wilford, "Knitting Needle Thrust Interrupts One Dissident," p. 8.

28. Boffey, "AAAS Conference," p. 46; Wilford, "Teller Deplores Secret Research."

29. Boffey, "AAAS Conference," p. 46; Wilford, "Teller Deplores Secret Research."

30. Science for the People, "An Indictment of Glenn T. Seaborg for the Crime of Science against the People," December 30, 1970, SftP Papers, Massachusetts Institute of Technology Archives and Special Collections, box 1, folder "AAAS–1969–1972."

31. "Teller Says Oppenheimer Talked Him Out of Opposing A-Bomb," *Washington Post*, December 28, 1970, p. A1.

32. Nancy Hicks, "3 Activists Score Scientists as Decadent," *New York Times*, December 30, 1970, p. 8.

33. "Teller Says Oppenheimer Talked Him Out of Opposing A-Bomb."

34. Philip H. Abelson, "The Chicago Meeting," *Science*, January 22, 1971, p. 239.

35. Peter Suedfeld, Letter to the Editor, *Science*, January 22, 1971, p. 230.

36. Boffey, "AAAS Conference," p. 48.

37. American Association for the Advancement of Science, Annual Meeting Program, 1970, AAAS Archives.

38. "American Association for the Advancement of Science 1970 Report of the Youth Council," pp. 1–2, Executive Office Files, Board of Directors, Council, and Committee Minutes, box 6, AAAS Archives.

39. Melanie J. Hunter, "Women in Science: Turmoil within the American Association for the Advancement of Science" (manuscript, July 1999), p. 7.

40. Steven Shapin and Simon Schaffer, *Leviathan and the Air Pump: Hobbes, Boyle, and the Experimental Life* (Princeton, NJ: Princeton University Press, 1986).

41. Les Fishbone, "Universities Review Think Tank Ties," *California Tech*, April 25, 1968, p. 1.

42. Students for a Democratic Society, "Who Defines the 'National Interest'?" *Princetonian*, October 17, 1967, p. 2.

43. *The Pentagon Papers: The Defense Department History of United States Decisionmaking in Vietnam*, Gravel edition (Boston: Beacon Press, 1971), book IV, p. 121.

44. John Morton Blum, *Years of Discord: American Politics and Society, 1961–1974* (New York: W. W. Norton, 1991), p. 387.

45. *The Pentagon Papers*, book IV, p. 121; Jason Group, "Explosively Produced Flechettes," Report no. 66–121 Institute for Defense Analysis, 1966.

46. *The Pentagon Papers*, book IV, p. 123; see also Seymour Melman, with Melvyn Baron and Dodge Ely, *In the Name of America: The Conduct of the War in Vietnam by the Armed Forces of the United States as Shown by Published Reports, Compared with the Laws of War Binding on the United States Government and on Its Citizens* (New York: Clergy and Laymen Concerned about Vietnam, 1968). Melman, a professor of industrial engineering at Columbia University, was a long-time critic of the federal government's military plans and activities. In 1958, he published the influential report *Inspection for Disarmament* (New York: Columbia University Press), in which he concluded that accurate inspections were almost impossible; four years later, he published *No Place to Hide: Fact and Fiction about Fallout Shelters* (New York: Grove Press, 1962), which was equally critical of the government's civil defense plan. At Berkeley, Charlie Schwartz was one of the main organizers of the anti-Jason campaign.

47. Berkeley Scientists and Engineers for Social and Political Action (contributors: Jan Brown, Martin Brown, Chandler Davis, Charlie Schwartz, Jeff Stokes, Honey Well, and Joe Woodard), *Science against the People* (Berkeley, CA: Berkeley Scientists and Engineers for Social and Political Action, 1972), p. 21.

48. Ibid., p. 20.

49. Ibid., p. 19.

50. Ibid., p. 24.

51. Ibid., pp. 33–37.

52. "Le Savant du Pentagone," *Le Novel Observateur*, June 26, 1972, reprinted in *The War Physicists: Documents about the Protests against the Physicists Working for the American Military through the JASON Division of the Institute for Defense Analysis*, ed. Bruno Vitale (Naples, Italy: B. Vitale, 1976), p. 78.

53. Vitale, *The War Physicists*, p. 7. The scientists listed on the poster were: Luis Alvarez, James D. Bjorken, Richard Blankenbecler, Lewis Branscomb, David Caldwell, Roger F. Dashen, Sydney Drell, Freeman Dyson, Val Fitch, Henry M. Foley, Edward Frieman, Richard Garwin, Murray Gell-Mann, Donald Glaser, Marvin Goldberger, Robert Gomer, Joseph Keller, Henry Kendall, George Kistiakowski, Robert Lelevier, Harold Lewis, Elliott Montroll, Walter H. Munk, William A. Nierenberg, Wolfgang Panofsky, Allen Peterson, Malvin Ruderman, Edwin Salpeter, Matthew Sands, Charles Townes, Steven Weinberg, John Wheeler, Eugene Wigner, C. Courtnay Wright, Frederick Zachariasen, and George Zweig (poster, "War Scientists," author's files). In addition, *Science against the People* listed Norman Christ, Henry Foley, Leon Lederman, Joseph Chamberlain, and Kenneth Case as members of Jason (*Science against the People*, p. 30). Not all of the scientists listed on the poster, in *Science against the People*, or in *The War*

Physicists were members of Jason at the time of the publication of these documents, but all had been members in the past.

54. Sidney Drell, oral history by Finn Aaserud, July 1, 1986, American Institute of Physics, College Park, MD, p. 6.

55. Charles Hard Townes, "A Life in Physics: Bell Telephone Laboratories and World War II, Columbia University and the Laser, MIT and Government Service, California and Research in Astrophysics," an oral history conducted in 1991–1992 by Suzanne B. Riess, Regional Oral History Office, Bancroft Library, University of California–Berkeley, 1994, p. 278.

56. Ibid.

57. Malvin Ruderman, in *Christianity and Crisis*, September 18, 1972, p. 6.

58. Richard Garwin, in *Christianity and Crisis*, September 18, 1972, p. 5.

59. Frank Baldwin, in *Christianity and Crisis*, September 18, 1972, p. 4.

60. Sidney Drell, oral history by Finn Aaserud, July 1, 1986, p. 10, American Institute of Physics.

61. Richard Blankenbecler, telephone interview by the author, August 30, 2006.

62. Richard Blankenbecler, oral history by Finn Aaserud, May 5, 1987, American Institute of Physics.

63. Richard Blankenbecler, telephone interview by the author, August 30, 2006.

64. Science for the People, *China: Science Walks on Two Legs* (New York: Discus Books / Avon, 1974).

65. Martin, "Science in a Citizen-Shaped World."

66. Northwestern University Science for the People, *AAAS in Mexico: Por Qué? Science and Technology in Latin America*, 1972, uncataloged documents, IP 156453; Science for the People, with the collaboration of Anna-Berta Al, Carlos, Christian, Esther, Gerardo, Manuel, Marco, and Mary, *Los Nuevos Conquistadores* (Jamaica Plain, MA: Science for the People, 1973).

67. Arthur R. Jensen, "How Much Can We Boost I.Q.?" *Harvard Educational Review* 39 (1969): 1–123.

68. Richard Herrnstein, "I.Q.," *Atlantic*, September 1971, p. 43–64.

69. Richard C. Lewontin, "Race and Intelligence," *Bulletin of the Atomic Scientists* 26 (March 1970): 2, 7.

70. Ullica Segerstråle, *Defenders of the Truth: The Sociobiology Debate* (New York: Oxford University Press, 2000).

71. Science for the People's membership was always in flux. Some chapters existed for nearly twenty years, others for only a year or two, and in some places chapters split or were revitalized after a period of dormancy. Through 1979, the organization's newsletter, *Science for the People*, lists local chapters. The longest-lasting were Boston, New York, Berkeley, Ann Arbor, Chicago, and Madison. Many of the chapters were centered around colleges or universities, although in some, such as New York and Boston, there were sometimes many science workers from outside academia, and different factions. I list here the chapters from 1973 only as an example of the dispersion of the group; the reader should consult *Science for the People* for the complete list of the locals from 1969 through 1979. The 1973 groups were located in Fayetteville, AR; Berkeley, Canoga Park, San

Diego, and Venice, CA; Boulder, CO; Storrs, CT; Washington, DC; Gainesville, FL; Atlanta, GA; Honolulu, HI; Chicago and Evanston, IL; Amherst, MA; Ann Arbor and Detroit, MI; St. Louis, MO; Endicott, Ithaca, New York, Schenectady, and Stonybrook, NY; Cincinnati, OH; Eugene, OR; Pittsburgh, Swarthmore, and University Park, PA; Nashville, TN; Burlington, VT; and Madison, WI.

72. Other choices, such as Ann Arbor, Berkeley, or Madison, would have produced similar variations in practices.

73. CPP was formed in 1969 by antiwar activists in New York City. They published *Interrupt*, a quarterly newsletter and magazine. Their early interest in the politics of political repression was reflected in their writings about the use of databanks to collect and distribute information about citizens in the United States. See for example, Jeanine Meyer, "Data Bank News," *Interrupt* 14 (1971): 3. One of the group's early projects was to provide political and financial support for the defense of Clarke Squire, a scientist and one of the "Panther 21" (members of the Black Panther Party) accused of conspiracy to murder police officers and bomb rail stations and retail stores. CPP, "Clarke Squire and the Panther 21: Acquitted," n.d., author's files. CPP had chapters in Boston, Chicago, Detroit, Houston, Los Angeles, Philadelphia, Poughkeepsie, San Francisco, Washington, DC, and Northern New Jersey in 1971 (*Interrupt*, February 1971, p. 15). See also Matthew Wisnioski, "Inside 'The System': Engineers, Scientists, and the Boundaries of Social Protest in the Long 1960s," *History and Technology* 19 (2003): 313–333.

74. CSRE was specifically interested in labor issues among engineers, including layoffs, and about the suppression or misuse of scientific information. The group encouraged engineers to uncover details about the contracts that engineers' employers held with the military and other groups. See *Spark*, the magazine of the CSRE, published beginning in spring 1971. CSRE, following the lead of SftP, focused attention on mobilizing engineers at professional association meetings, especially those of the Institute of Electrical and Electronics Engineers, but at other meetings as well. In 1971 alone, the group sponsored alternative sessions and a booth at the Spring Joint Computer Conference (with CPP), the Urban Technology Conference, the National Electronic Packaging and Production Conference, and the Semi-Conductor/IC Processing and Production Conference. CSRE also worked with industrial groups, including a group of radical engineers at General Electric, and the Stanford-based Technology and Society Committee.

75. Wisnioski, "Inside 'The System.'"

76. Barbara Ehrenreich, "Giving Power to the People: The Early Days of Health/PAC," *Health/PAC Bulletin* 18, no. 4 (1988): 5.

77. "New Engineering Conference," *Spark*, Fall 1971, p. 2.

78. Ann Rosenberg, *The Technological Warlords* (New York: Computer People for Peace, 1971).

79. American Committee on Africa, "I.B.M. in South Africa," November 1971, author's files; Science for the People, "Exporting ID-2—Instant Passbooks in South Africa," October 1971, author's files.

80. CPP, *Interrupt: Newsletter of Computer People for Peace* 10 (April 1970), author's files; CPP, *Project IBM*, n.d., author's files.

81. New York City CPP, *Data Banks: Privacy and Repression*, n.d. [1971?], author's files.

82. New York City CPP, *Health: Big Business for Computers*, May 1971, author's files.

83. Other New York members of NUC included: Borough of Manhattan Community College: Jim Perlstein, Ruth Misheloff, Nan Maglin, Bill Friedheim, Steve Cagan, and Beth Cagan (of whom the last two were political organizers previously based in Cleveland); Brooklyn College: Alyss Wolfe, Nanette Funk, Bill Zimmerman; Queens College: Bill Tabb and Mike Brown; Kingsborough Community College: Inez Martinez, Liz Diggs, and Daphne Joslin; SUNY–Old Westbury: John Ehrenreich, Paul Lauter, and Florence Howe; New York University: Bertell Ollman; Columbia University: Moe Levitt and Kate Ellis.

84. New York City SftP, *Hard Times: Employment, Unemployment, and Professionalism in the Sciences*, n.d. [1973?], pp. 33–34, 43, author's files. The New York City group also worked on the Medical Committee for Human Rights projects in Vietnam. "Medical Aid for Indochina," October 24, 1971, author's files.

85. Science for Vietnam, *Khoa học vì Viet Nâm/Science for Vietnam* (Chicago: Science for Viet Nam Chicago Collective, 1972); "People's Science Projects for Vietnam," *Science for the People* 3, no. 2 (1971): 12–13, both in Political Pamphlets Collection, Charles Deering McCormick Library Special Collections, Northwestern University.

86. Philip M. Boffey, "Herbicides in Vietnam: AAAS Study Finds Widespread Devastation," *Science*, January 8, 1971, pp. 43–47.

87. Science for Vietnam, "Herbicides in Vietnam," 1971; Science for Vietnam, "The Use of Chemical Defoliants in Vietnam," 1971, both in author's files.

88. Science for Vietnam, "SFV: Who We Are and What We Do," n.d., pp. E-4, E-6, Uncataloged Political Documents IP 185732, Northwestern University.

89. Richard Lewontin and Richard Levins were major proponents of a Marxist theory of biology. In 1968, Levins had published *Evolution in Changing Environments: Some Theoretical Considerations* (Princeton, NJ: Princeton University Press); in 1985, Levins and Lewontin coauthored *The Dialectical Biologist* (Cambridge, MA: Harvard University Press), which elaborated this theory.

90. Science for Vietnam, "SFV: Who We Are and What We Do," author's files.

91. SftP Boston, "Science for Vietnam Conference to Establish Programs to Provide Scientific and Technical Assistance and Material to the Vietnamese People," n.d. [1971?], author's files; Science for Vietnam, *Khoa học vì Viet Nâm/ Science for Vietnam*; "Science for Vietnam Conference," *Science for the People* 3, no. 3 (1971): 15.

92. Kathy Greeley and Sue Tafler, "Science for the People: A Ten-Year Retrospective," *Science for the People* 11, no. 1 (1979): 18; Science for the People, "The Technical Assistance Program," *Science for the People* 3, no. 3 (1971): 15.

93. Steven Rose and Hilary Rose were prolific and active members of the British Science for the People movement. They published an analysis of the relationships among science, technology, and class oppression, in *Science and Society* (London: Allen Lane, 1969). The British movement, never centered around a single organization, was influenced by the class-based analysis of science articulated by J. D. Bernal. In 1968, Steven Rose participated in a conference on chemical and biological welfare, and shortly afterward published, with David Paved, an edited volume of papers from the conference: *C.B.W.: Chemical and Biological Warfare, London Conference on Chemical and Biological Warfare* (London: Har-

rap, 1968). Rose also wrote, with Richard Lewontin and Leon J. Kamin, *Not in Our Genes: Biology, Ideology and Human Nature* (New York: Pantheon, 1984). Steven and Hilary Rose coedited *The Radicalisation of Science* (London: Macmillan, 1976), which contains an excellent historical analysis of the growth of science radicalism in Great Britain, including the British Association of Scientific Workers and the British Society for Social Responsibility in Science (BSSRS). Unlike the American group of the same name, the BSSRS maintained a class-based ideology, not a moral perfectionist ideology. Later, Hilary Rose would expand class-based analysis of science to include feminism, in *Love, Power, and Knowledge: Towards a Feminist Transformation of Science* (Bloomington: Indiana University Press, 1994).

94. Dave J. Boston General Meeting, "Complete Minutes of Principles of Unity Discussion Follows," October 30, 1974, p. 4, author's files.

95. The debates about the Principles of Unity were acrimonious, in part, because some members felt that members influenced by the Boston October League group were trying to force their principles on the group. For example, in 1975, six members wrote: "What does unity mean to the 'Unity Caucus'? It has become more and more clear the past year that to *them* "unity" means agreement with *their* carefully worked-out and detailed positions on a variety of positions which they deem crucial to the future of Science for the People. . . . Many people are being driven away—they rightfully are repulsed by the rhetoric and the absence of any connections to our constituency, as well as the heavy-handedness in evidence at many of our meetings." Frank Bove, Minna Goldfarb, Fred Gordon, Eric Entemann, Ted Goldfarb, and Mina Otmishi, "Science for the People and the Unity Caucus," 1975, author's files. For an overview of these debates, see Greeley and Tafler, "Science for the People," pp. 23–24; "The Northeast Regional Conference," *Science for the People* 6, no. 1 (1974): 17–18; "Northeast Regional Committee Membership Survey and Conference Call," *Science for the People* 6, no. 6 (1974): 22; "Northeast Regional Conference Report and Resolutions," *Science for the People* 7, no. 1 (1974): 23.

96. "Actions at NTSA," *Science for the People* 3, no. 3 (1971): 4; "Science Teaching: A Critique," *Science for the People* 3, no. 3 (1971): 5; *Science for the People* 4, no. 4 (1972) is a special issue on science teaching; "Science Teaching," *Science for the People* 6, no. 4 (1974): 32–33; "Science Teaching," *Science for the People* 7, no. 5 (1975): 29.

97. Rita, Michele, Sue T., Susanne, Brenda, and Anne S., "Alternative Draft of Principles of Unity," n.d., author's files.

98. *How Harvard Rules Women* (Cambridge, MA: New University Conference, 1970).

99. Rita Arditti, telephone interview by the author, May 4, 1999.

100. Ibid.

101. Letter from Paula to New York Science for the People, n.d. [1972?], author's files.

102. Carol Axelrod and Ruth Crocker, "Science for the People: The Natural Birth of a Woman's Group," *Science for the People* 6, no. 5 (1974): 15.

103. Robert Reinhold, "Scientists Isolate a Gene: Step in Heredity Control," *New York Times*, November 23, 1969, p. 1; James K. Glassman, "Harvard Genet-

ics Researcher Quits Science for Politics," *Science*, February 13, 1970, p. 963–964.

104. Glassman, "Harvard Genetics Researcher Quits Science for Politics."

105. Ibid.

106. Jacques R. Fresco, George P. Hess, Richard L. Russell, Jay C. Brown, Shyam Dube, John F. Wooten, James A. Spudich, Thomas A. Steitz, Francis A. Kallefelt, and Stephen J. Miller, "Shapiro's Defection," *Science*, March 27, 1970, pp. 1668–1669; Francis A. Kallefelt, Letter to the Editor, *Science*, March 27, 1970, p. 1668.

107. Glassman, "Harvard Genetics Researcher Quits Science for Politics."

108. Science for the People, Madison Collective, *The AMRC Papers: An Indictment of the Army Mathematics Research Center* (Madison, WI: Science for the People, Madison Collective, 1973).

109. Tom Bates, *Rads: The 1970 Bombing of the Army Math Research Center and Its Aftermath* (New York: Harper Collins, 1992), p. 404.

110. John Vandermeer, interview by the author, Ann Arbor, MI, May 16, 2001; Richard Lewontin, interview by the author, December 4, 1991; *Science for Nicaragua*, brochure, 2000, author's files; John H. Vandermeer, "The Ecological Basis of Alternative Agriculture," *Annual Review of Ecology and Systematics* 26 (1995): 201–224; John H. Vandermeer, *Reconstructing Biology: Genetics and Ecology in the New World Order* (New York: John Wiley and Sons, 1996); Peter Rosset and John Vandermeer, eds., *The Nicaragua Reader: Documents of a Revolution under Fire* (New York: Grove Press, 1983).

111. Ann Arbor Science for the People, *Biology as a Social Weapon* (Minneapolis: Burgess Publishing, 1977).

112. Because the network was so extensive and so many people were involved, it is not possible to capture the full range of the group's activities in one chapter, but I have highlighted what I think is typical of their work with other political groups. I have not, however, addressed a perhaps equally important aspect of the activities of SftP, the SSRS, or CNI, which is their effects on the intellectual and scientific activities of members of networks that overlapped with SftP. Examples include Donna Haraway, Evelyn Hammonds, Evelyn Fox Keller, and Ruth Hubbard. In her book *Has Feminism Changed Science?* Londa Schiebinger develops a method of understanding how feminist ideas have shaped the content of scientific knowledge; an adjunct to her method would be to use a biographical analysis of feminists and other activists to examine how their work—their questions and the range of answers they examine—have been shaped by political events and activities. Londa Schiebinger, *Has Feminism Changed Science?* (Cambridge, MA: Harvard University Press, 1999).

113. Martin, "Science in a Citizen-Shaped World."

CHAPTER 7: CONCLUSIONS

1. Yaron Ezrahi, *The Descent of Icarus: Science and the Transformation of Contemporary Democracy* (Cambridge, MA: Harvard University Press, 1990), pp. 274–275.

2. See, for example, Norman Levitt, *Prometheus Bedeviled: Science and the Contemporary Contradictions of Culture* (New Brunswick, NJ: Rutgers University Press 1999); Paul R. Gross and Norman Levitt, *Higher Superstition: The Academic Left and Its Quarrels with Science* (Baltimore: Johns Hopkins University Press, 1998); Alan Sokal and Jean Bricmon, *Fashionable Nonsense: Postmodern Intellectuals' Abuse of Science* (New York: Picador, 1998); Paul R. Gross, Norman Levitt, and Martin W. Lewis, eds., *The Flight from Science and Reason* (Baltimore: Johns Hopkins University Press, 1997); M. F. Ashley Montagu, ed., *Science and Creationism* (New York: Oxford University Press, 1994); National Academy of Sciences, *Science and Creationism: A View from the National Academy of Sciences* (Washington, DC: National Academies Press, 1999); Matt Young and Taner Edis, eds., *Why Intelligent Design Fails: A Scientific Critique of the New Creationism* (New Brunswick, NJ: Rutgers University Press, 2004); Jon D. Miller, "Public Understanding of, and Attitudes toward, Scientific Research: What We Know and What We Need to Know," *Public Understanding of Science* 13 (2004): 273–294; Gerald Holton, *Science and Anti-Science* (Cambridge, MA: Harvard University Press, 1993); and Nathan Newman, "Big Pharma, Bad Science," *The Nation*, July 25, 2002, pp. 4–8.

3. Jennifer Croissant, "Critical Legal Theory and Critical Science Studies: Engaging Institutions," *Cultural Dynamics* 12, no. 2 (2000): 225.

4. Peter J. Kuznick, *Beyond the Laboratory: Scientists as Political Activists in 1930s America* (Chicago: University of Chicago Press, 1987); Elizabeth Hodes, "Precedents for Social Responsibility among Scientists: The American Association of Scientific Workers and the Federation of American Scientists, 1938–1948" (Ph.D. dissertation, University of California–Santa Barbara, 1982); Edwin T. Layton Jr., *The Revolt of the Engineers: Social Responsibility and the American Engineering Profession* (Cleveland: Press of Case Western Reserve University, 1971); Charles E. Rosenberg, "Piety and Social Action: Some Origins of the American Public Health Movement," in *No Other Gods: On Science and American Social Thought*, rev. ed. (Baltimore: Johns Hopkins University Press, 1987), pp. 109–122; Garland Allen, "The Misuse of Biological Hierarchies: The American Eugenics Movement, 1900–1940," *History and Philosophy of the Life Sciences* 5 (1983): 105–128; Wendy Kline, *Building a Better Race: Gender, Sexuality, and Eugenics from the Turn of the Century to the Baby Boom* (Berkeley and Los Angeles: University of California Press, 2001).

5. Charles Thorpe, "Disciplining Experts: Scientific Authority and Liberal Democracy in the Oppenheimer Case," *Social Studies of Science* 32 (2002): 544.

6. Patrick Carroll, *Science, Culture, and Modern State Formation* (Berkeley and Los Angeles: University of California Press, 2006).

7. On political opportunities, see Doug McAdam, *Political Process and the Development of Black Insurgency, 1930–1970* (Chicago: University of Chicago Press, 1982).

8. Much of the analysis of scientists and politics is centered around the idea of the scientist as expert, with concerns about how to "bridge" the lay-expert divide. Scott Frickel, "Building an Interdiscipline: Collective Action Framing and the Rise of Genetic Toxicology," *Social Problems* 51 (2004): 269–287; Mike Michael and Nik Brown, "Switching between Science and Culture in Transpecies Transplanta-

tion," *Science, Technology, and Human Values* 26, no. 1 (2001): 3–22; Wolff-Michael Roth et al., "Those Who Get Hurt Aren't Always Being Heard: Scientist-Resident Interactions over Community Water," *Science, Technology, and Human Values* 29 (2004): 153–183; William R. Freudenberg, "Scientific Expertise and Natural Resource Decisions: Social Science Participation on Interdisciplinary Scientific Committees," *Social Science Quarterly* 83 (2002): 119–136; Heather Arksey, "Expert and Lay Participation in the Construction of Medical Knowledge," *Sociology of Health and Illness* 16 (1994): 448–468; Steven Epstein, *Impure Science: AIDS, Activism, and the Politics of Knowledge* (Berkeley and Los Angeles: University of California Press, 1996); Brian Wynne, "May the Sheep Safely Graze? A Reflexive View of the Expert-Lay Knowledge Divide," in *Risk, Environment, and Modernity: Towards a New Ecology*, ed. Scott Lash, Bronislaw Szerszynski, and Brian Wynne (London: Sage Publications, 1996), pp. 44–83; Edward J. Woodhouse and Dean A. Nieumsa, "Democratic Expertise: Integrating Knowledge, Power, and Participation," in *Knowledge, Power, and Participation in Environmental Policy Analysis*, ed. Matthjis Hisschemoller et al. (New Brunswick, NJ: Transaction, 2001), pp. 73–96.

9. Bruno Latour, *Politics of Nature: How to Bring the Sciences into Democracy* (Cambridge, MA: Harvard University Press, 2004).

10. By drawing attention to material, ideological, and organizational circumstances, I emphasize what Daniel Lee Kleinman, Scott Frickel, and I have called the "already constructed" systems of power that shape the distribution of power in sociotechnical debates. See Daniel Lee Kleinman, *Impure Cultures: University Biology and the World of Commerce* (Madison: University of Wisconsin Press, 2003), p. 62; and Scott Frickel and Kelly Moore, "Prospects and Challenges for a New Political Sociology of Science," in *The New Political Sociology of Science: Institutions, Networks, and Power*, ed. Scott Frickel and Kelly Moore (Madison: University of Wisconsin Press, 2006), pp. 3–34. This perspective draws on political sociology, which has paid close attention to how existing political systems shape possibilities for political innovation and transformation. Three exemplary books are Elisabeth S. Clemens, *The People's Lobby: Organizational Innovation and the Rise of Interest Group Politics in the United States, 1890–1925* (Chicago: University of Chicago Press, 1997); John D. Skrentny, *The Minority Rights Revolution* (Cambridge, MA: Belknap Press of Harvard University Press, 2002); and McAdam, *Political Process*. Although there has been a new turn in historical and laboratory-based studies toward an understanding of how larger-scale systems of power shape the activities of scientists and their knowledge claims, this understanding gives less priority—and power—to existing institutions, rules, and other relationships of power than I do here, emphasizing instead the "coproduction" of science. Exemplary of this turn toward coproduction is Sheila Jasanoff, ed., *States of Knowledge: The Co-production of Science and Social Order* (London: Routledge, 2004).

11. Maren Klawiter, "Racing for the Cure, Walking Women, and Toxic Touring: Mapping Cultures of Action within the Bay Area Terrain of Breast Cancer," *Social Problems* 46 (1999): 104–126.

12. Daniel J. Myers and Daniel M. Cress, "Authority in Contention," *Research in Social Movements, Conflicts and Change* 25 (2004): 279–293.

13. In Max Weber's words, "This whole process of rationalization . . . , especially in the bureaucratic state machine, parallels the centralization of the material implements of organization in the hands of the master. Thus, discipline inexorably takes over ever larger areas as the satisfaction of political and economic needs is increasingly rationalized. This universal phenomenon more and more restricts the importance of charisma and of individually differentiated conduct." *Max Weber on Law in Economy and Society*, ed. Max Rheinstein, trans. Edward Shils and Max Rheinstein (New York: Simon & Schuster, 1968), pp. 1151.

14. Charles Thorpe and Steven Shapin, "Who Was J. Robert Oppenheimer? Charisma and Complex Organization." *Social Studies of Science* 30 (2000): 545–590. Scientists' biographies and autobiographies nearly always emphasize the personal qualities of scientists, particularly their capacity to take intellectual risks beyond those their peers were taking. The narratives are normally structured to indicate that scientists had to overcome the skepticism of peers, but persevered as a result of their hunches and know-how.

15. Daniel Lee Kleinman, ed., *Science, Technology, and Democracy* (Albany: State University of New York Press, 2000).

16. David J. Hess, "Antiangiogenesis Research and Scientific Fields," in *The New Political Sociology of Science: Institutions, Networks, and Power*, ed. Scott Frickel and Kelly Moore (Madison: University of Wisconsin Press, 2006), pp. 122–147.

17. Sheila Jasanoff, "Ordering Knowledge, Ordering Society," in *States of Knowledge: The Co-production of Science and Social Order*, ed. Sheila Jasanoff (London: Routledge, 2004), pp. 3–45. Jasanoff's "coproduction" metaphor is a useful way of ensuring that we do not reify "science" and "society" but instead see them as end points that are created through interaction. The coproduction metaphor, however, does not capture the fact that playing fields are not level when it comes to negotiating power. Existing institutions shape who has which kinds of resources and what the rules are for rearranging them. This sense of power is more evident in Sheila Jasanoff, *Designs on Nature* (Princeton, NJ: Princeton University Press, 2005). As I hope I have shown, change is possible; it is just more difficult for those without power.

18. Steven Shapin, *A Social History of Truth: Civility and Science in Seventeenth-Century England* (Chicago: University of Chicago Press, 1994), especially chapter 6, "Knowing about People and Knowing about Things: A Moral History of Scientific Credibility."

19. Peter Galison and David J. Stump, eds., *The Disunity of Science: Boundaries, Contexts, and Power* (Stanford, CA: Stanford University Press, 1996); Karin Knorr Cetina, *Epistemic Cultures: How the Sciences Make Knowledge* (Cambridge, MA: Harvard University Press, 1999).

20. Scientists were not in any sense unified before 1945. All professions have intraprofessional networks and subunits that share knowledge practices and ideas about the proper codes of public conduct. These networks, however, were not routinely exposed to public scrutiny in the 1940s and 1950s. See Sydney A. Halpern, "Dynamics of Professional Control: Internal Coalitions and Crossprofessional Boundaries," *American Journal of Sociology* 97 (1992): 994–1021, on the importance of internal coalitions and changes in professions.

21. Stephen Hilgartner, *Science on Stage: Expert Advice as Public Drama* (Stanford, CA: Stanford University Press, 2000).

22. Brian Balogh, *Chain Reaction: Expert Debate and Public Participation in American Commercial Nuclear Power, 1945–1975* (New York: Cambridge University Press, 1991).

23. Jeff Goodwin, *No Other Way Out: States and Social Revolutions* (New York: Cambridge University Press, 2001); Arthur Stinchcombe, "Social Structure and Organizations," in *Handbook of Organizations*, ed. James G. March (Chicago: Rand McNally, 1965), pp. 142–193.

24. Daniel Sarewitz, *Frontiers of Illusion: Science, Technology, and the Politics of Progress* (Philadelphia: Temple University Press, 1996); Daniel S. Greenberg, *The Politics of Pure Science* (New York: New American Library, 1969); David Dickson, *The New Politics of Science* (New York: Pantheon, 1984).

25. Bruno Latour and Steve Woolgar, *Laboratory Life: The Construction of Scientific Facts* (Princeton, NJ: Princeton University Press, 1986).

26. Thomas F. Gieryn, "Boundary-Work and the Demarcation of Science from Non-Science: Strains and Interests in Professional Ideologies of Scientists," *American Sociological Review* 48 (1983): 782.

27. Kelly Moore, "Organizing Integrity: American Science and the Creation of Public Interest Organizations, 1955–1975," *American Journal of Sociology* 101 (1996): 1592–1627; Diane Vaughan, "Boundary Work: Levels of Analysis, the Macro-Micro Link, and the Social Control of Organizations," in *Social Science, Social Policy, and the Law*, ed. Patricia Ewick, Robert A. Kagun, and Austin Sarat (New York: Russell Sage Foundation, 1999), pp. 291–321; David H. Guston, "Stabilizing the Boundary between US Politics and Science: The Role of the Office of Technology Transfer as a Boundary Organization," *Social Studies of Science* 29 (1999): 87–111; Clark Miller, "Hybrid Management: Boundary Organizations, Science Policy, and Environmental Governance in the Climate Regime," *Science, Technology, and Human Values* 26 (2001): 478–501; Edmund Ramsden, "Carving Up Population Science: Eugenics, Demography and the Controversy over the 'Biological Law' of Population Growth," *Social Studies of Science* 32 (2002): 857–899; Tomas Hellstrom and Merle Jacob, "Boundary Organizations in Science: From Discourse to Construction," *Science and Public Policy* 30 (2003): 235–238; Felicity Mellor, "Between Fact and Fiction: Demarcating Science from Non-Science in Popular Physics Books," *Social Studies of Science* 33 (2003): 509–538; Susan E. Kelly, "Public Bioethics and Publics: Consensus, Boundaries, and Participation in Biomedical Science Policy," *Science, Technology, and Human Values* 28 (2003): 339–364; Juha Tuunainen, "Contesting a Hybrid Firm at a Traditional University," *Social Studies of Science* 35 (2005): 173–210. Two notable exceptions to this rule are Steven Peter Vallas, "Symbolic Boundaries and the New Division of Labor: Engineers, Workers, and the Restructuring of Factory Life," *Research in Social Stratification and Mobility* 18 (2001): 3–37; and Daniel Lee Kleinman and Abby J. Kinchy, "Boundaries in Science Policy Making: Bovine Growth Hormone in the European Union," *Sociological Quarterly* 44 (2003): 577–595.

28. Jason Owen-Smith, "From Separate Spheres to a Hybrid Order: Accumulative Advantage across Public and Private Science at Research One Universities,"

Research Policy 32 (2003): 1081–1104; Donna Haraway, "A Cyborg Manifesto: Science, Technology, and Socialist-Feminism in the Late Twentieth Century," in *Simians, Cyborgs and Women: The Reinvention of Nature* (London: Routledge, 1991), pp. 149–181; Daniel Lee Kleinman and Steven P. Vallas, "Science, Capitalism, and the Rise of the 'Knowledge Worker': The Changing Structure of Knowledge Production in the United States," *Theory and Society* 30, no. 4 (2001): 451–492; Matthias Gross, "Community by Experiment: Recursive Learning in Landscape Design and Ecological Restoration" (paper presented at the International Sociological Association's Mini-Conference on Community and Ecology, San Francisco, 2004).

29. Abby J. Kinchy and Daniel Lee Kleinman, "Organizing Credibility: Discursive and Organizational Orthodoxy on the Borders of Ecology and Politics," *Social Studies of Science* 33 (2003): 869–896.

30. See, for example, Londa Schiebinger, *Nature's Body: Gender in the Making of Modern Science* (Boston: Beacon Press, 1993); Sandra Harding, *The Science Question in Feminism* (Ithaca, NY: Cornell University Press, 1986); Sandra Harding, *Whose Science, Whose Knowledge?* (Ithaca, NY: Cornell University Press, 1991).

31. Sandra Morgen, *Into Our Own Hands: The Women's Health Movement in the United States, 1969–1990* (New Brunswick, NJ: Rutgers University Press, 2002); Carol S. Weisman, *Women's Health Care: Activist Traditions and Institutional Change* (Baltimore: Johns Hopkins University Press, 1998); Phil Brown, *Transfer of Care: Psychiatric Deinstitutionalization and Its Aftermath* (Boston: Routledge and Kegan Paul, 1985).

32. Rachel Morello-Frosch, Stephen Zavestoski, Phil Brown, Rebecca Gasior Altman, Sabrina McCormick, and Brian Mayer, "Embodied Health Movements: Responses to a 'Scientized' World," in *The New Political Sociology of Science: Networks, Institutions, and Power*, ed. Scott Frickel and Kelly Moore (Madison: University of Wisconsin Press, 2006), pp. 244–271.

33. See also Nick Crossley, *Contesting Psychiatry: Social Movements in Mental Health* (London: Routledge, 2005); Nick Crossley, "Mental Health, Resistance and Social Movements: The Collective-Confrontational Dimension," *Health Education Journal* 61, no. 2 (2002): 138–152.

34. Epstein, *Impure Science*.

35. See, for example, Barbara A. Israel, Eugenia Eng, Amy J. Schulz, Edith A. Parker, and David Satcher, *Methods in Community-Based Participatory Research for Health* (San Francisco: Jossey-Bass, 2005); Sabrina McCormick, Julia Brody, Phil Brown, and Ruth Polk, "Lay Involvement in Breast Cancer Research," *International Journal of Health Services* 34 (2004): 625–646; Stephen Zavestoski, Phil Brown, Sabrina McCormick, Brian Mayer, Maryhelen D'Ottavi, and Jaime Lucove, "Patient Activism and the Struggle for Diagnosis: Gulf War Illnesses and Other Medically Unexplained Physical Symptoms in the U.S.," *Social Science and Medicine* 58 (2004): 161–175; and Phil Brown, Sabrina McCormick, Brian Mayer, Stephen Zavestoski, Rachel Morello-Frosch, Rebecca Gasior Altman, and Laura Senie, "'A Lab of Our Own': Environmental Causation of Breast Cancer and Challenges to the Dominant Epidemiological Paradigm," *Science, Technology, and Human Values* 31, no. 5 (2006): 499–536. This discussion is based on

Kelly Moore, "Powered by the People: Scientific Authority in Participatory Science," in *The New Political Sociology of Science: Networks, Institutions, and Power*, ed. Scott Frickel and Kelly Moore (Madison: University of Wisconsin Press, 2006), pp. 299–326.

36. Jenny Reardon, *Race to the Finish: Identity in an Age of Genomics* (Princeton, NJ: Princeton University Press, 2005).

37. Laurel Smith-Doerr, "Learning to Reflect or Deflect? U.S. Policies and Graduate Programs' Ethics Training for Life Scientists," in *The New Political Sociology of Science: Institutions, Networks, and Power*, ed. Scott Frickel and Kelly Moore (Madison: University of Wisconsin Press, 2006), pp. 405–431. Sydney Halpern argues that current, formalized systems of risk regulation in medicine may not be as effective as harnessing more informal moral systems among researchers. Sydney A. Halpern, *Lesser Harms: The Morality of Risk in Medical Research* (Chicago: University of Chicago Press, 2004).

38. Vivian Weil, ed., *Trying Times: Science and Responsibilities after Daubert* (Chicago: Center for the Study of Ethics in the Professions, and Institute for Science, Law and Technology, Illinois Institute of Technology, 2001); *A Compilation of Federal Science Laws: As Amended through December 31, 2003*, prepared for the use of the House Committee on Science (Washington, DC: GPO, 2004); Paul Weindling, "The Origins of Informed Consent: The International Scientific Commission on Medical War Crimes, and the Nuremberg Code," *Bulletin of the History of Medicine* 75 (2001): 37–71.

39. Roger Smith and Brian Wynne, "Establishing the Rules of Laws: Constructing Expert Authority," in *Expert Evidence: Interpreting Science in the Law*, ed. Roger Smith and Bryan Wynne (London: Routledge, 1989), pp. 23–55; Kenneth Foster and Peter Huber, "Scientific Knowledge," in *Judging Science: Scientific Knowledge and the Federal Courts* (Cambridge, MA: MIT Press, 1999), pp. 1–22; Gary Edmond, "Legal Engineering: Contested Representations of Law, Science (and Non-science) and Society," *Social Studies of Science* 32 (2002): 371–412; Sheila Jasanoff, "Science and the Statistical Victim: Modernizing Knowledge in Breast Implant Litigation," *Social Studies of Science* 32 (2002): 37–69; Gary Edmond and David Mercer, "Litigation Life: Law-Science Knowledge Construction in (Bendectin) Mass Toxic Tort Litigation," *Social Studies of Science* 30 (2000): 265–316.

40. Sheila Jasanoff, *Science at the Bar: Law, Science, and Technology in America* (Cambridge, MA: Harvard University Press, 1995); Sheila Jasanoff, "The Eye of Everyman: Witnessing DNA in the Simpson Trial," *Social Studies of Science* 28 (1998): 713–740.

41. Michael Lynch, "Administrative Science: The Management of Objectivity and the Objectivity of Management [#1]" (paper presented at the Annual Meeting of the American Sociological Association, Philadelphia, 2005); see also Simon A. Cole, *Suspect Identities: A History of Fingerprinting and Criminal Identification* (Cambridge, MA: Harvard University Press, 2001).

42. Kleinman, *Impure Cultures*; Jennifer R. Fishman, "Manufacturing Desire: The Commodification of Female Sexual Dysfunction," *Social Studies of Science* 34 (2004): 187–218; Peter Conrad, "The Shifting Engines of Medicalization," *Journal of Health and Social Behavior* 46 (2005): 5–8; Walter W. Powell, Douglas

R. White, Kenneth W. Koput, and Jason Owen-Smith, "Network Dynamics and Field Evolution: The Growth of Inter-organizational Collaboration in the Life Sciences," *American Journal of Sociology* 110 (2004): 1132–1205; Jason Owen-Smith and Walter W. Powell, "Knowledge Networks as Channels and Conduits: The Effects of Spillovers in the Boston Biotechnology Community," *Organization Science* 15, no. 1 (2004): 5–21; Jennifer Croissant and Sal Restivo, eds., *Degrees of Compromise: Industrial Interests and Academic Values* (Albany: State University of New York Press, 2001).

43. Julie Bosman, "Reporters Find Science Journals Harder to Trust, but Not Easy to Verify," *New York Times*, February 13, 2006 (online version accessed June 15, 2006).

44. Mayer N. Zald and Roberta Ash, "Social Movement Organizations: Growth, Decay and Change," *Social Forces* 44 (1966): 61–72; D. VonEschen, D. Kirk, and Maurice Pinard, "The Organizational Substructure of Disorderly Politics," *Social Forces* 49 (1971): 529–544; Suzanne Staggenborg, *The Pro-Choice Movement: Organization and Activism in the Abortion Conflict* (New York: Oxford University Press, 1991); Debra C. Minkoff, "The Organization of Survival: Women's and Racial-Ethnic Voluntarist and Activist Organizations, 1955–1985," *Social Forces* 71 (1993): 887–908; John D. McCarthy and Mayer N. Zald, *The Trend of Social Movements in America: Professionalization and Resource Mobilization* (Morristown, NJ: General Learning Press, 1973); John D. McCarthy and Mayer N. Zald, "Resource Mobilization and Social Movements: A Partial Theory," *American Journal of Sociology* 82 (1977): 1212–1241; McAdam, *Political Process and the Development of Black Insurgency*; Jürgen Gerhards and Dieter Rucht, "Mesomobilization, Organizing, and Framing in Two Protest Campaigns in West Germany," *American Journal of Sociology* 98 (1992): 555–595, Debra C. Minkoff and John D. McCarthy, "Reinvigorating the Study of Organizational Processes in Social Movements," *Mobilization: An International Journal* 10 (2005): 289–308.

45. William A. Gamson, "Goffman's Legacy to Political Sociology," *Theory and Society* 14 (1985): 605–622.

46. Framing is now incorporated into a conceptual structure in the sociology of social movements that emphasizes the strategies that activists use to persuade others. Its logic is based in the sociology of rational choice theory, although most writers do not explicitly espouse this viewpoint. Most frame analysis is concerned with two issues: which ways of bounding a political claim are most persuasive, and how these claims function in social movements. What is to be explained is why some existing ideas are persuasive to others. What distinguishes "rim talk" and my perspective is that both emphasize the processes by which the "frames" are formed in the first place. Like frame analysts, scholars of social movements who study the "micromobilization" process examine how preexisting social networks and frames have led people to participate in social movements. In this view, society is organized like a layer cake, with "macro"-level systems being larger systems such as states, "meso" contexts being organizations, and "micro" contexts being face-to-face contexts. The central concern is to see how the "micro" aggregates to the "macro." See, for example, Karl-Dieter Opp and Wolfgang Roehl, "Repression, Micromobilization, and Political Protest," *Social Forces* 69

(1988): 521–547; Doug McAdam, "Micromobilization Contexts and Recruitment to Activism," *International Social Movement Research* 1 (1988): 125–154; Belinda Robnett, "African-American Women in the Civil Rights Movement, 1954–1965: Gender, Leadership, and Micromobilization," *American Journal of Sociology* 101 (1996): 1661–1693.

47. Francesca Polletta, *Freedom Is an Endless Meeting* (Chicago: University of Chicago Press, 2004); Paul Lichterman, *The Search for Political Community: American Activists Reinventing Commitment* (New York: Cambridge University Press, 1996).

48. Joseph R. Gusfield, *Symbolic Crusade: Status Politics and the American Temperance Movement* (Urbana: University of Illinois Press, 1963), p. 361.

49. Kelly Moore, "Political Protest and Institutional Change: The Anti–Viet Nam War Movement and American Science," in *How Social Movements Matter*, ed. Marco Guigni, Doug McAdam, and Charles Tilly (Minneapolis: University of Minnesota Press, 1999), pp. 97–118.

50. Rebecca Slayton, "Discursive Choices: Boycotting Star Wars between Science and Politics," *Social Studies of Science* 37 (2007): 27–66; Michael R. Nusbaumer, Judith A. DiIorio, and Robert D. Baller, "The Boycott of 'Star Wars' by Academic Scientists: The Relative Roles of Political and Technical Judgment," *Social Science Journal* 31 (1994): 375–388.

51. This oath is posted online at www.spusa.org/pledge/index.html (accessed June 1, 2006).

52. In 1971, a British group calling itself the Society for Social Responsibility in Science formed, but it had much more in common with the ideas of SftP than with those of the SSRS in the United States. Another group by the same name formed in Australia in the 1970s and collapsed in the 1980s. Neither group was connected to the American SSRS.

53. See www.ncsl.org/programs/health/conscienceclauses.htm (accessed June 18, 2006).

54. See the organization's Web site, at www.cbns.qc.edu/ (accessed June 18, 2006).

55. Gerald Holton and Gerhard Sonnert, *Ivory Bridges: Connecting Science and Society* (Cambridge, MA: MIT Press, 2002). Holton and Sonnert provide an appendix listing dozens of organizations in which "science" and "society" are bridged. The appendix is a welcome starting point, but the problem with counting such organizations is that they do not always survive for long periods, yet the networks of people who were linked by them sometimes do. What we need is more research on networks of scientists who are engaged in resolving political and social issues. At another level, one of the reasons that it is hard to sort out what constitutes a "bridge" or a "boundary" organization is that organizations that supposedly are apolitical, such as professional associations, also have political goals. The reproduction of the separate categories of political and scientific, then, may be less helpful than trying to understand how scientists and those with whom they work aim to change relations of power in networks of people and in the material world.

56. McCarthy and Zald, *The Trend of Social Movements in America*.

57. Barbara Ehrenreich and John Ehrenreich, "The New Left: A Case Study in Professional-Managerial Class Radicalism," *Radical America* 11 (1977): 7–22; David Wagner, *The Quest for a Radical Profession: Social Service Careers and Political Ideology* (Lanham, MD: University Press of America, 1990); Robert Perrucci, "In the Service of Man: Radical Movements in the Professions," *Sociological Review Monograph* 20 (1973): 179–194; Brian J. Heraud, "Professionalism, Radicalism and Social Change," *Sociological Review Monograph* 20 (1973): 85–101; Barbara Ehrenreich, "Giving Power to the People: The Early Days of Health/PAC," *Health/PAC Bulletin* 18, no. 4 (1988): 4–8.

58. See Walter W. Powell, "Neither Market nor Hierarchy: Network Forms of Organization," *Research in Organizational Behavior* 12 (1990): 295–336, on the importance of networks, rather than formal bureaucratic organizations, as an important form of organizing in contemporary life. Scott Frickel calls this new kind of activism among scientists "shadow mobilizing." In his research on the formation of the Environmental Mutagen Society in the 1960s and on contemporary environmental scientists, he has found that in their everyday scientific work, scientists carry out research projects with political consequences for people other than scientists. They do not normally call this work "political," but fold it into their everyday practices. Scott Frickel, "Scientific Authority and Expert Activism in Environmental Health and Justice Movements" (paper presented to the History and Sociology of Science Workshop, University of Pennsylvania, November 28, 2005).

59. Amy Harmon, "That Wild Streak? Maybe It Runs in the Family," *New York Times*, June 15, 2006, www.nytimes.com/2006/06/15/health/15gene.html?_r = 1&oref = slogin (accessed June 15, 2006).

60. Science education is one way to change this situation, but it must be the kind in which students do science, rather than learn about how scientists do it. Some may prefer to work on problems that are of interest in their own lives, as a way of sustaining interest. Most young children enjoy the everyday activity of exploring the patterns of nature in their world; by the time they are young adults, they are vastly less interested. One might, then, consider how to teach science in a way that allows people to solve problems of their everyday lives as a way to recapture this kind of enthusiasm.

61. Brian Wynne, "Reflexing Complexity: Post-Genomic Knowledge and Reductionist Returns in Public Science," *Theory, Culture and Society* 22 (2005): 69.

62. David J. Hess, "Medical Modernisation, Scientific Research Fields and the Epistemic Politics of Health Social Movements," *Sociology of Health and Illness* 24 (2004): 695–709.

63. Barbara Allen, *Uneasy Alchemy: Citizens and Experts in Louisiana's Chemical Corridor Disputes* (Cambridge, MA: MIT Press, 2003); Edward J. Woodhouse, "Change of State? The Greening of Chemistry," in *Synthetic Planet: Chemical Products in Modern Life*, ed. Monica J. Casper (New York: Routledge, 2003), pp. 177–193; Edward J. Woodhouse and Steve Breyman, "Green Chemistry as Social Movement?" *Science, Technology and Human Values* 30 (2005): 199–222; Weisman, *Women's Health Care*.

Bibliography

ORAL HISTORIES AND PERSONS INTERVIEWED

Interviews by the Author

Anonymous biologist, tape-recorded interview, Cambridge, Massachusetts, December 4, 1991

Anonymous member of SACC, tape-recorded interview, Greenfield, Massachusetts, March 20, 1991

Anonymous MIT professor of physics, tape-recorded interview, Cambridge, Massachusetts, June 13, 1991

Rita Arditti, telephone interview, May 4, 1999

Jonathan Beckwith, tape-recorded interview, Cambridge, Massachusetts, July 24, 1998

O. Theodore Benfey, telephone interview, June 11, 2004

Richard Blankenbecler, telephone interview, August 30, 2006

Alan Chodos, tape-recorded interview, New Haven, Connecticut, March 18, 1991

Barry Commoner, tape-recorded interview, Queens, New York, June 12, 1991

Joel Feigenbaum, tape-recorded interview, West Barnstable, Massachusetts, March 19, 1991

Britta Fischer, tape-recorded interview, Brookline, Massachusetts, December 5, 1991

Michael Goldhaber, tape-recorded interview, San Francisco, California, February 2, 1992

David Kotelchuck, tape-recorded interview, New York, New York, December 8, 1991

Franklin Miller, telephone interview, July 7, 2004

Stuart Newman, tape-recorded interviews, Valhalla, New York, February 3, 1991, and July 15, 1999

Charles Schwartz, tape-recorded interview, Berkeley, California, August 13, 1998

John Vandermeer, interview, Ann Arbor, Michigan, May 16, 2001

Al Weinrub, interview, Oakland, California, August 16, 2003

Oral Histories and Interviews by Other Researchers

AMERICAN INSTITUTE OF PHYSICS, COLLEGE PARK, MARYLAND

Richard Blankenbecler, oral history by Finn Aaserud, May 5, 1987

Sidney Drell, oral history by Finn Aaserud, July 1, 1986

Charles Schwartz, oral history by Finn Aaserud, May 15, 1987

WESTERN HISTORICAL MANUSCRIPT COLLECTION,
UNIVERSITY OF MISSOURI–ST. LOUIS

John Fowler, interview by William Sullivan, January 15, 1982

Joseph Klarman, Dan Bolef, and Gertrude Faust, interview by William Sullivan, February 4, 1982

CENTER FOR THE STUDY OF HISTORY AND MEMORY, INDIANA UNIVERSITY, BLOOMINGTON

Walter Bauer, interview by D. Scott Peterson, April 27, 1973

Virginia Brodine, interview by D. Scott Peterson, May 12, 1972

Barry Commoner, interview by D. Scott Peterson, April 24, 1973

REGIONAL ORAL HISTORY OFFICE, BANCROFT LIBRARY,
UNIVERSITY OF CALIFORNIA–BERKELEY

Charles Hard Townes. "A Life in Physics: Bell Telephone Laboratories and World War II, Columbia University and the Laser, MIT and Government Service, California and Astrophysics." 1994. An oral history conducted in 1991–1992 by Suzanne B. Reiss.

ARCHIVES AND SPECIAL COLLECTIONS

American Association for the Advancement of Science (AAAS) Archives, Washington, DC

American Institute of Physics, College Park, Maryland

American Physical Society (APS) Papers

American Philosophical Society, Philadelphia

E. G. Ramberg (ER) Papers

Charles Deering McCormick Library Special Collections, Northwestern University, Evanston, Illinois

Columbiana Archives, Columbia University, New York

Haverford College, Haverford, Pennsylvania

Otto Theodore Benfey (OTB) Papers

John M. Olin Library, Washington University, St. Louis, Missouri

Edna Gellhorn Papers

Library of Congress, Washington, DC
Barry Commoner Collection (BCC)

Massachusetts Institute of Technology Archives and Special Collections, Cambridge, Massachusetts
Science Action Coordinating Committee (SACC) Papers
Science for the People (SftP) Papers
Union of Concerned Scientists (UCS) Papers

Swarthmore College Peace Collection, Swarthmore, Pennsylvania
Papers of Victor Paschkis
Clergy and Laity Concerned Papers
Peace Groups—Misc. 1957

University of Chicago Archives and Special Collections
Atomic Scientists Collection

Western Historical Manuscript Collection, University of Missouri–St. Louis
Committee for Environmental Information Records (CEIR)

Secondary Sources

Abbott, Andrew. *The System of Professions.* Chicago: University of Chicago Press, 1982.
Ainley, Marianne Gosztonyi. "The Contribution of the Amateur to North American Ornithology: A Historical Perspective." *Living Bird* 18 (1979/80): 161–177.
Albert, Judith Clavir, and Stuart Albert, eds. *The Sixties Papers: Documents of a Rebellious Decade.* New York: Praeger, 1984.
Allen, Barbara. *Uneasy Alchemy: Citizens and Experts in Louisiana's Chemical Corridor Disputes.* Cambridge, MA: MIT Press, 2003.
Allen, Jonathan, ed. *March 4: Scientists, Students and Society.* Cambridge, MA: MIT Press, 1970.
Allen, Garland. "The Misuse of Biological Hierarchies: The American Eugenics Movement, 1900–1940." *History and Philosophy of the Life Sciences* 5 (1983): 105–128.
———. "Science as Moral Economy." *History and Philosophy of the Life Sciences* 18 (1996):129–134.
Amenta, Edwin, Neal Caren, Tina Fetner, and Michael P. Young. "Challengers and States: Toward a Political Sociology of Social Movements." *Research in Political Sociology* 10 (2002): 47–83.

Anonymous. "Witness at the Conspiracy Trial, 1969." In *The Sixties Papers: Documents of a Rebellious Decade*, ed. Judith Clavir Albert and Stuart Albert, p. 391. New York: Praeger, 1984.

Archibald, Sam. "The Early Years of the Freedom of Information Act, 1955–1975." *PS: Political Science and Politics* 26 (1993): 726–731.

Arksey, Heather. "Expert and Lay Participation in the Construction of Medical Knowledge. *Sociology of Health and Illness* 16 (1994): 448–468.

Atkinson, Richard C. "Science Advice at the Cabinet Level." In *Science and Technology Advice to the President, Congress, and Judiciary*, ed. William T. Golden, pp. 11–15. New York: Pergamon Books, 1993.

Avorn, Jerry. *Up against the Ivy Wall: A History of the Columbia University Crisis*. Ed. and with an introduction by Andrew Crane, with the editors of the *Columbia University Spectator*. New York: New Atheneum, 1969.

Badash, Lawrence. *Scientists and the Development of Nuclear Weapons: From Fission to the Limited Test Ban Treaty, 1939–1963*. Atlantic Highlands, NJ: Humanities Press, 1993.

Balogh, Brian. *Chain Reaction: Expert Debate and Public Participation in American Commercial Nuclear Power, 1945–1975*. New York: Cambridge University Press, 1991.

Baltzell, E. Digby. *Puritan Boston and Quaker Philadelphia*. Reprint edition. New Brunswick, NJ: Transaction Publishers, 1996.

Bates, Ralph S. *Scientific Societies in the United States*. Cambridge: Cambridge University Press, 1965.

Bates, Tom. *Rads: The 1970 Bombing of the Army Math Research Center and Its Aftermath*. New York: Harper Collins, 1992.

Bazerman, Charles. "Nuclear Information: One Rhetorical Moment in the Construction of the Information Age." *Written Communication* 18, no. 3 (2001): 259–295.

Beckwith, Jon. "The Origins of the Radical Science Movement." *Monthly Review* 3 (1986): 118–128.

Beisel, Nicola. "Constructing a Shifting Moral Boundary: Literature and Obscenity in Nineteenth-Century America." In *Cultivating Differences: Symbolic Boundaries and the Making of Inequality*, ed. Michèle Lamont and Marcel Fournier, pp. 104–128. Chicago: University of Chicago Press, 1992.

Bell, Daniel. *The End of Ideology: On the Exhaustion of Political Ideas in the Fifties*. New York: Free Press, 1962.

Ben-David, Joseph. *The Scientist's Role in Society: A Comparative Study*. Englewood Cliffs, NJ: Prentice-Hall, 1971.

Bennett, Scott H. *Radical Pacifism: The War Resisters League and Gandhian Nonviolence*. Syracuse, NY: Syracuse University Press, 2003.

Bloom, Alexander, and Wini Breines, eds. *Takin' It to the Streets: A Sixties Reader*. New York: Oxford University Press, 1995.

Blum, John Morton. *Years of Discord: American Politics and Society, 1961–1974*. New York: W. W. Norton, 1991.

Bourdieu, Pierre. "The Specificity of the Scientific Field and the Social Conditions of the Progress of Reason." *Social Science Information* 14 (1975): 19–47.

Boyer, Paul. *By the Bomb's Early Light: American Thought and Culture at the Dawn of the Atomic Age*. Chapel Hill: University of North Carolina Press, 1994.

Brinkley, Alan. *Liberalism and Its Discontents*. Cambridge, MA: Harvard University Press, 1998.

Brint, Steven G. *In an Age of Experts: The Changing Role of Professionals in Politics*. Princeton, NJ: Princeton University Press, 1994.

Brock, Peter, and Nigel Young. *Pacifism in the Twentieth Century*. Syracuse, NY: Syracuse University Press, 1999.

Brown, Phil. "Popular Epidemiology Revisited." *Current Sociology* 45 (1997): 137–156.

———. *Transfer of Care: Psychiatric Deinstitutionalization and Its Aftermath*. Boston: Routledge and Kegan Paul, 1985.

Brown, Phil, Sabrina McCormick, Brian Mayer, Stephen Zavestoski, Rachel Morello-Frosch, Rebecca Gasior Altman, and Laura Senie. "'A Lab of Our Own': Environmental Causation of Breast Cancer and Challenges to the Dominant Epidemiological Paradigm." *Science, Technology, and Human Values* 31, no. 5 (2006): 499–536.

Butcher, Sandra Ionno. "The Origins of the Russell-Einstein Manifesto." Manuscript, Pugwash Councils on World Affairs, May 2005.

Cahn, Anne Hessing. *Eggheads and Warheads: Scientists and the ABM*. Cambridge, MA: MIT Science and Public Policy Program, Department of Political Science and Center for International Studies, 1971.

Carlat, Louis E. "Partners with God: Arthur Holly Compton and the Moral Leadership of Science." A.B. thesis, Washington University, 1986.

Carroll, Patrick. *Science, Culture, and Modern State Formation*. Berkeley and Los Angeles: University of California Press, 2006.

———. "Science, Power, Bodies: The Mobilization of Nature as State Formation." *Journal of Historical Sociology* 9 (1996): 139–167.

Carson, Rachel. *Silent Spring*. Boston: Houghton Mifflin, 1962.

Caute, David. *The Great Fear: The Anti-Communist Purge under Truman and Eisenhower*. New York: Simon and Schuster, 1978.

Chomsky, Noam. "The Cold War and the University." In *The Cold War and the University: Toward an Intellectual History of the Postwar Years*, ed. André Schiffrin, pp. 171–194. New York: New Press, 1997.

Clarke, Adele E., Janet K. Shim, Laura Mamo, Jennifer Ruth Fosket, and Jennifer R. Fishman. "Biomedicalization: Technoscientific Transformations of Health, Illness, and U.S. Biomedicine." *American Sociological Review* 68 (2003): 161–194.

Clemens, Elisabeth S. "Organizational Form as Frame: Collective Identity and Political Strategy in the American Labor Movement, 1880–1920." In *Comparative Perspectives on Social Movements: Political Opportunities, Mobilizing Structures, and Cultural Framings*, ed. Doug McAdam, John D. McCarthy, and Mayer N. Zald, pp. 205–226. Cambridge: Cambridge University Press, 1996.

———. "Organizational Repertoires and Institutional Change: Women's Groups and the Transformation of U.S. Politics, 1890–1920." *American Journal of Sociology* 98 (1993): 755–798.

Clemens, Elisabeth S. *The People's Lobby: Organizational Innovation and the Rise of Interest Group Politics in the United States, 1890–1925.* Chicago: University of Chicago Press, 1997.

Clemens, Elisabeth S., and James M. Cook. "Politics and Institutionalism: Explaining Durability and Change." *Annual Review of Sociology* 25 (1999): 441–466.

Clemens, Elisabeth S., and Debra C. Minkoff. "Beyond the Iron Law: Rethinking the Place of Organizations in Social Movement Research." In *The Blackwell Companion to Social Movements*, ed. David A. Snow, Sarah A. Soule, and Hanspeter Kriesi, pp. 155–168. Malden, MA: Blackwell Publishing, 2004.

Clowse, Barbara Barksdale. "Education as an Instrument of National Security: The Cold War Campaign to 'Beat the Russians' from Sputnik to the National Defense Education Act of 1958." Ph.D. dissertation, University of North Carolina–Chapel Hill, 1977.

Cochrane, Rexmond C. *The National Academy of Sciences: The First Hundred Years, 1863–1963.* Washington, DC: National Academy of Sciences, 1978.

Cole, Simon A. *Suspect Identities: A History of Fingerprinting and Criminal Identification.* Cambridge, MA: Harvard University Press, 2001.

Commoner, Barry. *Science and Survival.* New York: Viking, 1966.

———. *Scientific Statesmanship in Air Pollution Control.* Washington, DC: United States Public Health Service, 1964.

A Compilation of Federal Science Laws: As Amended through 2003. Prepared for the use of the House Committee on Science. Washington, DC: GPO, 2004.

Conrad, Peter. "The Shifting Engines of Medicalization." *Journal of Health and Social Behavior* 46 (2005): 5–8.

Coulson, Charles A. *Science, Technology, and the Christian.* London: Abington Press, 1960.

Cowan, Wayne H., ed. *Witness to a Generation: Significant Writings from Christianity and Crisis, 1941–1966.* New York: Bobbs-Merrill, 1966.

Cress, Daniel M., and Daniel J. Myers. "Authority in Contention." *Research in Social Movements, Conflicts and Change* 25 (2004): 279–293.

Croissant, Jennifer. "Critical Legal Theory and Critical Science Studies: Engaging Institutions." *Cultural Dynamics* 12, no. 2 (2000): 223–226.

Croissant, Jennifer, and Sal Restivo, eds. *Degrees of Compromise: Industrial Interests and Academic Values.* Albany: State University of New York Press, 2001.

Crossley, Nick. *Contesting Psychiatry: Social Movements in Mental Health.* London: Routledge, 2005.

———. "Mental Health, Resistance and Social Movements: The Collective-Confrontational Dimension." *Health Education Journal* 61, no. 2 (2002): 138–152.

Culhane, Paul J. "Federal Organizational Change in Response to Environmentalism" *Humboldt Journal of Social Relations* 2 (1974): 31–44.

Daniels, George H., ed. *Nineteenth-Century American Science.* Evanston, IL: Northwestern University Press, 1972.

Davenport, Christian. "Introduction. Repression and Mobilization: Insights from Political Science and Sociology." In *Repression and Mobilization*, ed. Christian Davenport, Hank Johnston, and Carol Mueller, pp. vii–xli. Minneapolis: University of Minnesota Press, 2004.

Day, Dwayne A. "Cover Stories and Hidden Agendas: Early American Space and National Security Policy." In *Reconsidering Sputnik: Forty Years since the Soviet Satellite*, ed. Roger D. Launius, John M. Logsdon, and Robert W. Smith, pp. 161–196. London: Routledge, 2002.

DeBenedetti, Charles, with Charles Chatfield. *An American Ordeal: The Antiwar Movement of the Vietnam Era*. Syracuse, NY: Syracuse University Press, 1990.

D'Emilio, John. *Sexual Politics, Sexual Communities: The Making of a Homosexual Minority, 1940–1970*. Chicago: University of Chicago Press, 1983.

Dennis, Michael Aaron. "Reconstructing Sociotechnical Order: Vannevar Bush and U.S. Science Policy." In *States of Knowledge: The Co-production of Science and Social Order*, ed. Sheila Jasanoff, pp. 225–253. London: Routledge, 2004.

Dickson, David. *The New Politics of Science*. New York: Pantheon, 1984.

DiMaggio, Paul J., and Walter W. Powell. "The Iron Cage Revisited: Institutional Isomorphism and Collective Rationality in Organizational Fields." *American Sociological Review* 48 (1983): 147–160.

———, eds. *The New Institutionalism in Organizational Analysis*. Chicago: University of Chicago Press, 1991.

Divine, Robert A. *Blowing on the Wind: The Nuclear Test Ban Debate, 1954–1960*. New York: Oxford University Press, 1978.

Downey, Gary L. "Reproducing Cultural Identity in Negotiating Nuclear Power: The Union of Concerned Scientists and Emergency Core Cooling." *Social Studies of Science* 18 (1988): 231–264.

Du Bridge, Lee A. "Twenty-Five Years of the National Science Foundation." *Proceedings of the American Philosophical Society* 121 (1977): 191–194.

Duggan, Stephan, and Betty Drury. *The Rescue of Science and Learning*. New York: Macmillan, 1952.

Dupré, J. Stefan, and Sanford A. Lakoff. *Science and the Nation: Policy and Politics*. Englewood Cliffs, NJ: Prentice-Hall, 1962.

Dupree, A. Hunter. *Science in the Federal Government: A History of Policies and Activities to 1940*. Cambridge, MA: Belknap Press of Harvard University Press, 1957.

Earl, Jennifer. "Controlling Protest: New Directions for Research on the Social Control of Protest." *Research in Social Movements, Conflict and Change* 25 (2004): 55–83.

———. "Repression and the Control of Protest." *Mobilization* 11, no. 2 (2006): 129–143.

———. "Tanks, Tear Gas, and Taxes: Toward a Theory of Movement Repression." *Sociological Theory* 21 (2003): 44–68.

Eden, Murray. "Historical Introduction." In *March 4: Scientists, Students and Society*, ed. Jonathan Allen, pp. vii–xxi. Cambridge, MA: MIT Press, 1970.

Edmond, Gary. "Legal Engineering: Contested Representations of Law, Science (and Non-science) and Society." *Social Studies of Science* 32 (2002): 371–412.

Edmond, Gary, and David Mercer. "Litigation Life: Law-Science Knowledge Construction in (Bendectin) Mass Toxic Tort Litigation." *Social Studies of Science* 30 (2000): 265–316.

Ehrenreich, Barbara. "Giving Power to the People: The Early Days of Health/PAC." *Health/PAC Bulletin* 18, no. 4 (1988): 4–8.

Ehrenreich, Barbara, and John Ehrenreich. *Long March, Short Spring: The Student Uprisings at Home and Abroad.* New York: Monthly Review Press, 1969.

Ehrenreich, Barbara, and John Ehrenreich. "The New Left: A Case Study in Professional-Managerial Class Radicalism." *Radical America* 11 (1977): 7–22

Elbaum, Max. *Revolution in the Air: Sixties Radicals Turn to Lenin, Mao, and Che.* New York: Verso, 2002.

Epstein, Barbara. *Political Protest and Cultural Revolution: Nonviolent Direct Action in the 1970s and 1980s.* Berkeley and Los Angeles: University of California Press, 1991.

Epstein, Steven. *Impure Science: AIDS, Activism, and the Politics of Knowledge.* Berkeley and Los Angeles: University of California Press, 1996.

———. "Sexualizing Governance and Medicalizing Identities: The Emergence of 'State-Centered' LGBT Health Politics in the United States." *Sexualities* 6 (2003):131–171.

Ezrahi, Yaron. *The Descent of Icarus: Science and the Transformation of Contemporary Democracy.* Cambridge, MA: Harvard University Press, 1990.

———. "Science and the Political Imagination in Contemporary Democracies." In *States of Knowledge: The Co-production of Science and Social Order,* ed. Sheila Jasanoff, pp. 254–273. London: Routledge, 2004.

Farrell, James J. *The Spirit of the Sixties: The Making of Postwar Radicalism.* New York: Routledge, 1997.

Ferree, Myra Marx, and Patricia Yancey Martin. *Feminist Organizations: Harvest of the New Women's Movement.* Philadelphia: Temple University Press, 1991.

Finkbinder, Ann. *The Jasons: The Secret History of Science's Postwar Elite.* New York: Viking, 2006.

Fischer, Frank. *Citizens, Experts, and the Environment: The Politics of Local Knowledge.* Durham, NC: Duke University Press, 2000.

Fishman, Jennifer R. "Manufacturing Desire: The Commodification of Female Sexual Dysfunction." *Social Studies of Science* 34 (2004): 187–218.

Fleming, Arthur S. "The Philosophy and Objectives of the National Defense Education Act." *Annals of the American Academy of Political and Social Science* 327 (1962): 132–138.

Foster, Kenneth, and Peter Huber. "Scientific Knowledge." In *Judging Science: Scientific Knowledge and the Federal Courts,* pp. 1–22. Cambridge, MA: MIT Press, 1999.

Freidson, Eliot. *Professionalism Reborn: Theory, Prophecy, and Policy.* Chicago: University of Chicago Press, 1994.

Freudenberg, William. "Scientific Expertise and Natural Resource Decisions: Social Science Participation on Interdisciplinary Scientific Committees." *Social Science Quarterly* 83 (2002): 119–136.

Frickel, Scott. "Building an Interdiscipline: Collective Action Framing and the Rise of Genetic Toxicology." *Social Problems* 51 (2004): 269–287.

———. *Chemical Consequences: Environmental Mutagens, Scientist Activism, and the Rise of Genetic Toxicology.* New Brunswick, NJ: Rutgers University Press, 2004.

———. "Scientific Authority and Expert Activism in Environmental Health and Justice Movements." Paper presented to the History and Sociology of Science Workshop, University of Pennsylvania, November 28, 2005.

Frickel, Scott, and Kelly Moore. "Prospects and Challenges for a New Political Sociology of Science." In *The New Political Sociology of Science: Institutions, Networks, and Power*, ed. Scott Frickel and Kelly Moore, pp. 3–34. Madison: University of Wisconsin Press, 2006.

Futrell, Robert. "Technical Adversarialism and Participatory Collaboration in the U.S. Chemical Weapons Disposal Program." *Science, Technology, and Human Values* 28 (2003): 451–482.

Galison, Peter, and Bruce Hevly, eds. *Big Science: The Growth of Large-Scale Research*. Stanford, CA: Stanford University Press, 1992.

Galison, Peter, and David J. Stump, eds. *The Disunity of Science: Boundaries, Contexts, and Power*. Stanford, CA: Stanford University Press, 1996.

Gamson, William A. "Goffman's Legacy to Political Sociology." *Theory and Society* 14 (1985): 605–622.

Gaziano, Emanuel. "Ecological Metaphors as Scientific Boundary Work." *American Journal of Sociology* 101 (1996): 874–907.

Geiger, Roger. *Research and Relevant Knowledge: American Research Universities since World War II*. New York: Oxford University Press, 1993.

Gellhorn, Walter. *Security, Loyalty, Science*. Ithaca, NY: Cornell University Press, 1950.

Georgakas, Dan, and Marvin Surkin. *Detroit, I Do Mind Dying: A Study in Urban Revolution*. Boston: South End Press, 1998.

Gerhards, Jurgen, and Dieter Rucht. "Mesomobilization, Organizing, and Framing in Two Protest Campaigns in West Germany." *American Journal of Sociology* 98 (1992): 555–595.

Gieryn, Thomas F. "Boundary-Work and the Demarcation of Science from Non-Science: Strains and Interests in Professional Ideologies of Scientists." *American Sociological Review* 48 (1983): 781–795.

———. *Cultural Boundaries of Science: Credibility on the Line*. Chicago: University of Chicago Press, 1999.

Gieryn, Thomas F., George M. Bevins, and Stephen C. Zehr. "Professionalization of American Scientists: Public Science in Creation/Evolution Trials." *American Sociological Review* 50 (1985): 392–409.

Gilpin, Robert, and Christopher Wright, eds. *Scientists and National Policy-Making*. New York: Columbia University Press, 1964.

Gitlin, Todd. *The Sixties: Years of Hope, Days of Rage*. New York: Bantam, 1987.

Goffman, Erving. *Encounters: Two Studies in the Sociology of Interaction*. 1961; reprint, London: Allen Lane, 1972.

Golden, William T., ed. *Science and Technology Advice to the President, Congress, and Judiciary*, 2nd ed. New Brunswick, NJ: Transaction Publishers, 1995.

Goodman, Paul. *Growing Up Absurd*. New York: Random House, 1964.

Goodwin, Jeff. *No Other Way Out: States and Social Revolutions*. New York: Cambridge University Press, 2001.

Goodwin, Jeff, and James M. Jasper. "Caught in a Winding, Snarling Vine: The Structural Bias of Political Process Theory." *Sociological Forum* 14 (1999): 27–54.

Greenberg, Daniel S. *The Politics of Pure Science*. New York: New American Library, 1969.

Gross, Matthias. "Community by Experiment: Recursive Learning in Landscape Design and Ecological Restoration." Paper presented at the International Sociological Association's Mini-Conference on Community and Ecology, San Francisco, 2004.

Gross, Paul R., and Norman Levitt. *Higher Superstition: The Academic Left and Its Quarrels with Science*. Baltimore: Johns Hopkins University Press, 1998.

Gross, Paul R., Norman Levitt, and Martin W. Lewis, eds. *The Flight from Science and Reason*. Baltimore: Johns Hopkins University Press, 1997.

Gusfield, Joseph R. *Symbolic Crusade: Status Politics and the American Temperance Movement*. Urbana: University of Illinois Press, 1963.

Guston, David H. "Stabilizing the Boundary between US Politics and Science: The Role of the Office of Technology Transfer as a Boundary Organization." *Social Studies of Science* 29 (1999): 87–111.

Guston, David H., and Daniel Sarewitz. "Real-Time Technology Assessment." *Technology in Society* 24 (2002): 93–109.

Haber, Barbara, and Al Haber. "Getting By with a Little Help from Our Friends." In *Radicals in Professions: Selected Papers*, pp. 44–62. Ann Arbor, MI: Radical Education Project, 1967.

Hagen, Thomas. *Force of Nature: The Life of Linus Pauling*. New York: Simon and Schuster, 1995.

Halpern, Sydney A. "Dynamics of Professional Control: Internal Coalitions and Crossprofessional Boundaries." *American Journal of Sociology* 97 (1992): 994–1021.

———. *Lesser Harms: The Morality of Risk in Medical Research*. Chicago: University of Chicago Press, 2004.

Haraway, Donna. "A Cyborg Manifesto: Science, Technology, and Socialist-Feminism in the Late Twentieth Century." In *Simians, Cyborgs and Women: The Reinvention of Nature*, pp. 149–181. London: Routledge, 1991.

———. "The Transformation of the Left in Science: Radical Associations in Britain in the 30's and the U.S.A. in the 60s's." *Soundings* 58 (1973): 441–462.

Harding, Sandra. *The Science Question in Feminism*. Ithaca, NY: Cornell University Press, 1986.

———. *Whose Science, Whose Knowledge?* Ithaca, NY: Cornell University Press, 1991.

Heineman, Kenneth J. *Campus Wars: The Peace Movement at American State Universities in the Vietnam Era*. New York: New York University Press, 1993.

Hellstrom, Tomas, and Merle Jacob. "Boundary Organizations in Science: From Discourse to Construction." *Science and Public Policy* 30 (2003): 235–238.

Heraud, Brian J. "Professionalism, Radicalism and Social Change." *Sociological Review Monograph* 20 (1973): 85–101.

Hess, David J. "Antiangiogenesis Research and Scientific Fields." In *The New Political Sociology of Science: Institutions, Networks, and Power*, ed. Scott

Frickel and Kelly Moore, pp. 122–147. Madison: University of Wisconsin Press, 2006.

———. "Medical Modernisation, Scientific Research Fields and the Epistemic Politics of Health Social Movements." *Sociology of Health and Illness* 24 (2004): 695–709.

Hewlett, Richard G., and Oscar E. Anderson. *A History of the United States Atomic Energy Commission.* University Park: Pennsylvania State University Press, 1962–1969.

———. *The New World, 1939–1946.* University Park: Pennsylvania State University Press, 1962.

Hewlett, Richard G., and Jack Hall. *Atoms for Peace and War: Eisenhower and the Atomic Energy Commission.* Berkeley and Los Angeles: University of California Press, 1983.

Hilgartner, Stephen. *Science on Stage: Expert Advice as Public Drama.* Stanford, CA: Stanford University Press, 2000.

Hirsch, Paul M., and Michael Lounsbury. "Ending the Family Quarrel: Towards a Reconciliation of 'Old' and 'New' Institutionalism." *American Behavioral Scientist* 40 (1997): 406–418.

Hodes, Elizabeth. "Precedents for Social Responsibility among Scientists: The American Association of Scientific Workers and the Federation of American Scientists, 1938–1948." Ph.D. dissertation, University of California–Santa Barbara, 1982.

Hoffman, Lily. *The Politics of Knowledge: Activist Movements in Medicine and Planning.* Albany: State University of New York Press, 1989.

Hollinger, David A. "The Defense of Democracy and Robert K. Merton's Formulation of the Scientific Ethos." In *Science, Jews, and Secular Culture: Studies in Mid-Twentieth-Century Intellectual History*, ed. David A. Hollinger, pp. 80–96. Princeton, NJ: Princeton University Press, 1996.

———. "Free Enterprise and Free Inquiry: The Emergence of Laissez-Faire Communitarianism in the Ideology of Science in the United States." In *Science, Jews, and Secular Culture: Studies in Mid-Twentieth-Century Intellectual History*, ed. David A. Hollinger, pp. 97–120. Princeton, NJ: Princeton University Press, 1996.

———. "Jewish Intellectuals and the De-Christianization of American Public Culture in the Twentieth Century." In *Science, Jews, and Secular Culture: Studies in Mid-Twentieth-Century American Intellectual History*, ed. David A. Hollinger, pp. 17–41. Princeton, NJ: Princeton University Press, 1996.

Holton, Gerald. *Science and Anti-Science.* Cambridge, MA: Harvard University Press, 1993.

Holton, Gerald, and Gerhard Sonnert. *Ivory Bridges: Connecting Science and Society.* Cambridge, MA: MIT Press, 2002.

Indyk, David, and David Rier. "Grassroots AIDS Knowledge: Implications for the Boundaries of Science and Collective Action." *Knowledge* 15 (1993): 3–43.

Ingle, H. Larry. "The American Friends Service Committee, 1947–1949: The Cold War's Effect." *Peace and Change* 23 (1998): 27–48.

Irwin, Alan. *Citizen Science: A Study of People, Expertise, and Sustainable Development.* New York: Routledge, 1995.

Israel, Barbara A., Eugenia Eng, Amy J. Schulz, Edith A. Parker, and David Satcher. *Methods in Community-Based Participatory Research for Health*. San Francisco: Jossey-Bass, 2005.

Isserman, Maurice. *If I Had a Hammer: The Death of the Old Left and the Birth of the New Left*. New York: Basic Books, 1987.

Jasanoff, Sheila. *Designs on Nature*. Princeton, NJ: Princeton University Press, 2005.

———. "The Eye of Everyman: Witnessing DNA in the Simpson Trial." *Social Studies of Science* 28 (1998): 713–740.

———. "Ordering Knowledge, Ordering Society." In *States of Knowledge: The Co-production of Science and Social Order*, ed. Sheila Jasanoff, pp. 3–45. London: Routledge, 2004.

———. *Science at the Bar: Law, Science, and Technology in America*. Cambridge, MA: Harvard University Press, 1995.

———. "Science and the Statistical Victim: Modernizing Knowledge in Breast Implant Litigation." *Social Studies of Science* 32 (2002): 37–69.

———, ed. *States of Knowledge: The Co-production of Science and Social Order*. London: Routledge, 2004.

Jasper, James M. *The Art of Moral Protest: Culture, Biography, and Creativity in Social Movements*. Chicago: University of Chicago Press, 1997.

Jensen, Arthur R. "How Much Can We Boost I.Q.?" *Harvard Educational Review* 39 (1969): 1–123.

Kahn, Herman. *On Thermonuclear War*. New York: Macmillan, 1962.

Kaiser, David. "Cold War Requisitions, Scientific Manpower, and the Production of American Physicists after World War II." *Historical Studies in the Physical and Biological Sciences* 33 (2002): 131–159.

Kamen, Martin D. *Radiant Science, Dark Politics: A Memoir of the Nuclear Age*. Berkeley and Los Angeles: University of California Press, 1985.

Katz, Milton. *Ban the Bomb: A History of SANE, the Committee for a Sane Nuclear Policy, 1957–1985*. Westport, CT: Greenwood, 1987.

Kay, Lily E. *Who Wrote the Book of Life? A History of the Genetic Code*. Stanford, CA: Stanford University Press, 2000.

Keller, Evelyn Fox. "Gender and Science: Origin, History, and Politics." *Osiris* 10 (1995): 27–38.

———. *Secrets of Life, Secrets of Death: Essays on Gender, Language, and Science*. Cambridge, MA: MIT Press, 2000.

Kelly, Susan E. "Public Bioethics and Publics: Consensus, Boundaries, and Participation in Biomedical Science Policy." *Science, Technology, and Human Values* 28 (2003): 339–364.

Kerr, Anne, Sarah Cunningham-Burley, and Amanda Amos. "The New Genetics: Professionals' Discursive Boundaries." *Sociological Review* 45 (1997): 279–303.

Kevles, Daniel J. *The Physicists: The History of a Scientific Community in Modern America*. Cambridge, MA: Harvard University Press, 1995.

———. "Scientists, the Military, and the Control of Postwar Defense Research: The Case of the Research Board for National Security, 1944–46." *Technology and Culture* 16 (1975): 20–47.

Kinchy, Abby J., and Daniel Lee Kleinman. "Organizing Credibility: Discursive and Organizational Orthodoxy on the Borders of Ecology and Politics." *Social Studies of Science* 33 (2003): 869–896.

Klaw, Spencer. *The New Brahmins: Scientific Life in America*. New York: William Morrow, 1968.

Klawiter, Maren. "Racing for the Cure, Walking Women, and Toxic Touring: Mapping Cultures of Action within the Bay Area Terrain of Breast Cancer." *Social Problems* 46 (1999): 104–126.

Kleinman, Daniel Lee. *Impure Cultures: University Biology and the World of Commerce*. Madison: University of Wisconsin Press, 2003.

———. *Politics on the Endless Frontier: Postwar Research Policy in the United States*. Durham, NC: Duke University Press, 1995.

———, ed. *Science, Technology, and Democracy*. Albany: State University of New York Press, 2000.

Kleinman, Daniel Lee, and Abby J. Kinchy. "Boundaries in Science Policy Making: Bovine Growth Hormone in the European Union." *Sociological Quarterly* 44 (2003): 577–595.

Kleinman, Daniel Lee, and Mark Solovey. "Hot Science/Cold War: The National Science Foundation after World War II." *Radical History Review* 63 (1995): 110–139.

Kleinman, Daniel Lee, and Steven P. Vallas. "Science, Capitalism, and the Rise of the 'Knowledge Worker': The Changing Structure of Knowledge Production in the United States." *Theory and Society* 30, no. 4 (2001): 451–492.

Kline, Wendy. *Building a Better Race: Gender, Sexuality, and Eugenics from the Turn of the Century to the Baby Boom*. Berkeley and Los Angeles: University of California Press, 2001.

Knorr Cetina, Karin. *Epistemic Cultures: How the Sciences Make Knowledge*. Cambridge, MA: Harvard University Press, 1999.

Kopp, Carolyn. "The Origins of the American Scientific Debate over Fallout Hazards." *Social Studies of Science* 9 (1979): 403–422.

Kuhn, Thomas. *The Structure of Scientific Revolutions*. Chicago: University of Chicago Press, 1962.

Kurlansky, Mark. *1968: The Year That Rocked the World*. New York: Ballantine Books, 2004.

Kuznick, Peter J. *Beyond the Laboratory: Scientists as Political Activists in 1930s America*. Chicago: University of Chicago Press, 1987.

———. "The Ethical and Political Crisis of Science: The AAAS Confronts the War in Vietnam." Paper presented at the Annual Meeting of the History of Science Society, New Orleans, October 1994.

Laird, Frank N. "Participatory Analysis, Democracy, and Technical Decision Making." *Science, Technology, and Human Values* 18 (1993): 341–361.

Lapp, Ralph. *The New Priesthood: The Scientific Elite and the Uses of Power*. New York: Harper and Row, 1965.

Larson, Magali Sarfatti. *The Rise of Professionalism*. Berkeley and Los Angeles: University of California Press, 1977.

Lassmann, Thomas Charles. "Compton's Effect on Washington University." A.B. thesis, Washington University, 1991.

Latour, Bruno. "From Realpolitik to Dingpolitik: Or How to Make Things Public." In *Making Things Public: Atmospheres of Democracy*, ed. Bruno Latour and Peter Weibel, pp. 14–41. Karslruhe, Germany: Center for Art and Media, and Cambridge, MA: MIT Press, 2005.

———. *Politics of Nature: How to Bring the Sciences into Democracy*. Cambridge, MA: Harvard University Press, 2004.

Latour, Bruno, and Steve Woolgar. *Laboratory Life: The Construction of Scientific Facts*. Princeton, NJ: Princeton University Press, 1986.

Launius, Roger D. "Sputnik and Its Repercussions: A Historical Analysis." *Aerospace Historian* 17 (1970): 89.

Layton, Edwin T., Jr. *The Revolt of the Engineers: Social Responsibility and the American Engineering Profession*. Cleveland: Press of Case Western Reserve University, 1971.

Leslie, Stuart W. *The Cold War and American Science: The Military-Industrial-Academic Complex at MIT and Stanford*. New York: Columbia University Press, 1993.

Levins, Richard. *Evolution in Changing Environments: Some Theoretical Considerations*. Princeton, NJ: Princeton University Press, 1968.

Levins, Richard, and Richard Lewontin. *The Dialectical Biologist*. Cambridge, MA: Harvard University Press, 1985.

Levitt, Norman. *Prometheus Bedeviled: Science and the Contemporary Contradictions of Culture*. New Brunswick, NJ: Rutgers University Press, 1999.

Lewenstein, Bruce V. "Magazine Publishing and Popular Science after World War II." *American Journalism* 6 (1989): 220.

———. "The Meaning of 'Public Understanding of Science' in the United States after World War II." *Public Understanding of Science* 1 (1992): 47–48.

———. "Shifting Science from People to Programs: AAAS in the Postwar Years." In *The Establishment of Science in America: 150 Years of the American Association for the Advancement of Science*, ed. Sally Gregory Kohlstedt, Michael M. Sokal, and Bruce V. Lewenstein, pp. 103–166. New Brunswick, NJ: Rutgers University Press, 1999.

Lewontin, Richard, Steven Rose, and Leon J. Kamin. *Not in Our Genes: Biology, Ideology and Human Nature*. New York: Pantheon, 1984.

Lichterman, Paul. *The Search for Political Community: American Activists Reinventing Commitment*. New York: Cambridge University Press, 1996.

Lounsbury, Michael. "Institutional Sources of Practice Variation: Staffing College and University Recycling Programs." *Administrative Science Quarterly* 46 (2001): 29–56.

Lounsbury, Michael, and Marc J. Ventresca. "'Social Structure and Organizations' Revisited." *Research in the Sociology of Organizations* 19 (2002): 3–38.

Luker, Kristen. *Abortion and the Politics of Motherhood*. Berkeley and Los Angeles: University of California Press, 1984.

Lynch, Michael. "Administrative Science: The Management of Objectivity and the Objectivity of Management [#1]." Paper presented at the Annual Meeting of the American Sociological Association, Philadelphia, 2005.

Lyons, Eugene M., and John W. Masland. *Education and Military Leadership: A Study of the R.O.T.C.* Princeton, NJ: Princeton University Press, 1959.

Lyttle, Bradford. *The Chicago Anti–Vietnam War Movement*. Chicago: Midwest Pacifist Center, 1988.

Marcuse, Herbert. *One-Dimensional Man: Studies in the Ideology of Advanced Industrial Society*. Boston: Beacon Press, 1964.

Martin, Brian. "Science in a Citizen-Shaped World." In *The New Political Sociology of Science: Institutions, Networks, and Power*, ed. Scott Frickel and Kelly Moore, pp. 272–298. Madison: University of Wisconsin Press, 2006.

Matusow, Allen J. *The Unraveling of America: A History of Liberalism in the 1960s*. New York: Harper & Row, 1984.

Mayer, Robert N. *The Consumer Movement: Guardians of the Marketplace*. Boston: Twayne, 1989.

Mazuzan, George T., and J. Samuel Walker. *Controlling the Atom: The Beginnings of Nuclear Regulation, 1946–1962*. Berkeley and Los Angeles: University of California Press, 1985.

McAdam, Doug. "Micromobilization Contexts and Recruitment to Activism." *International Social Movement Research* 1 (1988): 125–154.

———. *Political Process and the Development of Black Insurgency, 1930–1970*. Chicago: University of Chicago Press, 1982.

———. "Tactical Innovation and the Pace of Insurgency." *American Sociological Review* 48 (1983): 735–754.

McAdam, Doug, Sidney Tarrow, and Charles Tilly. *Dynamics of Contention*. New York: Cambridge University Press, 2001.

McCarthy, John D., and Mayer N. Zald. "Resource Mobilization and Social Movements: A Partial Theory." *American Journal of Sociology* 82 (1977): 1212–1241.

———. *The Trend of Social Movements in America: Professionalization and Resource Mobilization*. Morristown, NJ: General Learning Press, 1973.

McCormick, Sabrina, Julia Brody, Phil Brown, and Ruth Polk. "Lay Involvement in Breast Cancer Research." *International Journal of Health Services* 34 (2004): 625–646.

Mellor, Felicity. "Between Fact and Fiction: Demarcating Science from Non-Science in Popular Physics Books." *Social Studies of Science* 33 (2003): 509–538.

Melman, Seymour, with Melvyn Baron and Dodge Ely. *In the Name of America: The Conduct of the War in Vietnam by the Armed Forces of the United States as Shown by Published Reports, Compared with the Laws of War Binding on the United States Government and on Its Citizens*. New York: Clergy and Laymen Concerned about Vietnam, 1968.

———. *Inspection for Disarmament*. New York: Columbia University Press, 1958.

———. *No Place to Hide: Fact and Fiction about Fallout Shelters*. New York: Grove Press, 1962.

Mendelsohn, Everett. "The Politics of Pessimism: Science and Technology Circa 1968." In *Technology, Pessimism, and Postmodernism*, ed. Yaron Ezrahi, Everett Mendelsohn, and Howard Segal, pp. 151–274. Amherst: University of Massachusetts Press, 1994.

Merton, Robert K. "The Normative Structure of Science." In *The Sociology of Science*, ed. Norman W. Storer, pp. 267–278. Chicago: University of Chicago Press, 1973.

Meyer, John W., and Brian Rowan. "Institutionalized Organizations: Formal Structure as Myth and Ceremony." *American Journal of Sociology* 83 (1977): 340–363.

Mezerik, A. G. *The Pursuit of Plenty: The Story of Man's Expanding Domain.* New York: Harper, 1950.

Michael, Mike, and Nik Brown. "Switching between Science and Culture in Transpecies Transplantation." *Science, Technology, and Human Values* 26, no. 1 (2001): 3–22.

Miller, Clarke. "Hybrid Management: Boundary Organizations, Science Policy, and Environmental Governance in the Climate Regime." *Science, Technology, and Human Values* 26 (2001): 478–501.

Miller, Jon D. "Public Understanding of, and Attitudes toward, Scientific Research: What We Know and What We Need to Know." *Public Understanding of Science* 13 (2004): 273–294.

Mills, C. Wright. *The Causes of World War Three.* New York: Ballantine Books, 1958.

———. *The Power Elite.* Oxford: Oxford University Press, 1956.

Minkler, Meredith, Angela Glover Blackwell, Mildred Thompson, and Heather Tamir. "Community-Based Participatory Research: Implications for Public Health Funding." *American Journal of Public Health* 8 (2003): 1210–1214.

Minkoff, Debra C. "The Organization of Survival: Women's and Racial-Ethnic Voluntarist and Activist Organizations, 1955–1985." *Social Forces* 71 (1993): 887–908.

Minkoff, Debra C., and John D. McCarthy. "Reinvigorating the Study of Organizational Processes in Social Movements." *Mobilization: An International Journal* 10 (2005): 289–308.

Mische, Ann. *Partisan Publics: Activist Trajectories and Communicative Styles in Brazilian Youth Politics, 1977–1997.* Princeton, NJ: Princeton University Press, 2006.

Montagu, M. F. Ashley, ed. *Science and Creationism.* New York: Oxford University Press, 1984.

Moore, Kelly. 2006. "Organizing Integrity: American Science and the Creation of Public Interest Organizations, 1955–1975." *American Journal of Sociology* 101 (1996): 1592–1627.

———. "Political Protest and Institutional Change: The Anti–Viet Nam War Movement and American Science." In *How Social Movements Matter*, ed. Marco Guigni, Doug McAdam, and Charles Tilly, pp. 97–118. Minneapolis: University of Minnesota Press, 1999.

———. "Powered by the People: Scientific Authority in Participatory Science." In *The New Political Sociology of Science: Networks, Institutions, and Power*, ed. Scott Frickel and Kelly Moore, pp. 299–326. Madison: University of Wisconsin Press, 2006.

Moore, Kelly, and Nicole Hala. "Organizing Identity: The Creation of Science for the People." *Research in the Sociology of Organizations* 19 (2002): 309–355.

Morello-Frosch, Rachel, Stephen Zavestoski, Phil Brown, Rebecca Gasior Altman, Sabrina McCormick, and Brian Mayer. "Embodied Health Movements:
Responses to a 'Scientized' World." In *The New Political Sociology of Science:
Networks, Institutions, and Power*, ed. Scott Frickel and Kelly Moore, pp. 244–
271. Madison: University of Wisconsin Press, 2006.

Morgen, Sandra. *Into Our Own Hands: The Women's Health Movement in the
United States, 1969–1990*. New Brunswick, NJ: Rutgers University Press,
2002.

Mukerji, Chandra. *A Fragile Power: Scientists and the State*. Princeton, NJ:
Princeton University Press, 1989.

———. *Territorial Ambitions and the Gardens of Versailles*. New York: Cambridge University Press, 1997.

National Academy of Sciences. *Science and Creationism: A View from the National Academy of Sciences*. Washington, DC: National Academies Press, 1999.

Nelkin, Dorothy. *The University and Military Research: Moral Politics at MIT*.
Ithaca, NY: Cornell University Press, 1972.

Nepstad, Sharon Erickson. *Convictions of the Soul: Religion, Culture, and
Agency in the U.S. Central America Solidarity Movement*. New York: Oxford
University Press, 2004.

Nichols, David. "The Associational Interest Groups of American Science." In *Scientists in Public Affairs*, ed. Albert H. Teich, pp. 123–167. Cambridge, MA:
MIT Press, 1974.

Niebuhr, Reinhold. "The Christian Faith and World Crisis." *Christianity and Crisis* 1, no. 1 (1941): 4–6.

Nusbaumer, Michael R., Judith A. DiLorio, and Robert D. Baller. "The Boycott
of 'Star Wars' by Academic Scientists: The Relative Roles of Political and Technical Judgment." *Social Science Journal* 31 (1994): 375–388.

Oltman, Ruth M. "Women in the Professional Caucuses." *American Behavioral
Scientist* 18 (1972): 281–302.

Opp, Karl-Dieter, and Wolfgang Roehl. "Repression, Micromobilization, and Political Protest." *Social Forces* 69 (1988): 521–547.

Owen-Smith, Jason. "From Separate Spheres to a Hybrid Order: Accumulative
Advantage across Public and Private Science at Research One Universities."
Research Policy 32 (2003): 1081–1104.

Owen-Smith, Jason, and Walter W. Powell. "Knowledge Networks as Channels
and Conduits: The Effects of Spillovers in the Boston Biotechnology Community." *Organization Science* 15, no. 1 (2004): 5–21.

Pasley, Jeffrey L. *"The Tyranny of Printers": Newspaper Politics in the Early
American Republic*. Richmond: University of Virginia Press, 2000.

Pauling, Linus. *No More War!* New York: Dodd, Mead, 1958.

*The Pentagon Papers: The Defense Department History of United States Deci-
sionmaking in Vietnam*. Gravel edition. Boston: Beacon Press, 1971.

Perrucci, Robert. "In the Service of Man: Radical Movements in the Professions."
Sociological Review Monograph 20 (1973): 179–194.

Polanyi, Michael. *Science, Faith, and Society*. London: Oxford University Press,
1946.

Polletta, Francesca. *Freedom Is an Endless Meeting: Democracy in American Social Movements*. Chicago: University of Chicago Press, 2004.

———. *It Was Like a Fever: Storytelling in Protest and Politics*. Chicago: University of Chicago Press, 2006.

Popper, Karl. *The Logic of Scientific Discovery*. New York: Routledge Classics, 2002.

Porter, Jack Nusan. *Student Protest and the Technocratic Society: The Case of ROTC*. Chicago: Adams Press, 1973.

Powell, Walter W. "Neither Market nor Hierarchy: Network Forms of Organization." *Research in Organizational Behavior* 12 (1990): 295–336.

Powell, Walter W., Douglas R. White, Kenneth W. Koput, and Jason Owen-Smith. "Network Dynamics and Field Evolution: The Growth of Inter-organizational Collaboration in the Life Sciences." *American Journal of Sociology* 110 (2004): 1132–1205.

Powers, Richard Gid. *Not without Honor: The History of American Anticommunism*. New Haven, CT: Yale University Press, 1995.

Price, Don K. *Government and Science: Their Dynamic Relation in American Democracy*. New York: New York University Press, 1954.

———. *The Scientific Estate*. Cambridge, MA: Harvard University Press, 1965.

Ramsden, Edmund. "Carving Up Population Science: Eugenics, Demography and the Controversy over the 'Biological Law' of Population Growth." *Social Studies of Science* 32 (2002): 857–899.

Reagan, Michael D. *Science and the Federal Patron*. New York: Oxford University Press, 1969.

Reardon, Jenny. *Race to the Finish: Identity in an Age of Genomics*. Princeton, NJ: Princeton University Press, 2005.

Reingold, Nathan. "Refugee Mathematicians in the United States of America, 1933–1941: Reception and Reaction." *Annals of Science* 38 (1981): 313–338.

———. "Vannevar Bush's New Deal for Research: Or the Triumph of the Old Order." *Historical Studies in the Physical and Biological Sciences* 17 (1987): 299–344.

Robinson, Jo Ann. "A. J. Muste and Ways to Peace." In *Peace Movements in America*, ed. Charles Chatfield, pp. 81–94. New York: Schocken, 1973.

———. *Abraham Went Out: A Biography of A. J. Muste*. Philadelphia: Temple University Press, 1981.

Robnett, Belinda. "African-American Women in the Civil Rights Movement, 1954–1965: Gender, Leadership, and Micromobilization." *American Journal of Sociology* 101 (1996): 1661–1693.

Rose, Hilary. *Love, Power, and Knowledge: Toward a Feminist Transformation of Science*. Bloomington: Indiana University Press, 1994.

Rose, Hilary, and Steven Rose, eds. *The Radicalisation of Science*. London: Macmillan, 1976.

———. *Science and Society*. London: Allen Lane, 1969.

Rose, Steven, and David Paved, eds. *C.B.W.: Chemical and Biological Warfare, London Conference on Chemical and Biological Warfare*. London: Harrap, 1968.

Rosenberg, Ann. *The Technological Warlords*. New York: Computer People for Peace, 1971.

Rosenberg, Charles. "Piety and Social Action: Some Origins of the American Public Health Movement." In *No Other Gods: On Science and American Social Thought*, pp. 109–122. Revised edition. Baltimore: Johns Hopkins University Press, 1997.

Rosset, Peter, and John Vandermeer, eds. *The Nicaragua Reader: Documents of a Revolution under Fire*. New York: Grove Press, 1983.

Rossinow, Doug. *The Politics of Authenticity: Liberalism, Christianity, and the New Left in America*. New York: Columbia University Press, 1998.

Rotblat, Joseph. *History of the Pugwash Conferences*. London: Taylor and Francis, 1962.

———. *Pugwash, the First Ten Years: History of the Conferences of Science and World Affairs*. London: Heinemann, 1967.

Roth, Wolff-Michael, Janet Riecken, Lilian Pozzer-Ardenghi, Robin McMillan, Brenda Storr, Donna Tait, Gail Bradshaw, and Trudy Pauluth Penner. "Those Who Get Hurt Aren't Always Being Heard: Scientist-Resident Interactions over Community Water." *Science, Technology, and Human Values* 29 (2004): 153–183.

Rudolph, John L. *Scientists in the Classroom: The Cold War Reconstruction of American Science Education*. New York: Palgrave, 2002.

Russell, Bertrand. "Man's Peril." In *Portraits from Memory*, pp. 233–238. New York: Simon & Schuster, 1951.

Sale, Kirkpatrick. *SDS*. New York: Random House, 1973.

Sarewitz, Daniel. *Frontiers of Illusion: Science, Technology, and the Politics of Progress*. Philadelphia: Temple University Press, 1996.

Schelling, Thomas C. "The Future of the Academic Community." In *March 4: Scientists, Students and Society*, ed. Jonathan Allen, pp. 83–87. Cambridge, MA: MIT Press, 1970.

Schiebinger, Londa. *Has Feminism Changed Science?* Cambridge, MA: Harvard University Press, 1999.

———. *Nature's Body: Gender in the Making of Modern Science*. Boston: Beacon Press, 1993.

Science for the People. *China: Science Walks on Two Legs*. New York: Discus Books / Avon, 1974.

Science for the People, Madison Collective. *The AMRC Papers: An Indictment of the Army Mathematics Research Center*. Madison, WI: Science for the People, Madison Collective, 1973.

Sclove, Richard. "Research by the People, for the People." *Futures* 29 (1997): 541–551.

Segerstråle, Ullica. *Defenders of the Truth: The Sociobiology Debate*. New York: Oxford University Press, 2000.

Shapin, Steven. "Science and the Public." In *Companion to the History of Modern Science*, ed. R. C. Olby, G. N. Cantor, J.R.R. Christie, and M.J.S. Hodge, pp. 990–1007. London: Routledge, 1990.

———. *A Social History of Truth: Civility and Science in Seventeenth-Century England*. Chicago: University of Chicago Press, 1994.

Shapin, Steven, and Simon Schaffer. *Leviathan and the Air Pump: Hobbes, Boyle, and the Experimental Life*. Princeton, NJ: Princeton University Press, 1986.

Shapley, Harlow. *Through Rugged Ways to the Stars.* New York: Charles Scribner and Sons, 1969.

Shils, Edward. "The Autonomy of Science." In *The Torment of Secrecy: The Background and Consequences of American Security Policies,* chapter 7. Chicago: Ivan R. Dee, 1956.

Shute, Nevil. *On the Beach.* New York: William Morrow, 1957.

Siddiqi, Asif A. "Korolev, Sputnik, and the International Geophysical Year." In *Reconsidering Sputnik: Forty Years since the Soviet Satellite,* ed. Roger D. Launius, John M. Logsdon, and Robert W. Smith, pp. 43–72. London: Routledge, 2002.

Simon, Herbert A. "Theories of Decision-Making in Economics and Behavioral Science." *American Economic Review* 49 (1959): 253–283.

Skocpol, Theda. *Protecting Soldiers and Mothers: The Political Origins of Social Policy in the United States.* Cambridge, MA: Harvard University Press, 1992.

Skrentny, John D. *The Minority Rights Revolution.* Cambridge, MA: Belknap Press of Harvard University Press, 2002.

Slayton, Rebecca. "Discursive Choices: Boycotting Star Wars between Science and Politics." *Social Studies of Science* 37 (2007): 27–66.

Smith, Alice Kimball. *A Peril and a Hope: The Scientists' Movement in America, 1945–1947.* Chicago: University of Chicago Press, 1965.

Smith, Allen. "Converting America: Three Community Efforts to End the Cold War, 1956–1973." Ph.D. dissertation, American University, 1995.

———. "Democracy and the Politics of Information: The St. Louis Committee for Nuclear Information." *Gateway Heritage* 17 (1996): 2–13.

———. "The Renewal Movement: The Peace Testimony and Modern Quakerism." *Quaker History* 85 (1996): 1–23.

Smith, Bruce L. R. *The Advisers: Scientists in the Policy Process.* Washington, DC: Brookings Institution Press, 1992.

———. *American Science Policy since World War II.* Washington, DC: Brookings Institution Press, 1990.

Smith, Christian, ed. *Disruptive Religion: The Force of Faith in Social Movement Activism.* New York: Routledge, 1996.

———. *The Emergence of Liberation Theology: Radical Religion and Social Movement Theory.* Chicago: University of Chicago Press, 1991.

Smith, R. Allen. "Mass Society and the Bomb: The Discourse of Pacifism in the 1950s." *Peace and Change* 18 (1993): 347–372.

Smith, Roger, and Brian Wynne. "Establishing the Rules of Laws: Constructing Expert Authority." In *Expert Evidence: Interpreting Science in the Law,* ed. Roger Smith and Bryan Wynne, pp. 23–55. London: Routledge, 1989.

Smith-Doerr, Laurel. "Learning to Reflect or Deflect? U.S. Policies and Graduate Programs' Ethics Training for Life Scientists." In *The New Political Sociology of Science: Institutions, Networks, and Power,* ed. Scott Frickel and Kelly Moore, pp. 405–431. Madison: University of Wisconsin Press, 2006.

Snow, C. P. *The Two Cultures and the Scientific Revolution.* Cambridge: Cambridge University Press, 1959.

Snow, David A. "Social Movements." In *Handbook of Symbolic Interactionism,* ed. Larry T. Reynolds and Nancy J. Herman-Kinney, pp. 811–834. Walnut Creek, CA: Altamira Press, 2003.

———. "Social Movements as Challenges to Authority: Resistance to an Emerging Conceptual Hegemony." *Research in Social Movements, Conflicts and Change* 25 (2004): 3–25.

Sokal, Alan, and Jean Bricmon. *Fashionable Nonsense: Postmodern Intellectuals' Abuse of Science*. New York: Picador, 1998.

Staggenborg, Suzanne. *The Pro-Choice Movement: Organization and Activism in the Abortion Conflict*. New York: Oxford University Press, 1991.

Steege, Ted. "Introduction." In *Radicals in Professions: Selected Papers*, pp. 3–7. Ann Arbor, MI: Radical Education Project, 1967.

Stewart, George R. *The Year of the Oath: The Fight for Academic Freedom at the University of California*. Garden City, NY: Doubleday, 1950.

Stinchcombe, Arthur. "Social Structure and Organizations." In *Handbook of Organizations*, ed. James G. March, pp. 142–193. Chicago: Rand McNally, 1965.

Strickland, Donald A. *Scientists in Politics: The Atomic Scientists Movement, 1945–46*. Lafayette, IN: Purdue University Studies, 1968.

Sullivan, William Cuyler, Jr. *Nuclear Democracy: A History of the Greater St. Louis Citizen's Committee on Nuclear Information*. University College Occasional Papers, no. 1. St. Louis: Washington University, 1982.

Swerdlow, Amy. *Women Strike for Peace: Traditional Motherhood and Radical Politics in the 1960s*. Chicago: University of Chicago Press, 1993.

Sylves, Richard T. *Nuclear Oracles: A Political History of the General Advisory Committee of the Atomic Energy Commission, 1947–1977*. Ames: Iowa State University Press, 1987.

Theoharis, Athan. *Seeds of Repression: Harry S. Truman and the Origins of McCarthyism*. New York: Quadrangle, 1971.

Thorpe, Charles. "Disciplining Experts: Scientific Authority and Liberal Democracy in the Oppenheimer Case." *Social Studies of Science* 32 (2002): 525–562.

———. *Oppenheimer: The Tragic Intellect*. Chicago: University of Chicago Press, 2006.

Thorpe, Charles, and Steven Shapin. "Who Was J. Robert Oppenheimer? Charisma and Complex Organization." *Social Studies of Science* 30 (2000): 545–590.

Tilly, Charles. "Major Forms of Collective Action in Western Europe." *Theory and Society* 3 (1976): 365–375.

Turner, Ralph H., and Lewis M. Killian. *Collective Behavior*. 3rd edition. Englewood Cliffs, NJ: Prentice-Hall, 1987.

Tuunainen, Juha. "Contesting a Hybrid Firm at a Traditional University." *Social Studies of Science* 35 (2005): 173–210.

Vallas, Steven Peter. "Symbolic Boundaries and the New Division of Labor: Engineers, Workers, and the Restructuring of Factory Life." *Research in Social Stratification and Mobility* 18 (2001): 3–37.

Vandermeer, John H. *Reconstructing Biology: Genetics and Ecology in the New World Order*. New York: John Wiley and Sons, 1996.

Vaughan, Diane. "Boundary Work: Levels of Analysis, the Macro-Micro Link, and the Social Control of Organizations." In *Social Science, Social Policy, and the Law*, ed. Patricia Ewick, Robert A. Kagun, and Austin Sarat, pp. 291–321. New York: Russell Sage Foundation, 1999.

Vitale, Bruno, ed. *The War Physicists: Documents about the Protests against the Physicists Working for the American Military through the JASON Division of the Institute for Defense Analysis*. Naples, Italy: B. Vitale, 1976.

Von Eschen, Donald, Jerome Kirk, and Maurice Pinard. "The Organizational Substructure of Disorderly Politics." *Social Forces* 49 (1971): 529–544.

Wagner, David. *The Quest for a Radical Profession: Social Service Careers and Political Ideology*. Lanham, MD: University Press of America, 1990.

Wang, Jessica. *American Scientists in an Age of Anxiety: Scientists, Anticommunism, and the Cold War*. Chapel Hill: University of North Carolina Press, 1999.

———. "Liberals, the Progressive Left, and the Political Economy of Postwar American Science: The National Science Foundation Debate Revisited." *Historical Studies in the Physical and Biological Sciences* 26 (1995): 139–166.

———. "Merton's Shadow: Perspectives on Science and Democracy since 1940." *Historical Studies in the Physical and Biological Sciences* 30 (1999): 279–306.

———. "Science, Security and the Cold War: The Case of E. U. Condon." *Isis* 83 (1992): 246–248.

———. "Scientists and the Problem of the Public in Cold War America, 1945–1960." *Osiris* 17 (2002): 323–347.

War Resisters League. *History of the War Resisters League*. New York: War Resisters League, 1980.

Weber, Max. *Max Weber on Law in Economy and Society*. Ed. Max Rheinstein, trans. Edward Shils and Max Rheinstein. New York: Simon & Schuster, 1968.

———. "Science as a Vocation." In *From Max Weber: Essays in Sociology*, ed. Hans Gerth and C. Wright Mills, pp. 129–156. New York: Oxford University Press, 1946.

———. "The Social Psychology of the World's Religions." In *From Max Weber: Essays in Sociology*, ed. Hans Gerth and C. Wright Mills, pp. 267–301. New York: Oxford University Press, 1946.

Weil, Vivian, ed. *Trying Times: Science and Responsibilities after Daubert*. Chicago: Center for the Study of Ethics in the Professions, and Institute for Science, Law, and Technology, Illinois Institute of Technology, 2001.

Weindling. Paul. "The Origins of Informed Consent: The International Scientific Commission on Medical War Crimes, and the Nuremberg Code." *Bulletin of the History of Medicine* 75 (2001): 37–71.

Weiner, Charles. "A New Site for the Seminar: The Refugees and American Physics in the Thirties." In *The Intellectual Migration: Europe and America, 1930–1960*, ed. Donald Fleming and Bernard Bailyn, pp. 190–234. Cambridge, MA: Belknap Press of Harvard University Press, 1969.

Weiner, Douglas R. *Models of Nature: Ecology, Conservation, and Cultural Revolution in Soviet Russia*. Pittsburgh, PA: University of Pittsburgh Press, 2000.

Weinstein, Sandra. *Personnel Security Programs of the Federal Government*. New York: Fund for the Republic, 1954.

Weisman, Carol S. *Women's Health Care: Activist Traditions and Institutional Change*. Baltimore: Johns Hopkins University Press, 1998.

Wisnioski, Matthew H. "Engineers and the Intellectual Crisis of Technology, 1957–1973." Ph.D. dissertation, Princeton University, 2005.

———. "Inside 'The System': Engineers, Scientists, and the Boundaries of Social Protest in the Long 1960s." *History and Technology* 19 (2003): 313–333.

Wittner, Lawrence. *One World or None: A History of the World Nuclear Disarmament Movement through 1953*. Stanford, CA: Stanford University Press, 1993.

———. *Resisting the Bomb: A History of the World Nuclear Disarmament Movement, 1954–1970*. Stanford, CA: Stanford University Press, 1997.

Wolfle, Dael. *Renewing a Scientific Society: The American Association for the Advancement of Science from World War II to 1970*. Washington, DC: American Association for the Advancement of Science, 1989.

Wood, James C. "Scientists and Politics: The Rise of an Apolitical Elite." In *Scientists and National Policy-Making*, ed. Robert Gilpin and Christopher Wright, pp. 41–72. New York: Columbia University Press, 1964.

Woodhouse, Edward J. "Change of State? The Greening of Chemistry." In *Synthetic Planet Chemical Products in Modern Life*, ed. Monica J. Casper, pp. 177–193. New York: Routledge, 2003.

Woodhouse, Edward J., and Steve Breyman. "Green Chemistry as Social Movement?" *Science, Technology and Human Values* 30 (2005): 199–222.

Woodhouse, Edward J., and Dean A. Nieumsa. "Democratic Expertise: Integrating Knowledge, Power, and Participation." In *Knowledge, Power, and Participation in Environmental Policy Analysis*, ed. Matthjis Hisschemoller, Rob Hoppe, William N. Dunn, and Jerry R. Ravetz, pp. 73–96. New Brunswick, NJ: Transaction, 2001.

Wynne, Brian. "May the Sheep Safely Graze? A Reflexive View of the Expert-Lay Knowledge Divide." In *Risk, Environment, and Modernity: Towards a New Ecology*, ed. Scott Lash, Bronislaw Szerszynski, and Brian Wynne, pp. 44–83. London: Sage Publications, 1996.

———. "Reflexing Complexity: Post-Genomic Knowledge and Reductionist Returns in Public Science." *Theory, Culture and Society* 22 (2005): 67–94.

York, Herbert F. *Making Weapons, Talking Peace: A Physicist's Odyssey from Hiroshima to Geneva*. New York: Basic Books, 1987.

Young, Matt, and Taner Edis, eds. *Why Intelligent Design Fails: A Scientific Critique of the New Creationism*. New Brunswick, NJ: Rutgers University Press, 2004.

Young, Michael P. *Bearing Witness against Sin: The Evangelical Birth of the American Social Movement*. Chicago: University of Chicago Press, 2006.

Zald, Mayer N., and Roberta Ash. "Social Movement Organizations: Growth, Decay and Change." *Social Forces* 44 (1966): 61–72.

Zaroulis, Nancy, and Gerald Sullivan. *Who Spoke Up? American Protest against the War in Vietnam, 1963–1975*. Garden City, NY: Doubleday, 1984.

Zavestoski, Stephen, Phil Brown, Sabrina McCormick, Brian Mayer, Maryhelen D'Ottavi, and Jaime Lucove. "Patient Activism and the Struggle for Diagnosis: Gulf War Illnesses and Other Medically Unexplained Physical Symptoms in the U.S." *Social Science and Medicine* 58 (2004): 161–175.

Zwerman, Gilda, and Patricia Steinhoff. "When Activists Ask for Trouble: State-Dissident Interactions and the New Left Cycle of Resistance in the United States and Japan." In *Repression and Mobilization*, ed. Christian Davenport, Hank Johnston, and Carol Mueller, pp. 87–105. Minneapolis: University of Minnesota Press, 2004.

Index

Abele, Ralph, 107, 115
Abelson, Philip, 1, 168
abolitionism, 39
Abrams, Henry, 120
academic freedom, 110, 118, 172
Acheson-Lilienthal report, 58
Advanced Research Projects Agency
  (ARPA), 28–29, 30, 31
African Americans, 7, 19, 103, 134, 140,
  156, 205. *See also* civil rights movement
agriculture, 159, 185
AIDS movement, 206
Alaskan Inuit, 122
Alternative Agriculture Group (AAG), 186
American Association for the Advancement of Science (AAAS), 60, 100, 101,
  104; and anti–Vietnam War movement,
  91, 148, 164; Arden House statement,
  108; Committee on Science in the Promotion of Human Welfare, 79, 123; and
  Commoner, 108, 123, 126; and Condon,
  92; "Crime, Science, and Technology"
  panel, 148; and education and communication, 123; and fallout information
  groups, 124; and information, 108–10;
  Interim Committee on the Social Aspects
  of Science, 78, 108–10, 113, 123; "Is Defense against Ballistic Missiles Possible?"
  panel, 148; and militarism, 110; and military, 194; 1969 meeting of, 161–64;
  1970 meeting of, 165–69; organization
  of, 209; political discussion at, 148,
  161–62; and Science for the People, 155,
  158–59, 161–64, 165–69, 176; Scientist's Committee on Chemical and Biological Warfare, 161–62; and security concerns, 71; and social problems, 148; and
  Society for Social Responsibility in Science, 64, 68, 69, 76, 80, 81; "The Sorry
  State of Science" symposium, 161; subunits of, 210–11; "Unanticipated Environmental Hazards Resulting from Technological Intrusions" symposia, 148;
  Youth Council, 168

American Association of Scientific Workers
  (AASW), 59, 91, 110, 148
American Chemical Society, 80, 176
American Committee for Democracy and
  Intellectual Freedom, 59
American Friends Service Committee
  (AFSC), 40, 55, 61, 62, 66, 107, 119,
  133; *Speak Truth to Power,* 40. *See also*
  Quakers
American Historical Association, 147
American Institute of Chemical Engin-
  eers, 63
American Legion, 80, 120
American Philosophical Association, 147
American Physical Society (APS), 60, 80,
  92, 130, 149–51, 152, 153, 176
American Political Science Associa-
  tion, 147
American Psychiatric Association, 165
American Society for Microbiologists, 147
American Sociological Association, 147
Amory, Cleveland, 110
antiballistic missile (ABM) program, 36,
  51, 132, 135–37, 139, 144, 152, 153,
  154–55, 161, 175
anti–Vietnam War movement, 18, 20, 42,
  50; and American Association for the Advancement of Science, 91, 148, 164; and
  American Physical Society, 146; and antiballistic missile (ABM) program, 137; in
  Bay area, 149–50, 151; beginnings of,
  88; and biologists, 91; and Blankenbecler, 175; on campuses, 132; and destruction of property, 52; development
  of, 133–34; and development of radicalism, 156; and European activism, 172;
  and herbicide use, 161–62, 164, 180–81;
  and March 4 movement, 138–39, 144–
  45; and professional associations, 146,
  148; and radicalism, 132, 154; and Science for the People, 161, 178; and scientists, 128–29, 134; and Society for Social
  Responsibility in Science, 57, 89, 92;
  and spring 1970 student demonstrations,
  165. *See also* war

PRINCETON STUDIES IN CULTURAL SOCIOLOGY